The past 20 years have seen a steady increase in the number of viruses, of both man and animals, which predispose to the development of cancer. The mechanisms involved are now being elucidated, and the means of intervention seems, at least in some instances, close at hand. This volume reviews the current knowledge of oncogenic viruses, their mechanism of action and the prospect for vaccination or therapy. Whilst the emphasis is on viruses of man: papillomaviruses, Epstein–Barr virus, hepatitis B virus and human T cell leukaemia virus, the impact of animal studies is recognized by the inclusion of five chapters on oncogenic viruses of animals.

VIRUSES AND CANCER

SYMPOSIA OF THE
SOCIETY FOR GENERAL MICROBIOLOGY

Series editor (1991–1996): Dr Martin Collins, Department of Food Microbiology, The Queen's University of Belfast
Volumes currently available:

VIRUSES AND CANCER

EDITED BY

A. MINSON, J. NEIL AND M. McCRAE

FIFTY-FIRST SYMPOSIUM OF THE
SOCIETY FOR GENERAL MICROBIOLOGY
HELD AT THE UNIVERSITY OF CAMBRIDGE
MARCH 1994

Published for the Society of General Microbiology

CAMBRIDGE
UNIVERSITY PRESS

Published by the Press Syndicate of the University of Cambridge
The Pitt Building, Trumpington Street, Cambridge CB2 1RP
40 West 20th Street, New York, NY 10011-4211, USA
10 Stamford Road, Oakleigh, Melbourne 3166, Australia

First published 1994

Printed in Great Britain at the University Press, Cambridge

A catalogue record for this book is available from the British Library

Library of Congress cataloguing in publication data

Society for General Microbiology. Symposium (51st : 1994 : University
  of Cambridge)
    Viruses and cancer : Fifty-first Symposium for the Society for
  General Microbiology, held at the University of Cambridge, March
  1994 / edited by A. Minson, J. Neil, and M. McCrae.
        p.   cm.
    Includes bibliographical references and index.
    ISBN 0-521-45472-7
    1. Oncogenic viruses—Congresses. I. Minson, A. C. II. Neil,
  J., 1953–   . III. McCreae, M. IV. Title.
  QR372.06S63   1994
  6.6.99′4071—dc20   93-46636
                                                                  CIP

ISBN 0 521 45472 7 hardback

PN

CONTENTS

CONTRIBUTORS

ADAM, E. Department of Molecular Biology, University of Brussels, B1640 Rhode-St-Genèse, Belgium and National Institute for Veterinary Research, B 1180 Brussels, Belgium

ALTMANN, A. Deutsches Krebsforschungszentrum, Angewandte Tumorvirologie, Forschungsschwerpunkt Genomveränderungen und Carcinogenese, 69121 Heidelberg, Germany

ANDERSON, R. The Beatson Institute for Cancer Research, CRC Beatson Laboratories, Garscube Estate, Glasgow G61 1BD, UK

BEAR, S. E. Fox Chase Cancer Center, Philadelphia, PA 19111, USA

BECKER, S. A. Division of Molecular Virology, Baylor College of Medicine, Houston, TX 77030, USA

BELLACOSA, A. Fox Cancer Center, Philadelphia, PA 19111, USA. *Present address*: Institute of Human Genetics, Catholic University Medical School, L.goF.Vito 00168 Roma, Italy

BUENDIA, M. A. Unité de Recombinaison et Expression Génétique (INSERM U163), Institut Pasteur, 28 rue du Dr Roux, 75724 Paris Cedex 15, France

BURNY, A. Molecular Biology and Animal Physiology Unit, Faculty of Agronomy, 13, avenue Maréchal Juin, B5030 Gembloux, Belgium

BUTEL, J. S. Division of Molecular Virology, Baylor College of Medicine, Houston, TX 77030,USA

CAIRNEY, M. The Beatson Institute for Cancer Research, CRC Beatson Laboratories, Garscube Estate, Glasgow G61 1BD, UK

CALLEBAUT, I. Molecular Biology and Animal Physiology Unit, Faculty of Agronomy, 13, avenue Maréchal Juin, B5030 Gembloux, Belgium and Department of Biological Macromolecules, University of Paris 6, F75252 Paris Cedex 05

CAMPO, M. S. Beatson Institute for Cancer Research, CRC Beatson Laboratories, Garscube Estate, Glasgow G61 1BD, UK

CHATTOPADHYAY, S. K. Laboratory of Immunopathology, National Institute of Allergy and Infectious Diseases, and Registry of Experimental Cancers, National Cancer Institute, National Institutes of Health, Bethesda, MD 20892, USA

CHEN, I. S. Y. Division of Hematology–Oncology, Departments of Medicine and Microbiology & Immunology, UCLA School of Medicine and Jonsson Comprehensive Cancer Center, Los Angeles, CA 90024-1678, USA

CLUDTS, I. Department of Molecular Biology, University of Brussels, B1640 Rhode-St-Genèse, Belgium

DEQUIEDT, F. Faculty of Agronomy, 13, avenue Maréchal Juin, B5030 Gembloux, Belgium

DROOGMANS, L. Department of Molecular Biology, University of Brussels, B1640 Rhode-St-Genèse, Belgium

FARRELL, P. J. Ludwig Institute for Cancer Research, St Mary's Hospital Medical School, Norfolk Place, London W2 1PG, UK

FREDRICKSON, T. N. Laboratory of Immunopathology, National Institute of Allergy and Infectious Diseases, and Registry of Experimental Cancers, National Cancer Institute, National Institutes of Health, Bethesda, MD 20892, USA

GIESE, N. Laboratory of Immunopathology, National Institute of Allergy and Infectious Diseases, and Registry of Experimental Cancers, National Cancer Institute, National Institutes of Health, Bethesda, MD 20892, USA

GILKS, C. B. Fox Chase Cancer Center, Philadelphia, PA 19111, USA. *Present address*: Department of Pathology, University of British Columbia, 2211 Wesbrook Mall, Room G227, Vancouver, BC, V6T-2B5, Canada

GISSMANN, L. Loyola University Medical Center, Department of Obstetrics and Gynecology, Maywood, Illinois 60153, USA

GRIMONPONT, C. Faculty of Agronomy, B5030 Gembloux, Belgium

HARTLEY, J. W. Laboratory of Immunopathology, National Institute of Allergy and Infectious Diseases, and Registry of Experimental Cancers, National Cancer Institute, National Institutes of Health, Bethesda, MD 20892, USA

IZUMI, K. Program in Virology and Department of Microbiology and Molecular Genetics and Medicine, Harvard University, 75 Francis St, Boston, MA 02115, USA

JACKSON, M. E. The Beatson Institute for Cancer Research, CRC Beatson Laboratories, Garscube Estate, Glasgow G61 1BD, UK

JARRETT, O. University of Glasgow, Department of Veterinary Pathology, Bearsden, Glasgow G61 1QH, UK

JOCHMUS, I. Deutsches Krebsforschungszentrum, Angewandte Tumorvirologie, Forschungsschwerpunkt Genomveränderungen und Carcinogenese, 69121 Heidelberg, Germany

KAYE, K. Program in Virology and Department of Microbiology and Molecular Genetics and Medicine, Harvard University, 75 Francis St, Boston, MA 02115, USA

KERKHOFS, P. National Institute for Veterinary Research, B1180 Brussels, Belgium

KETTMANN, R. Faculty of Agronomy, B5030 Gembloux, Belgium

KIEFF, E. Program in Virology and Department of Microbiology and Molecular Genetics and Medicine, Harvard University, 75 Francis St, Boston, MA 02115, USA

KLEIN, G. Department of Tumour Biology, Karolinska Institutet, S-171 77 Stockholm, Sweden

LANE, D. P. CRC Laboratories, Department of Biochemistry, University of Dundee, Dundee DD1 4HN, UK

LONGNECKER, J. Program in Virology and Department of Microbiology and Molecular Genetics and Medicine, Harvard University, 75 Francis St, Boston, MA 02115, USA

MCCRAE, M. Department of Biological Sciences, University of Warwick, Coventry CV4 7AL, UK

MAKRIS, A. Fox Chase Cancer Center, Philadelphia, PA 19111, USA

MAMMERICKX, M. National Institute for Veterinary Research, B1180 Brussels, Belgium

MANNICK, J. Program in Virology and Department of Microbiology and Molecular Genetics and Medicine, Harvard University, 75 Francis St, Boston, MA 02115, USA

MILLER, C. Program in Virology and Department of Microbiology and Molecular Genetics and Medicine, Harvard University, 75 Francis St, Boston, MA 02115, USA

MINSON, A. Department of Pathology, University of Cambridge, Tennis Court Road, Cambridge CB2 1QP, UK

MORSE, H. C. Laboratory of Immunopathology, National Institute of Allergy and Infectious Diseases, and Registry of Experimental Cancers, National Cancer Institute, National Institutes of Health, Bethesda, MD 20892, USA

NEIL, J. Department of Veterinary Pathology, University of Glasgow Veterinary School, Bearsden, Glasgow G61 1QH, UK

PATRIOTIS, C. Fox Chase Cancer Center, Philadelphia, PA 19111, USA

POON, B. Division of Hematology–Oncology, Departments of Medicine and Microbiology & Immunology, UCLA School of Medicine and Jonsson Comprehensive Cancer Center, Los Angeles, CA 90024-1678, USA

PORTETELLE, D. Faculty of Agronomy, B5030 Gembloux, Belgium

RICKINSON, A. B. CRC Laboratories, Department of Cancer Studies, University of Birmingham, The Medical School, Birmingham B15 2TJ, UK

ROBERTSON, E. Program in Virology and Department of Microbiology and Molecular Genetics and Medicine, Harvard University, 75 Francis St, Boston, MA 02115, USA

SINCLAIR, A. J. Ludwig Institute for Cancer Research, St Mary's Hospital Medical School, Norfolk Place, London W2 1PG, UK

SLAGLE, E. L. Division of Molecular Virology, Baylor College of Medicine, Houston, TX 77030, USA

STEWART, S. Division of Hematology-Oncology, Departments of Medicine and Microbiology & Immunology, UCLA School of Medicine and

Jonsson Comprehensive Cancer Center, Los Angeles, CA 90024-1678, USA

SWAMINATHAN, S. Program in Virology and Department of Microbiology and Molecular Genetics and Medicine, Harvard University, 75 Francis St, Boston, MA 02115, USA

TANG, Y. Laboratory of Immunopathology, National Institute of Allergy and Infectious Diseases, and Registry of Experimental Cancers, National Cancer Institute, National Institutes of Health, Bethesda, MD 20892, USA

TOMKINSON, B. Program in Virology and Department of Microbiology and Molecular Genetics and Medicine, Harvard University, 75 Francis St, Boston, MA 02115, USA

TSICHLIS, P. N. Fox Chase Cancer Center, Philadelphia, PA 19111, USA

TUNG, X. Program in Virology and Department of Microbiology and Molecular Genetics and Medicine, Harvard University, 75 Francis St, Boston, MA 02115, USA

VOUSDEN, K. H. Ludwig Institute for Cancer Research, Norfolk Place, London W2 1PG, UK

WELLER, I. V. D. Academic Department of Genito Urinary Medicine, University College London Medical School, London, UK

WILLEMS, L. Faculty of Agronomy, B5030 Gembloux, Belgium

YALAMANCHILI, R. Program in Virology and Department of Microbiology and Molecular Genetics and Medicine, Harvard University, 75 Francis St, Boston, MA 02115, USA

VIRUSES AND CANCER

G. KLEIN

*Dept of Tumour Biology, Karolinska Institutet, S-171 77
Stockholm, Sweden.*

INTRODUCTION

The birth of viral oncology at the beginning of this century came from the work of a few pioneers, with Peyton Rous in the lead. Hopes were high in the anticipation that the understanding of tumour aetiology was imminent and would open the way towards immunological prevention of cancer. These hopes evaporated quickly when it appeared that mammalian tumours were not readily transmitted by filtrates like chicken tumours.

After a stagnation of several decades, expectations rose again in the 1950s and reached a totally unprecedented peak, triggered by the discovery of murine leukaemia viruses by Ludwik Gross and of polyoma virus by Gross, Stewart and Eddy. Many cancer researchers and virologists turned to the field of viral oncology. The majority of these researchers were propelled by the hope that many, and perhaps most, human tumours would be found to have a viral aetiology. The subsequent development has failed to fulfil these high expectations. Nevertheless, it had an enormous spin-off effect. We owe several of the important early developments in molecular biology to it, such as the discovery of reverse transcription, the molecular dissection of the DNA tumour viruses, much of the pioneering work on recombinant DNA technology, and the discovery of the oncogenes. While the significance of these discoveries is beyond debate, they were hardly in line with original hopes. What has remained of them, and to what extent do viruses contribute to the incidence of naturally occurring cancers in humans and wild animal populations?

The current impression is that the oncogene-transducing, directly trans-forming class I retroviruses do not play any major causative role for naturally occurring cancers. Are the class I viruses freaks or perhaps even monsters generated by laboratory experimentations? Probably so. The acquisition of cell-derived oncogenes by illegitimate recombination may be the unavoid-able, but rare, consequence of the retroviral lifestyle, with its cycles of forward and reverse transcription. The pick-up of a cellular oncogene provides no advantage for viral replication. On the contrary, it is an unnecessary and often crippling burden for the virus. It is therefore likely that the oncogene-transducing recombinants are readily eliminated under natural circumstances. In order to be perpetuated, they require the intervention of the investigator who is in the need of regularly reproducible *in vitro* transformation and/or *in vivo* tumourigenicity systems. Serial passage and re-isolation from tumours or transformed cells may selectively enrich the transforming particles, provide them with appropriate helpers, and end up with the established highly oncogenic laboratory strains that we know. Some of them have even managed to pick up two unlinked oncogenes that have totally different mechanisms of action, but are complementary in their transforming effect. While they are a source of great delight for the investigator, it is hard to see how they could have survived in nature.

All RNA tumour viruses that are known to contribute to the aetiology of naturally occurring animal or human tumours belong to the chronic or class II retroviruses that do not carry cell-derived oncogene sequences. They are exemplified by the feline and bovine leukemia viruses and by the human adult T-cell leukemia virus (HTLV-1). All three viruses are of exogenous origin; the species for which they are pathogenic lack homologous endogenous sequences. The mechanism for their leukemogenic action is obscure. Insertion within the immediate neighbourhood of a cellular oncogene is one of the currently favoured models.

The DNA tumour viruses involved in the causation of naturally occurring tumours are highly diversified. They range from the small papilloma viruses to the large herpes viruses. Their tumourigenic action can be rapid and direct as in the case of Marek's disease virus, or chronic and probably indirect, as in the case of hepatitis virus B. Some of them like Epstein–Barr virus (EBV) or the papilloma viruses may merely increase the target cell population that runs the risk of undergoing malignant transformation. In the case of EBV, the combined effect of viral immortalization and chronic infection probably creates a pre-neoplastic population of virally immortalized B lymphocytes, partially blocked in their differentiation. Due to their inability to leave the B-cell pool and chronically stimulated to divide, they are risking the crucial chromosomal translocation accident that triggers the growth of the autonomous tumour clone. The papilloma viruses are probably also responsible for the pre-neoplastic rather than for the final carcinomatous change.

The presently known viruses that cause tumours in nature represent a small but firmly established group. Some of them may eventually satisfy the original aspiration of the tumour virologist; immunological prevention. Tumours that appear in congenitally or iatrogenically immunodefective individuals are sometimes, and perhaps always, associated with potentially oncogenic and often highly transforming viruses that do not cause tumours in hosts with an intact immune system. Ubiquitous tumour viruses of this category such as polyoma in mice or EBV in humans have a long symbiotic history with their natural host. Their selective impact has led to the evolution of a tight, multi-effector surveillance system that prevents the growth of the transformed cells. In this situation, tumour development is always a biological accident.

The interactions between potentially oncogenic viruses, their host cells and concomitant but causally unrelated genetic changes in the latter may contribute to the neoplastic development in complex ways. Studies on Burkitt's lymphoma (BL) are particularly relevant in this respect. They will be reviewed below.

THE ROLE OF IG/MYC TRANSLOCATION AND EBV IN BL

Reciprocal translocations between c-*myc* and an Ig-locus represent the common denominator of all BLs (for review see Klein, 1993). They are universally associated with both the high endemic and the sporadic form, whether EBV carrying or EBV negative. The juxtaposition of c-*myc* and Ig-sequences deregulates the former, by bringing it under the control of the latter. Ig-loci are constitutively active in B-cells. The c-*myc* gene is regularly expressed in dividing but not in resting cells. Ig-juxtaposed *myc* sequences cannot be down-regulated when the B-cell leaves the cycling compartment and goes to the resting (G0) stage, in contrast to the normal, non-translocated allele. This happens at least three times during normal B-cell development: (i) following the rearrangement of the immunoglobulin genes that leads to the short-lived virgin B-cell, awaiting the activating stimulus; (ii) following antigen induced clonal B-cell expansion, after the waning of antigenic stimulus, when long lived memory B-cells are generated; (iii) upon terminal differentiation into plasma cells. Since they cannot down regulate their Ig-juxtaposed *myc* gene, the translocation carrying cells are unable to leave the cycling compartment and may generate tumours. BL-cells may represent virgin or memory B-cells that have escaped control, whereas the Ig/*myc* translocation carrying mouse and rat tumours are plasmacytomas.

What causes the translocations? The large variability of the breakpoints in and around the c-*myc* gene suggests that the chromosome breaks at random, as an accident of prolonged cell division in the target cell at risk. Chronic stimulation of cell division is a notable and probably essential feature within the pre-neoplastic history of all three Ig/*myc* carrying tumours, BL, mouse plasmacytoma (MPC) and rat immunocytoma (RIC). The Ig-locus carrying chromosome breaks at sites where the recombinases involved in physiological Ig rearrangement are known to act, indicating that these enzymes may play a role in the illegitimate joining of the translocated chromosome pieces.

It is not known whether the probability of an Ig/*myc* translocation is influenced by genetic factors in humans. There is some circumstantial evidence to suggest that the MPC-associated translocation is. The Balb/c mouse strain is eminently susceptible to MPC induction. Most other mouse strains are resistant. Resistance is dominant over susceptibility (for review see Wiener & Potter, 1993). Using reciprocal chimeras between Balb/c and the resistant DBA/2 strain, we have shown that at least part of the genetic susceptibility to MPC induction is determined at the target cell level (Silva *et al.*, 1991). The probability of the Ig/*myc* translocation may be a rate-limiting factor. Overnight incubation of Balb/c spleen cells with Abelson virus, followed by transplantation into pristane treated Balb/c recipients with a different chromosomal marker has led to the rapid appearance of donor type MPCs, indicating that translocation carrying cells were already present in the untreated Balb/c donor (Sugiyama *et al.*, 1989). This is further supported by the development of translocation carrying spontaneous MPCs in a Balb/c substrain, selected for high MPC susceptibility (Silva *et al.*, to be published). A genetically determined high translocation proneness of the Balb/c strain is also indicated by the findings of Kishimoto's group on Emu-IL-6 transgenic mice (Suematsu *et al.*, 1992). As long as the transgene was expressed on the C57B1 background, the mice developed benign plasmacytosis only. Introduction of the BALB/c background has led to the appearance of malignant plasmacytomas that carried rearranged *myc* genes.

Can the Ig/*myc* translocation fully account for the genesis of the B-cell derived tumours that carry it? Facsimile experiments with transgenic mice have conclusively proven the ability of immunoglobulin enhancer activated myc genes to induce B-cell tumours (Adams *et al.*, 1985; Wang *et al.*, 1992). The tumours are oligo- or monoclonal, in spite of the fact that the Emu-*myc* construct is potentially active in every B-cell. This suggests that additional changes are required for autonomous tumour development.

Is p53 mutation one of these additional changes? Is the expression of mutated p53 important for the continuous proliferation of the cell? Wild-type p53 can form heterodimers with its mutated counterpart. This can suppress tumourigenicity in some mutant carrying cells and can cause complete growth arrest in others (for review see Zambetti & Levine, 1993). We have expressed wtp53 from a temperature sensitive conformational mutant p53 gene in BL-cells that carried a mutant p53 gene (Ramqvist *et al.*, 1993). At 37 °C where the introduced gene expresses mutant p53, the transfected cells grew as well as the controls. Following the shift of the temperature to 32 °C where a major part of the expressed exogenous protein assumes wild-type conformation, the cells died by massive apoptosis. We have suggested that the apoptotic mechanism is triggered by the contradicting signals from the activated, proliferation stimulating *myc*-gene and the growth inhibiting wtp53. We could reproduce the same phenomenon in a v-*myc* induced, originally p53-negative murine T-lymphoma line, transfected with the same tsp53 construct (Wang *et al.*, 1993).

Does EBNA1, the single EBV-encoded protein product expressed in the EBV-carrying BL-cell, play a positive role in the induction and/or maintenance of the tumourigenic phenotype? Recent evidence from Ian Magrath's laboratory indicates that this may be the case (Magrath, 1993). EBNA1 was found to transactivate the c-*myc*-promoter, but only in the presence of a *cis*-positioned immunoglobulin enhancer. The IgH-enhancer is often included within the fused IgH/*myc* sequence in the high endemic BL tumours that regularly carry EBV. The situation is quite different in the sporadic, largely EBV-negative BLs, where the *myc*-gene breaks in the first exon or intron, as a rule, bringing the coding *myc* sequences, located in the second and third exon, into close proximity with the juxtaposed Ig-sequences (Magrath, 1990). In the sporadic BLs that carry a typical 8;14 translocation, *myc* is usually linked to an IgH switch region. This excludes the IgH enhancer from the fused gene. The tight proximity between the active IgH chromatin and the coding *myc* exons may conceivably obviate the need for the IgH enhancer and for the activating effect of EBNA1 on the IgH-enhancer/*myc* promoter complex. The contribution of EBNA1 may become important in the high endemic form with its greater distances between IgH and *myc* sequences. The activating effect of EBNA1 on the IgH enhancer–*myc* promoter complex may permit the survival of rearranged configurations that would have been eliminated in its absence.

Recently, Magrath has obtained direct evidence in support of the postu-lated effect of EBNA1 by expressing an EBNA1 antisense construct in EBV-carrying BL-cells under an inducible promoter (Magrath, 1993). Induction of the antisense message inhibited the growth of the cells *in vitro* and *in vivo*.

EBV AND IMMUNOSURVEILLANCE

EBV is a highly successful virus. It infects most members of our species and establishes a latent, non-pathogenic interaction in the vast majority. Dis-eases develop only as the result of accidents. This double success of viral and host survival can be attributed, at least in part, to the fortunate resolution of two seemingly paradoxical problems that are related to each other. EBV has an outstanding transforming ability, second to no known virus, but causes normally no disease at all, or only a self-limiting disease, mononucleosis. Even this benign disease can be seen as a disturbance of the normal virus–host equilibrium, caused by the modern hygiene-related postponement of the non-pathogenic early childhood infection, promoted by poor salivary hygiene, to the time of intensified salivary exchanges between adolescents and young adults. The second paradox concerns the immunological behav-iour of the virally infected cells. EBV-transformed B-blasts are highly immunogenic for autologous T-cells. They induce them to proliferate in mixed lymphocyte cultures and to generate EBV-specific cytotoxic T-cells (CTLs). The strength of these autologous responses is comparable to the strongest alloresponses across MHC-barriers. Similarly strong reactions occur *in vivo* during the convalescent phase of mononucleosis. But, in spite of this, the virus is not eliminated.

The following scenario is consistent with the currently available evidence.

On primary infection, the virus attaches to, and enters, resting B-cells by using the B-cell specific CR2 complement receptor as its receptor. It transforms part of the infected cells into proliferating immunoblasts that express six virally encoded nuclear proteins (EBNA1-6) and two membrane antigens (LMP1-2). The virus acts as a polyclonal B-cell activator. It switches on B-cell growth factors and growth factor receptors. EBNA-positive blasts can be readily detected in the peripheral blood and lymphoid tissues during acute mononucleosis. Many of them are in mitosis. EBV specific CTLs appear after a couple of weeks. Subsequently, the EBNA positive blasts are cleared away and become undetectable by ordinary staining, although EBV-carrying B-cells persist in the blood, as shown by PCR. The CTL response is directed against the MHC class I associated peptide derivatives of the EBNAs and LMPs, with the notable exception of EBNA1. The targeting of the CTLs depends on the HLA class I genotype of the host (for review see Masucci *et al.*, 1993). Some of the immunodominant peptides that serve as the targets of the CTL response have been identified (Gavioli *et al.*, 1992; Khanna *et al.*, 1992; Murray *et al.*, 1992; Zhang *et al.*, 1993).

The non-immunogenicity of EBNA1 for T-cells appears to be more than a coincidence of experimentation. It has been confirmed in a transfected murine tumour cell system (Trivedi *et al.*, 1991) and it may reflect an important aspect of the viral strategy. Consider the following:

In contrast to the other EBNAs, EBNA1 is expressed in a cell type independent fashion. It is produced by all presently known cells that carry viral DNA. This includes B-cells, epithelial (nasopharyngeal carcinoma) cells and EBV-carrying T-cell and Hodgkins lymphomas. In contrast, EBNA2-6 are only expressed in activated B-blasts and their *in vitro* counterparts, the EBV-immortalized lymphoblastoid cell lines (LCLs). EBV-carrying BL-cells express only EBNA1 but not EBNA2-6 *in vivo*. Derived cell lines behave in the same way, as long as they maintain the original (group I) BL phenotype. When the BL-cells 'drift' to a more immunoblastic (group II or III) phenotype in the course of serial *in vitro* passage, EBNA2-6 are turned on (Rowe *et al.*, 1987). EBV-carrying nasopharyngal carcinoma cells, T-cell lymphomas and Hodgkins lymphomas express EBNA1 but not EBNA2-6 (for review see Pallesen *et al.*, in press).

The cell phenotype dependent differences in the EBV expression pattern have been also brought out by somatic cell hybridization experiments. When BL type I cells that only express EBNA1 are fused with LCLs that express all six EBNAs, the LCL immunoblastic phenotype and the corresponding EBNA expression is dominant in the hybrids. When LCLs are fused with non-B (epithelial, fibroblastic or myeloid) cells, hybrids with a dominating non-B phenotype express only EBNA1 (Contreras-Brodin *et al.*, 1991).

The regulation of LMP expression is more complex, and will not be discussed here. Some of the main facts can be found in the papers of Fåhraeus *et al.* (1990, 1993).

Table 1. *Promoter usage and EBV encoded protein expression in growth transformed cells*[a]

Cell	Promoter usage		Antigen expression
	Fp	Wp/Cp	
LCL[b]	−	+	E1-6[c],LMP
BLI[d]	+	−	E1 only
BLI+5azaC[c]	+	+	E1-6, LMP induced in a fraction of the cells
BLIII[f]	+	+	E1-6, LMP
NPC[g]	+	−	E1, no E2-6, LMP in 60%
Somatic hybrids			
LCL × BL[h]	+	+	E1-6, LMP
LCL × dominant non-B[i]	+	−	E1 only

[a] For details and references, see Sample *et al.*, 1991; Schaefer *et al.*, 1991; Rowe *et al.*, 1992; Hu, L.-F., Chen, F., Altiok, E., Winberg, G., Klein, G. & Ernberg, I.: Cell phenotype dependent alternative splicing of EBNA mRNAs in the EBV-carrying cells, subm for publ.
[b] EBV-immortalized lymphoblastoid cell line.
[c] E1–6: EBNA1–6.
[d] BL group I, representative of the *in vivo* phenotype.
[e] BL group I, after exposure to 5aza-cytidine.
[f] EBV-carrying BL-lines *in vitro*, after drift to a more lymphoblastoid (group III) phenotype.
[g] Nasopharyngeal carcinoma.
[h] The LCL phenotype dominates in these hybrids.
[i] Hybrids generated by fusing LCL with epithelial, fibroblastic or myeloid cells. The non-B phenotype dominates in these hybrids.

The phenotype dependent differential expression of the EBNAs in latently infected cells is due to the differential utilization of viral promoters. The main patterns are indicated in Table 1. The EBNA1 message can be generated in two basically different ways. In the EBV-carrying immuno-blasts (LCLs and BL group II/III cells), an approximately 85 kb giant message is generated from two alternative promoters, in the BamHI C or W region. They are collectively symbolized as Wp/Cp. In BL type I cells, a different promoter is used, located in the F region. It generates a message that encodes EBNA1 only.

Interaction of EBNA5 with RB and p53

In view of the remarkable interactions between the transforming proteins of SV40, the adenoviruses and the human tumour associated papilloma viruses with the retinoblastoma (RB) and p53 proteins, we looked for similar interactions with EBV encoded, growth transformation proteins. In a first immunohistochemical study, we have found that one of the EBV encoded, growth transformation associated nuclear proteins, EBNA5, colocalizes with the RB protein (Jiang *et al.*, 1991). Subsequently, we have shown that the repetitive (W) part of EBNA5 can form complexes with both RB and p53 *in vitro* (Szekely *et al.*, 1993). The details of these reactions differ from the previously known interactions with other DNA tumour viruses, since p53 and RB compete with each other for EBNA5 binding, and since mutants of both RB and p53 that no longer bind SV40 LT can still bind EBNA5. The possible functional significance of EBNA5 complexing is not yet known, but it is noteworthy that EBNA5 is essential for the B-lymphocyte immortalizing function of EBV (Mannick *et al.*, 1991).

No p53 mutations have been found in EBV transformed lymphoblastoid cell lines (LCLs) of non-neoplastic origin (Wiman *et al.*, 1991), nor would they be expected to occur as a transformation requirement if EBNA5, a regularly expressed protein in EBV–LCLs, complexes with and possibly inactivates RB and p53. The situation is different in EBV carrying Burkitt's lymphoma (BL), where the virus uses a different programme, as already mentioned (see also Table 1).

ESCAPES OF EBV-CARRYING CELLS FROM REJECTION

Viral escape.

Exposure of T-cells from HLA-A11 positive individuals to autologous EBV-transformed B-cells was found to generate a predominantly EBNA4-specific, A11-restricted response. The immunodominant EBNA4 peptide has been identified. It can sensitize A11 positive PHA-blasts to specific CTL-mediated lysis in dilutions up to 10^{-12} (Zhang *et al.*, 1993).

All Caucasian and African EBV-isolates so far studied were found to encode the immunodominant EBNA4 peptide. All six EBV isolates originating from Papua-New Guinea had the same point mutation in one of the A11-anchorage sites, however, cancelling the ability of the peptide to serve as a CTL target (de Campos-Lima *et al.*, 1993). This may be related to the fact that A11 is a more frequent HLA class I allele in Papua-New Guinea than in Caucasian or African populations.

If the point mutation in the immunodominant EBNA4 peptide has increased the chances of viral survival in the New Guinea population, as these data indicate, it suggests that the CTL response directed against EBNA4 positive cells may hamper the chances of viral survival. EBNA4 is only expressed in proliferating immunoblasts, however. This further emphasizes the importance of the early lymphoproliferative phase for viral survival. Virus production is probably largely, if not entirely confined to epithelial cells, mainly in the oropharynx. Since no viral genomes have been detected in the basal, proliferating epithelial layers and, in view of the fact that lytic viral replication takes place in the keratinizing epithelium, this suggests that the virus may reach the epithelium via the lymphocytes, by contact or by emperipolesis.

Lymphoproliferative diseases in immunodefectives.
EBV-carrying immunoblastomas are frequent complications in organ transplant recipients. They also appear in some congenital immunodeficiences, XLP (X-linked lymphoproliferative syndrome) in particular. The lymphoproliferative disease can take the form of fatal mononucleosis or is more lymphoma-like. It is often polyclonal in the beginning, but may progress to oligo- and monoclonal. Part of the various lymphoproliferative diseases seen in AIDS patients also belong to the category of immunoblastomas while others are quite different.

Phenotypically, the immunoblastomas resemble the lymphoblastoid cell lines (LCLs) *in vitro*. They express EBNA1-6 and LMP. Obviously, these cells proliferate because the surveillance of the host has broken down, rather than due to any cellular escape. This is also in line with the clinical features of the disease. The immunoblastomas of kidney transplant recipients may regress without any treatment after the immunosuppressive regimen has been lifted and the patient was allowed to reject his kidney. No second lymphoma arose, as a rule, following a second kidney transplantation. This makes a good case for boosting the EBV-specific immune response of organ and bone marrow transplant recipients preventively, prior to transplantation. This particular risk group would provide the best justified target for a vaccination programme, in our opinion. It would aim at the prevention of immunoblast proliferation, rather than protection from viral infection–most prospective graft recipients would have been infected earlier. The HLA genotype of the patients would allow rational predictions about the most likely EBV-encoded protein targets of the CTL response.

Burkitt's lymphoma (BL).
This develops in immunocompetent patients. The EBV carrying form provides an example of cellular, rather than viral escape. Several features of the BL-cell contribute to this escape. They are related to specific aspects of the 'quasi-resting' BL-phenotype, fixed by the Ig/*myc* translocation. As already mentioned, the constitutive expression of c-*myc* prevents the cells from leaving the cycling compartment, even though they have switched their phenotype to what corresponds to a resting B-cell or, in the case of the Ig-*myc* carrying mouse and rat plasmacytomas, to a plasma cell.

Three phenotypic features of the BL-cell concur in facilitating the cellular escape: (i) the exclusive use of the F-promoter and the corresponding lack of EBNA2-6 and LMP expression removes the virus-specific targets of an EBV-specific CTL-attack; (ii) in contrast to the EBV-carrying LCLs, BL-cells express little or no adhesion molecules, known to be required for efficient effector/target cell interactions; (iii) allele specific down-regulation of certain MHC class I antigens that is another important feature of the BL phenotype (Andersson *et al.*, 1991). A11 is one of the most constantly down-regulated specificities.

The powerful penetrance of the Ig/*myc* translocation as the unifying feature of the BL-cell phenotype can be seen against these pleiotropic features of the phenotypic window where 'the resting cell that is not resting' stays fixed. It has the double capacity of preventing the cells from leaving the cycling compartment and of facilitating immune escape.

Nasopharyngeal carcinoma

This provides another example of EBV-carrying tumour cell escape in immunocompetent hosts. All NPCs express EBNA1. They do not express the immunogenic EBNA2-6 proteins, but 60% of them express LMP. We have recently introduced a Chinese NPC-derived LMP1-gene into a non-immunogenic mouse mammary carcinoma cell. Whereas the B-LMP1 (B95-8 virus-derived LMP1) gene rendered the target cells highly immunogenic (rejectable) in a mouse model system (Trivedi *et al.*, 1991), the NPC-derived LMP1 gene failed to do so (Trivedi P. *et al.*, in press). If confirmed for additional LMP1-positive NPCs, it would indicate that NPC-cells may escape rejection by a mutation in the potentially immunogenic LMP1 protein.

SUMMARY OF BL DEVELOPMENT

At least three genetic changes are known to contribute to the genesis of Burkitt's lymphoma (BL): the Ig/*myc* translocation, the presence of EBV in the vast majority of the endemic and a minority of sporadic tumours, and a p53 mutation, present in approximately 60% of the BL derived lines. Activation of c-*myc* by juxtaposition to Ig sequences is a universal common denominator of endemic and sporadic, EBV positive and negative BLs. It acts by preventing the cell from leaving the cycling compartment and by facilitating immune escape. EBV probably acts by expanding the target cell population at risk and prolonging its life-span. This, together with the malaria co-factor, would increase the risk of the translocation accident. The p53 mutation may be essential for the continued growth of the tumours where it occurs, since introduction of wild-type p53 leads to their apoptotic death.

REFERENCES

Adams, J. M., Harris, A. W., Pinkert, C. A. *et al.* (1985). The c-myc oncogene driven by immunoglobulin enhancers induces lymphoid malignancy in transgenic mice. *Nature,* **318**, 533–8.

Alfieri, C., Birkenbach, M. & Kieff, E. (1991). Early events in Epstein–Barr virus infection of human B lymphocytes. *Virology,* **181**, 595–608.

Andersson, M. L., Stam, N., Klein, G., Ploegh, H. & Masucci, M. G. (1991). Aberrant expression of HLA class I antigens in BL cells. *International Journal of Cancer*, **47**, 544–50.

Collins, V. P. & James, C. D. (1993). Gene alterations in the origin and progression of human gliomas. *FASEB Journal* (Thematic issue on Tumor Suppressor Genes), in press.

Contreras-Brodin, B. A., Anvret, M., Imre, S., Altiok, E., Klein, G. & Masucci, M. G. (1991), B-cell phenotype dependent expression of the Epstein–Barr virus nuclear antigens EBNA2 to EBNA6: studies with somatic cell hybrids. *Journal of General Virology*, **72**, 3025–33.

de Campos-Lima, P. O., Gavioli, R., Zhang, Q.-J. *et al.* (1993). Epstein–Barr virus isolates from a highly ELA A11 positive population lack the dominant A11 restricted T cell epitope. *Science*, **260**, 98–100.

Farrell, P. J., Allan, G. J., Shanahan, F., Vousden, K. H. & Crook, T. (1991), p53 is frequently mutated in Burkitt's lymphoma cell lines. *EMBO Journal*, **10**, 2879–87.

Fouds, L. (1958). The natural history of cancer. *Journal of Chronic Diseases*, **8**, 2–37.

Fåhraeus, R., Jansson, A., Ricksten, A., Sjöblom, A. & Rymo, L. (1990). Epstein–Barr virus-encoded nuclear antigen 2 activates the viral latent membrane protein promoter by modulating the activity of a negative regulatory element. *Proceedings of the National Academy of Sciences, USA*, 7390–4.

Fåhraeus, R., Jansson, A., Sjöblom, A., Nilsson, T., Klein, G. & Rymo, L. (1993). Cell phenotype dependent control of Epstein–Barr virus latent membrane protein 1 (LMP1) gene regulatory sequences. *Virology*, **195**, 71–80.

Gaidano, G., Ballerini, P., Gong, J. Z. *et al.* (1991). p53 mutations in human lymphoid malignancies: association with Burkitt's lymphoma and chronic lymphocytic leukemia. *Proceedings of the National Academy of Sciences, USA*, **88**, 5413–17.

Gavioli, R., De Campos-Lima, P. O., Kurilla, M. G., Kieff, E., Klein, G. & Masucci, M. G. (1992). Recognition of the Epstein–Barr virus-encoded nuclear antigens EBNA-4 and EBNA-6 by HLA-A11-restricted cytotoxic T-lymphocytes: implications for down-regulation of HLA-A11 in Burkitt's lymphoma. *Proceedings of the National Academy of Sciences USA*, **89**, 5862–6.

Gratama, J. W., Oosterveer, M. A. P., Zwaan, F. E., Lepoutre, J., Klein, G. & Ernberg, I. (1988). Eradication of Epstein–Barr virus by allogeneic bone marrow transplantation: implications for sites of viral latency. *Proceedings of the National Academy of Sciences, USA*, **85**, 8693–6.

Jiang, W.-Q., Szekely, L., Wendel-Hansen, V., Ringertz, N., Klein, G. & Rosén, A. (1991). Co-localization of the retinoblastoma protein and the Epstein-Barr virus-encoded nuclear antigen EBNA-5. *Experimental Cell Research*, **197**, 314–18.

Khanna, R., Borrows, S., Kurilla, M. *et al.* (1992). Localization of Epstein–Barr virus cytotoxic T-cell epitopes using recombinant vaccinia: Implication for vaccine development. *Journal of Experimental Medicine*, **176**, 169–76.

Klein, G. (1993). The road to the Ig/myc translocation. In *The Causes and Consequences of Chromosomal Aberrations*, ed. I. R. Kirsch pp. 481–5032, Oxford: CRC Press.

Klein, G. & Klein, E. (1985). Evolution of tumours and the impact of molecular oncology. *Nature*, **315**, 67–75.

Lewin, N., Åman, P., Masucci, M. G. *et al.* (1987). Characterization of EBV-carrying B-cell populations in healthy seropositive individuals with regard to density, release of transforming virus and spontaneous outgrowth. *International Journal of Cancer*, **39**, 472–6.

Magrath, I. (1990). The pathogenesis of Burkitt's lymphoma. *Advances in Cancer Research,* **55**, 133–270.

Magrath, I. (1993). Lecture presented at Keystone Symp on B- and T-cell lymphomas.

Mannick, J. B., Cohen, J. I., Birkenbach, M., Marchini, A. & Kieff, E. (1991). The Epstein–Barr virus nuclear protein encoded by the leader of the EBNA RNAs is important in B-lymphocyte transformation. *Journal of Virology,* **65**, 6826–37.

Masucci, M. G., Gavioli, R., de Campos-Lima, P. O., Zhang, Q.-J., Trivedi, P. & Dolcetti, R. (1993). Transformation associated Epstein–Barr virus antigens as targets for immune attack. In *Annals of the New York Academy of Sciences,* **690**, 86–100.

Murray, R., Kurilla, M., Brooks, J. *et al.* (1992). Identification of target antigens for the human cytotoxic T-cell responses to Epstein–Barr virus (EBV): Implications for the immune control of EBV positive malignancies. *Journal of Experimental Medicine,* **176**, 157–68.

Pallesen, G., Hamilton-Dutoit, S. J. & Zhou, Z. (1993). The association of Epstein–Barr virus (EBV) with T cell lymphoproliferations and Hodgkin's disease: two new developments in the EBV field. *Advances in Cancer Research,* in press.

Ramqvist, T., Magnusson, K. P., Wang, Y., Szekely, L., Klein, G. & Wiman, K. G. (1993). Wild-type p53 induces apoptosis in a Burkitt lymphoma (BL) line that carries mutant p53. *Oncogene,* **8**, 1495–500.

Rowe, M., Lear., A. L., Croom-Carter, D., Davies, A. H. & Rickinson, A. B. (1992). Three pathways of Epstein–Barr virus gene activation from EBNA-1 positive latency in B-lymphocytes. *Journal of Virology,* **66**, 122–31.

Rowe, M., Rowe, D. T., Gregory, C. D. *et al.* (1987). Differences in B-cell growth phenotype reflect novel patterns of Epstein-Barr virus latent gene expression in Burkitt's lymphoma cells. *EMBO Journal,* **6**, 2743–51.

Sample, J., Brooks, L., Sample, C. *et al.* (1991). Restricted Epstein–Barr virus protein expression in Burkitt lymphoma is reflected in a novel, EBNA1 mRNA and transcriptional initiation site. *Proceedings of the National Academy of Sciences, USA,* **88**, 6343–7.

Schaefer, B. C., Woisetschläger, M., Strominger, J. L. & Speck, S. H. (1991). Exclusive expression of Epstein–Barr virus nuclear antigen 1 in Burkitt's lymphoma arises from a third promoter, distinct from the promoters utilized in latently infected lymphocytes. *Proceedings of the National Academy of Sciences, USA,* **88**, 6550–4.

Silva, S., Sugiiyama, H., Babonits, M., Wiener, F. & Klein, G. (1991) Differential susceptibility of BALB/c and DBA/2 cells to plasmacytoma induction in reciprocal chimeras. *International Journal of Cancer,* **49**, 224–8.

Suematsu, S., Matsusaka, T., Matsuda, T. *et al.* (1992). Generation of plasmacytomas with the chromosomal translocation t(12;15) in interleukin 6 transgenic mice. *Proceedings of the National Academy of Sciences, USA,* **89**, 232–5.

Sugiyama, H., Weber, G., Silva, S., Babonits, M., Wiener, F. & Klein, G.: (1989). The accelerating role of Abelson murine leukemia virus in murine plasmacytoma development: *in vitro* infection of spleen cells generates donor-type tumours after transfer to pristane-treated Balb/c mice. *International Journal of Cancer,* **44**, 348–52.

Szekely, L., Selivanova, G., Magnusson, K. P., Klein, G. & Wiman, K. G.: (1993). EBNA-5, an EBV encoded nuclear antigen, binds to the RB and p53 protein. *Proceedings of the National Academy of Sciences, USA,* **90**, 5455–9.

Trivedi, P., Masucci, M. G., Winberg, G. & Klein, G.: (1991). The Epstein–Barr-virus-encoded membrane protein LMP but not the nuclear antigen EBNA1 induces rejection of transfected murine mammary carcinoma cells. *International Journal of Cancer,* **48**, 794–800.

Trivedi, P., Hu, L.-F., Chen, F. *et al.* (1993). The EBV encoded membrane protein LMP1 from a nasopharyngeal carcinoma is non-immunogenic in a murine model system, in contrast to a B-cell derived homologue. *European Journal of Cancer,* in press.

Wang, J.-Y., Ramqvist, T., Szekely, L., Axelson, H., Klein, G. & Wiman, K. G. (1993). Reconstitution of wild-type p53 expression triggers apoptosis in a p53 negative v-myc retrovirus-induced T cell lymphoma line. *Cell Growth and Differentiation,* **4**, 467–73.

Wang, Y., Sugiyama, H., Axelson, H. *et al.* (1992). Functional homology between N-myc and c-myc in murine plasmacytomagenesis: plasmacytoma development in N-myc transgenic mice. *Oncogene,* **7**, 1241–7.

Wiener, F. & Potter, M. (1993). Myc-associated chromosomal translocations and rearrangements in plasmacytomagenesis in mice. In *The Causes and Consequences of Chromosomal Aberrations.* ed. I. R. Kirsch pp. 91–124, Oxford: CRC Press.

Wiman, K. G., Magnusson, K. P., Ramqvist, T. & Klein, G. (1991). Mutant p53 detected in a majority of Burkitt lymphoma cell lines by monoclonal antibody PAb240. *Oncogene,* **6**, 1633–9.

Yokota, J. & Sugimura, T. (1993). Multiple steps in carcinogenesis involving alterations of multiple tumour suppressor genes. *FASEB Journal* (Thematic issue on Tumour Suppressor Genes), in press.

Zambetti, G. P. & Levine, A. J. (1993). A comparison of the biological activities of wild-type and mutant p53. *FASEB Journal,* (Thematic Issue on Tumour Suppressor Genes), **7**, 855–63.

Zhang, Q.-J., Gavioli, R., Klein, G. & Masucci, M. G. (1993). An HLA-A11-specific motif in nonamer peptides derived from viral and cellular proteins. *Proceedings of the National Academy of Sciences, USA,* **90**, 2217–21.

TUMOUR SUPPRESSOR GENES AND p53

D. P. LANE

Cancer Research Campaign Laboratories, Department of Biochemistry, University of Dundee, Dundee DD1 4HN, Scotland, UK.

INTRODUCTION AND DISCUSSION

The revolution in methods brought about by the development of modern molecular biology is allowing enormous progress to be made in the analysis of human cancer. It is clear that many separate genetic changes are required before a normal cell will behave as a malignant cell. Principal among these changes are the mutational activation of the growth promoting functions of the proto oncogenes and the mutational inactivation of the function of the tumour suppressor genes whose existence was first deduced by the cell fusion studies of Henry Harris (for reviews, see Klein, 1987; Bishop, 1991). Tumours in different tissues are typically associated with certain specific changes to defined oncogenes and suppressor genes. The number of identified tumour suppressor genes is growing quite fast and it is clear that like the oncogenes they will act at many different points in the cancer process. Studies of inherited predispositions to malignancy suggest that the action of many suppressor genes is of special importance in a particular tissue and at a particular developmental stage. The classic example of this is the retinoblastoma (Rb) gene. Inheritance of a defective allele of this gene confers a predominantly tissue specific and developmental stage specific predisposition to malignant disease. However, when the Rb gene was isolated, it became clear that loss of this gene function was a common occurrence in many somatic cancers that people who carried a defective allele were not apparently particularly susceptible to. While at first sight this seems paradoxical, it can be explained by the multistep nature of genetic change in somatic cancers and the discovery of a family of proteins whose biochemical activities overlap with that of Rb. Since a suppressor gene can only act if it is expressed, the tissue specific regulation of these genes is of particular importance in discovering their role in the incidence of neoplasia of a particular cell type. One gene, however, the p53 gene, stands out because of its almost universal involvement in the development of the common tumours in man (Vogelstein, 1990; Lane & Benchimol, 1990; Vogelstein & Kinzler, 1992; Levine, Momand & Finlay, 1991). The existence of a common step in the development of cancer compels us to determine the precise function(s) of p53 because that knowledge should deepen our understanding of the whole neoplastic process.

The properties of the p53 protein

The p53 protein was first discovered through its interaction with large T antigen the oncogene product of the SV40 small DNA tumour virus (Lane & Crawford, 1979; Linzer & Levine, 1979). Subsequent work has shown that p53 also forms physical complexes with the oncogene products of two other groups of DNA tumour virus, the E1B product of Adenovirus 5 (Sarnow *et al.*, 1982) and the E6 product of Human Papilloma virus 16 (Werness, Levine & Howley, 1990). The p53 gene is on chromosome 17p and consists of 11 exons (Soussi, Caron de Fromentel & May, 1990). It has been conserved in evolution among the vertebrates but has not so far been found in the invertebrates. The gene, in man, encodes a 393 amino acid phospho-protein that is commonly located in the cell nucleus. The p53 protein binds specifically to double stranded DNA and can act as a regulator of transcription both to promote the transcription of some genes, for example, the GADD 45 gene (Kastan *et al.*, 1992), but also to reduce the transcription of several other genes, for example, the myc gene. High levels of the protein can inhibit both viral and cellular DNA synthesis *in vivo*. The p53 protein is normally present at very low levels in the tissues of the adult organism due to its very short half-life (for review see Vogelstein, 1990; Lane & Benchimol, 1990; Vogelstein & Kinzler, 1992; Levine, Momand & Finlay, 1991). The protein accumulates to high levels in cells exposed to DNA damaging agents (Maltzman & Czyzyk, 1984; Lu *et al.*, 1992; Hall *et al.*, 1993; Kastan *et al.*, 1991; Lu & Lane, 1993) which suggests, as will be discussed later, that p53 may act as a checkpoint control to prevent DNA synthesis in cells with damaged template. The breakdown of this control pathway, brought about by mutations in p53 or other components of the pathway, may be a key step in the progression of cancer, as it will permit the proliferation of cells that contain aberrant DNA and give rise to a pool of variants from which malignant or drug resistant clones of cancer cells will arise (Livingstone *et al.*, 1992; Yin *et al.*, 1992; Lane, 1992).

p53 tumour suppressor gene and oncogene?

Mice in which both alleles of the p53 gene have been inactivated develop a wide variety of tumours at a very young age (Donehower *et al.*, 1992). This establishes that loss of p53 function is sufficient to predispose to the development of neoplasia. In humans the germ line inheritance of one mutant allele of the p53 gene predisposes the affected individual to a wide range of neoplasias sometimes manifest as the Li–Fraumeni syndrome (Malkin *et al.*, 1990; Srivastava *et al.*, 1990; Santibanez-Koref *et al.*, 1991). The mutant p53 proteins found in human tumour cells have lost their growth suppressive function, however several lines of evidence suggest that they might also have gained a growth promoting function. Transfection of mutant p53 with an activated ras gene will transform primary rodent cells. This function depends on the conformation and continued expression of the p53 protein since temperature sensitive point mutants in p53 have been identified that will act as transforming genes at 30 °C in this assay but will suppress transformation and growth of the same cells at 32 °C (Michalovitz, Halvey & Oren, 1990). Part of the mechanism of this dominant transforming activity resides in the capacity of mutant p53 proteins to bind to, and inactivate the function, of the wild-type protein. This function has now been localized to the C terminal oligomerization domain of p53 (Shaulian et al., 1992). This classic dominant negative mutant behaviour is not sufficient however to explain all the characteristics of the p53 system. The mutant proteins may also act on other targets to promote growth since, at least in some systems, introduction of a mutant p53 gene into cells that have lost both wild type alleles enhances their neoplastic properties (Wolf et al., 1984). In human tumours it is very apparent that there is a strong selection for loss of the normal allele and retention of the mutant allele. There also seems to be a selection for high level expression of the mutant protein in the tumour (Barnes et al., 1993).

Expression of p53 protein in human tumours

The production of recombinant p53 protein in bacteria has allowed the isolation of polyclonal and monoclonal antibodies that efficiently detect the protein in routine histopathology samples (Midgley et al., 1992; Vojtesek et al., 1992). This has confirmed earlier work that identified high levels of p53 protein in 20% to 40% of breast cancers (Crawford, Pim & Lamb, 1984; Cattoretti et al., 1988) but has also extended the observation to many other tumour types including, for example, lung, colon, stomach, brain, bladder, skin, cervical and ovarian cancer (Bartek et al., 1990a,b, 1991; Bennett et al., 1991; Eccles et al., 1992; Gusterson et al., 1991; Hanski et al., 1992; Iggo et al., 1990; Pignatelli et al., 1991; Rodrigues et al., 1990; Wrede et al., 1991; Wright et al., 1991). High levels of p53 are frequently found in tumours of all major cell lineages including sarcomas, carcinomas, glioblastomas and haemopoetic malignancies. The molecular basis for the accumulation of high levels of p53 specifically in tumours seems to be an increase in the post-translational stability of the protein, as no consistent increase in mRNA levels is found. Direct sequencing of the p53 gene in immunohistochemically strongly positive tumours has shown that the protein is usually the product of a p53 gene carrying a point mutation within the coding region. The tumours are typically homozygous at the p53 locus having lost the wild type allele. High levels of the p53 protein are clearly associated with a poor prognosis in breast (Isola et al., 1992) lung and gastric cancers (Martin et al., 1992). In the case of breast cancer, p53 levels are fast emerging as a key indicator of disease free interval and may soon be used to identify a subset of women requiring differential post operative therapy (Barnes *et al.*, 1993). The patterns of expression of p53 seen by immunohistochemistry are of great interest. The most striking, which seems to be closely associated with a point mutation of the protein is an intense staining of all cells within the tumour. This pattern is seen in about 20% of primary breast cancers and in 50–70% of colon, gastric, oesophageal and lung cancers (Midgley *et al.*, 1992; Bennett *et al.*, 1991). In these tumours the events that lead to the morphological appearance of malignancy must be closely coincident with the events that lead to the high levels of p53. In other cases more focal expression of p53 immunoreactivity is apparent suggesting that the appearance of immunoreactivity may mark a stage in tumour progression. Molecular studies and the high levels of p53 apparent in the metastatic sites of heterogeneously positive primary tumours support this view. In other cases just a few scattered positive cells are seen and finally some tumours show no enhancement of immunoreactivity over background (Midgley *et al.*, 1992).

Regulation of p53 protein levels

The association between high levels of p53 protein and poor prognosis and the variable expression patterns of immunoreactivity seen in individual tumours has prompted a detailed investigation of the molecular basis of the accumulation of p53 protein in human cancer. A series of initial studies from my group and others established that in a series of cell lines and primary tumours high levels of p53 where associated with point missense mutations in the p53 gene (Bartek *et al.*, 1990a,b; Iggo *et al.*, 1990; Rodrigues *et al.*, 1990). It is high levels of mutant protein that are being detected. Many of the point mutations in p53 change the conformation of the protein and this conformational change can be detected with monoclonal antibodies (Gannon *et al.*, 1990; Stephen & Lane, 1992). Several studies have shown that this conformational change renders the protein more stable metabolically and may, in part, account for its accumulation. However, this effect whilst important is not sufficient to explain all cases. Mutation in the p53 protein is not the only way that it can become stable in tumour cells. We recently found a cancer prone family (Barnes *et al.*, 1992) in which the proband and her affected mother showed a high level of p53 constitutively expressed in normal epithelial, endothelial and stromal cells. The p53 open reading frame was normal in sequence in these individuals suggesting that mutation in another gene was causing both the high level of p53 and the associated increased risk of neoplasia. By contrast in families with germline p53 mutations high levels of p53 are not found in normal tissues though they are often readily apparent in the tumours of these individuals. This suggests that other events may be required to stabilize even mutant p53 proteins. Certainly wild type p53 can be stabilized dramatically by treating cells with DNA damaging agents, and by the action of SV40 large T and Adenovirus

E1B protein. In some tumours containing high levels of p53 the gene appears to be wild type. In a mouse model of prostate cancer (Lu *et al.*, 1992) we found high levels of wild-type p53 in tumour cells derived from precursors expressing an activated ras and an activated myc gene. The pattern of expression was sporadic and was clearly associated with cells within the population that had nuclear abnormalities consistent with cell cycle failures. It is thus possible that the stabilization of both mutant and wild-type p53 in tumours is a direct result of their genetic instability which is itself caused by a breakdown in the functioning of the p53 pathway (Fig. 1). Failure in the p53 pathway could be due to mutations in p53 itself or in the gene products responsible for inducing the increase in p53 levels following DNA damage or in the gene products required for the cellular response to high levels of p53. To understand this in detail, we will need to identify these other genes on the p53 pathway. This process is already under way; we have found human cells that do not produce stable p53 in the presence of UV radiation or SV40 T antigen (Lu & Lane, 1993). Their p53 gene is however wild type in sequence and will respond to UV when the defective cells are fused to responsive mouse cells. Similarly Kastan's laboratory have shown that the p53 response to ionizing radiation is defective in some cases of Ataxia telangiectasia (Kastan *et al.*, 1992); however we have not been able to confirm this (Lu & Lane 1993). Intense efforts are now under way to determine how p53 is degraded and what blocks this process following DNA damage. Identification of the gene products involved in this pathway will allow an investigation of their function in normal and tumour cells and may identify new cancer susceptibility syndromes. A very promising avenue for investigation in this area has again come from studies of viral systems. The human papilloma virus oncogene product produced by the E6 open reading frame is able to bind to p53 and promote its rapid degradation apparently by targeting it for ubiquitination. Recently an essential cellular factor for this process the E6 AP (E6 associated protein) was purified and the gene encoding it cloned (Huibregtse, Scheffner & Howley, 1992). The E6 protein forms a complex with the E6 AP and this complex then binds to p53 and targets its destruction. This effect seems to be very important since it is only a property of the E6 proteins of high risk HPV serotypes, i.e. those associated with cervical cancer and not those that are associated with benign warts. This viral model of the degradation of p53 may provide a route in to understanding the normal regulation of p53 stability and it will be of great interest to search for a cellular protein than can interact with the E6 AP and p53.

Regulation of p53 protein function

If p53 can act as a potent suppressor of cell growth in cells exposed to DNA damaging agents, or transfected with extra copies of the p53 gene, how is it

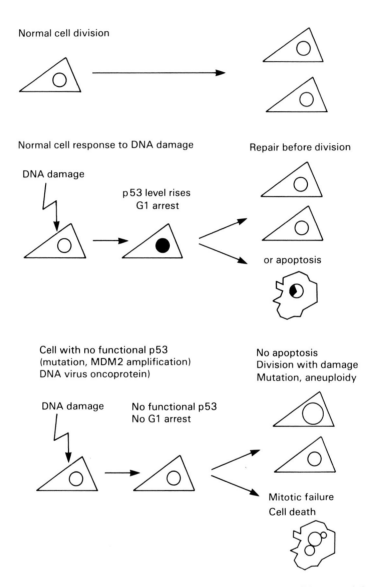

Fig. 1. A model for the function of p53. In normal cells where p53 is wild type and the p53 pathway is intact DNA damage, and perhaps other stresses, lead to the stabilization of p53 which accumulates to high levels. This induces a cell cycle arrest. In some cells (stem cells?), this arrest will lead to apoptosis, in others the growth arrest is transitory and once the inducing signal is gone p53 levels return to normal and the cell can go back into cycle. In tumour cells, the p53 pathway is often inactivated by mutation in p53 itself or in other genes that act upstream or downstream of p53 and affect its function. When these cells are exposed to DNA damaging agents, DNA synthesis continues leading to genetic errors and instability. This may result in cell death but also gives rise to aneuploidy. From this pool of damaged cells, malignant clones will arise at elevated frequencies.

possible for normal growth and development to take place? Three models could explain this, first that normal levels of p53 in undamaged cells are too low to restrict growth, second that p53 function is normally ablated by the activity of other gene products and third that p53 is normally produced in a latent form and only becomes activated in the presence of signals induced by DNA damage. All three models are probably correct leading to a very tight control on the growth suppressive function of p53. The level of p53 is very low in normal cells and a gene product MDM2 that binds and inactivates p53 has recently been discovered (Momand, et al., 1992; Wu et al., 1993). Work in my own laboratory has shown that p53 is normally present in cells in a form that cannot bind to ds DNA with sequence specificity. To do so the latent protein must be activated by conformational modifications at the C terminus of the protein. These modifications can be brought about by phosphorylation with casein kinase 2, by the action of members of the hsp70 heat shock protein family and finally by the binding of specific antibodies (Hupp et al., 1992). This allosteric regulation of p53 function is exciting because it suggests that small molecules may be discovered or designed that will modulate p53 function. Support for this idea has come from very recent experiments in which we were able to reactivate the DNA binding activity of some mutant p53 proteins by treating them with specific antibodies to the C terminus or with the molecular chaperonin dnaK in the presence of ATP. However these same mutants could not be activated by phosphorylation which may explain why they are normally inactive in the tumour cell (Hupp et al., 1993).

CONCLUSIONS

We are now at an immensely exciting stage in cancer research. We have identified a common step in the molecular progression of the disease and novel findings about the structure and function of p53 are being made at breakneck speed as the full resources of modern molecular and cellular biology are brought to bear on this one protein. One can only be optimistic for the future. I am convinced that novel therapeutics targeted to the p53 pathway will be discovered soon. They will exploit the fundamental differ- ence in genetic stability and cell cycle control found in cells that lack the p53 pathway. They will exploit the tumour specific accumulation and confor- mational alteration in p53 protein. They will therefore strike selectively at the fundamental differences between the normal cell and the tumour cell. We can hope that they will be more effective and kinder to the patient than any current agent. If that dream comes true then all our work will have been worthwhile.

SUMMARY

Mutation of the p53 tumour suppressor gene and the accumulation of the p53 protein are amongst the most common specific changes found in human cancer. An understanding of the precise biochemical and biological effects of these changes will provide new insights into the fundamental nature of the neoplastic process. Recent results show that the p53 protein is a critical part of a damaged induced cell cycle checkpoint control that maintains genetic stability. Biochemical analysis of the activity of wild-type and mutant p53 proteins have shown that the wild-type protein can bind specifically to DNA and activate transcription of genes that contain p53 binding sites. Most mutations in p53 found in human tumours inactivate this DNA binding function of p53. Recently, however, we have found that the DNA binding function of p53 is regulated by a C terminal domain that can both activate and inactivate p53 function. The DNA binding activity of some mutant proteins can be 'rescued' by molecules that bind to this region of p53. The p53 protein thus provides an attractive target for the development of new diagnostic and therapeutic agents that will hopefully reduce the impact of cancer on society.

ACKNOWLEDGMENTS

Our work is currently supported by the Cancer Research Campaign.

REFERENCES

Barnes, D. M., Hanby, A. M., Gillett, C. E. *et al.* (1992). Abnormal expression of wild type p53 protein in normal cells of a cancer family patient. *Lancet,* **340**, 259–63.

Barnes, D. M., Dublin E. A., Fisher, C. J., Levinson, D. A. & Millis, R. R. (1993). Immunohistochemical detection of p53 in mammary carcinoma: an important new independent indicator of prognosis. *Human Pathology,* **24**, 469–76.

Bartek, J., Bartkova, J., Vojtesek, B. *et al.* (1991). Aberrant expression of the p53 oncoprotein is a common feature of a wide spectrum of human malignancies. *Oncogene,* **6**, 1699–703.

Bartek, J., Bartkova, J., Vojtesek, B. *et al.* (1990a). Patterns of expression of the p53 tumour suppressor in human breast tissues and tumours *in situ* and *in vitro. International Journal of Cancer,* **46**, 839–44.

Bartek, J., Iggo, R., Gannon, J. & Lane, D. P. (1990b). Genetic and immunochemical analysis of mutant p53 in human breast cancer cell lines. *Oncogene,* **5**, 893–9.

Bennett, W. P., Hollstein, M. C., He, A. *et al.* (1991). Archival analysis of p53 genetic and protein alterations in Chinese esophageal cancer. *Oncogene,* **6**, 1779–84.

Bishop, J. M. (1991). Molecular themes in oncogenesis. *Cell,* **64**, 235–48.

Cattoretti, G., Rilke, F., Andreola, S., D'Amato, L. & Delia, D. (1988). p53 expression in breast cancer. *International Journal of Cancer,* **41**, 178–83.

Crawford, L. V., Pim, D. C. & Lamb, P. (1984). The cellular protein p53 in human tumours. *Molecular and Biological Medicine,* **2**, 261–72.

Donehower, L. A., Harvey, M., Slagle, B. L. *et al.* (1992). Mice deficient for p53 are developmentally normal but susceptible to spontaneous tumours. *Nature,* **356**, 215–21.

Eccles, D. M., Brett, L., Lessells, A. *et al.* (1992). Overexpression of the p53 protein and allele loss at 17p13 in ovarian carcinoma. *British Journal of Cancer,* **65**, 40–4.

Gannon, J. V., Greaves, R., Iggo, R. & Lane, D. P. (1990). Activating mutations in p53 produce a common conformational effect. A monoclonal antibody specific for the mutant form. *EMBO Journal,* **9**, 1595–602.

Gusterson, B. A., Anbazhagan, R., Warren, W. *et al.* (1991). Expression of p53 in premalignant and malignant squamous epithelium. *Oncogene,* **6**, 1785–9.

Hall, P. A., McKee, P. H., Menage, H. D., Dover, R. & Lane, D. P. (1993). High levels of p53 protein in UV irradiated human skin. *Oncogene,* **8**, 203–7.

Hanski, C., Bornhoeft, G., Shimoda, T. *et al.* (1992). Expression of p53 protein in invasive colorectal carcinomas of different histological type. In press.

Huibregtse, J. M., Scheffner, M. & Howley, P. M. (1992). Cloning and expression of the cDNA for E6-AP: a protein that mediates the interaction of the human papillomavirus E6 oncoprotein with p53. *Molecular and Cellular Biology,* **13**, 775–84.

Hupp, T. R., Meek, D. W., Midgley, C. A. & Lane, D. P. (1992). Regulation of the specific DNA binding function of p53. *Cell,* **71**, 875–86.

Hupp, T. R., Meek, D. W., Midgley, C. A. & Lane, D. P. (1993). Activation of the DNA binding function of mutant forms of p53. *Nucleic Acids Research,* **21**, 3167–74.

Iggo, R., Gatter, K., Bartek, J., Lane, D. & Harris, A. L. (1990). Increased expression of mutant forms of p53 oncogene in primary lung cancer. *Lancet,* **335**, 675–9.

Isola, J., Visakorpi, T., Holli, K. & Kallioniemi, O.-P. (1992). Association of overexpression of tumour suppressor protein p53 with rapid cell proliferation and poor prognosis in node-negative breast cancer patients. *Journal of the National Cancer Institute,* **84**, 1109–14.

Kastan, M. B., Onyekwere, O., Sidransky, D., Vogelstein, B. & Craig, R. W. (1991). Participation of p53 protein in the cellular response to DNA damage. *Cancer Research,* **51**, 6304–11.

Kastan, M. B., Zhan, Q., El-Deiry, W. K. *et al.* (1992). A mammalian cell cycle checkpoint pathway utilising p53 and gadd45 is defective in ataxia–telangiectasia. *Cell,* **71**, 587–97.

Klein, G. (1987). The approaching era of the tumour suppressor genes. *Science,* **238**, 1539–44.

Lane, D. & Benchimol, S. (1990). p53: oncogene or anti-oncogene. *Genes and Development,* **4**, 1–8.

Lane, D. P. (1992). p53, guardian of the genome. *Nature,* **358**, 15–16.

Lane, D. P. & Crawford, L. V. (1979). T-antigen is bound to host protein in SV40-transformed cells. *Nature,* **278**, 261–3.

Levine, A. J., Momand, J. & Finlay, C. A. (1991). The p53 tumour suppressor gene. *Nature,* **351**, 453–6.

Linzer, D. I. H. & Levine, A. J. (1979). Characterization of a 54K Dalton cellular SV40 tumour antigen present in SV40 transformed cells and uninfected embryonal carcinoma cells. *Cell,* **17**, 43–52.

Livingstone, L. R., White, A., Sprouse, J., Livanos, E., Jacks, T. & Tlsty, T. D. (1992). Altered cell cycle arrest and gene amplification potential accompany loss of wild type p53. *Cell,* **70**, 923–35.

Lu, X., Park, S. H., Thompson, T. C. & Lane, D. P. (1992). ras-induced hyperplasia occurs with mutation of p53, but an activated ras and myc together can induce carcinoma without p53 mutation. *Cell,* **70**, 153–61.

Lu, X. & Lane, D. P. (1993). Differential induction of transcriptionally active p53 following UV or ionising radiation: defects in chromosome instability syndromes. *Cell*, **75**, 765–78.

Malkin, D., Li, F. P., Strong, L. C. *et al.* (1990) Germ line p53 mutations in a familial syndrome of breast cancer, sarcomas, and other neoplasms. *Science*, **250**, 1233–8.

Maltzman, W. & Czyzyk, L. (1984). UV irradiation stimulates levels of p53 cellular tumour antigen in nontransformed mouse cells. *Molecular and Cellular Biology*, **4**, 1689–94.

Martin, H. M., Filipe, M. I., Morris, R. W., Lane, D. P. & Silvestre, F. (1992). p53 expression and prognosis in gastric carcinoma. *International Journal of Cancer*, **50**, 859–62.

Michalovitz, D., Halvey, O. & Oren, M. (1990). Conditional inhibition of transformation and of cell proliferation by a temperature-sensitive mutant of p53. *Cell*, **62**, 671–80.

Midgley, C. A., Fisher, C. J., Bartek, J., Vojtesek, B., Lane, D. P. & Barnes, D. M. (1992). Analysis of p53 expression in human tumours: an antibody raised against human p53 expressed in *E. coli. Journal of Cell Science*, **101**, 183–9.

Momand, J., Zambetti, G. P., Olson, D. C., George, D. & Levine, A. J. (1992). The mdm-2 oncogene product forms a complex with the p53 protein and inhibits p53-mediated transactivation. *Cell*, **69**, 1237–45.

Pignatelli, M., Stamp, G. W. H., Kafiri, G., Lane, D. P. & Bodmer, W. F. (1991). Overexpression of p53 nuclear oncoprotein in colorectal adenomas. *International Journal of Cancer*, **50**, 683–8.

Rodrigues, N. R., Rowan, A., Smith, M. E. F. *et al.* (1990). p53 mutations in colorectal cancer. *Proceedings of the National Academy of Sciences, USA*, **87**, 7555–9.

Santibanez-Koref, M. F., Birch, J. M., Hartley, A. L. *et al.* (1991). p53 germline mutations in Li–Fraumeni syndrome. *Lancet*, **338**, 1490–1.

Sarnow, P., Ho, Y. S., Williams, J. & Levine, A. J. (1982). Adenovirus E1B-58 kd tumour antigen and SV40 large tumor antigen are physically associated with the same 54 kd cellular protein in transformed cells. *Cell*, **28**, 387–94.

Shaulian, E., Zauberman, A., Ginsberg, D. & Oren, M. (1992). Identification of a minimal transforming domain of p53: negative dominance through abrogation of sequence-specific DNA binding. *Molecular and Cellular Biology*, **12**, 5581–92.

Soussi, T., Caron de Fromentel, C. & May, P. (1990). Structural aspects of the p53 protein in relation to gene evolution. *Oncogene*, **5**, 945–52.

Srivastava, S., Zou, Z., Pirollo, K., Blattner, W. & Chang, E. H. (1990). Germ-line transmission of a mutated p53 gene in a cancer prone family with Li–Fraumeni syndrome. *Nature*, **348**, 747–9.

Stephen, C. W. & Lane, D. P. (1992). Mutant conformation of p53: precise epitope mapping using a filamentous phage epitope library. *Journal of Molecular Biology*, **225**, 577–83.

Vogelstein, B. (1990). A deadly inheritance. *Nature*, **348**, 681–2.

Vogelstein, B. and Kinzler, K. W. (1992). p53 Function and dysfunction. *Cell*, **70**, 523–6.

Vojtesek, B., Bartek, J., Midgley, C. A. & Lane, D. P. (1992). An immunochemical analysis of human p53: New monoclonal antibodies and epitope mapping using recombinant p53. *Journal of Immunology Methods*, **151**, 237–44.

Werness, B. A., Levine, A. J. & Howley, P. M. (1990). Association of human papillomavirus types 16 and 18 E6 proteins with p53. *Science*, **248**, 76–9.

Wolf, D., Harris, N., Goldfinger, N. & Rotter, V. (1984). Reconstitution of p53 expression in nonproducer Ab-MuLV-transformed cell line by transfectional of a functional p53 gene. *Cell*, **38**, 119–26.

Wrede, D., Tidy, J. A., Crook, T., Lane, D. & Vousden, K. H. (1991). Expression of abnormal p53 and RB protein detectable only in HPV negative cervical carcinoma cell lines. *Molecular Carcinogenesis,* **4**, 171–5.

Wright, C., Mellon, K., Johnston, P. *et al*. (1991). Expression of mutant p53, c-erbB-2 and the epidermal growth factor receptor in transitional cell carcinoma of the human urinary bladder. *British Journal of Cancer,* **63**, 967–70.

Wu, X., Bayle, J. H., Olson, D. & Levine, A. J. (1993). The p53-mdm-2 autoregulatory feedback loop. *Genes and Development,* **7**, 1126–32.

Yin, Y., Tainsky, M. A., Bischoff, F. Z., Strong, L. C. & Wahl, G. M. (1992). Wild-type p53 restores cell cycle control and inhibits gene amplification in cells with mutant p53 alleles. *Cell,* **70**, 937–48.

CELL TRANSFORMATION BY HUMAN PAPILLOMAVIRUSES

K. H. VOUSDEN

Ludwig Institute for Cancer Research, St Mary's Hospital Medical School, Norfolk Place, London W2 1PG, UK.

INTRODUCTION

The human papillomaviruses (HPVs) comprise a large group of small DNA viruses which infect many types of epithelial tissue (von Knebel Doeberitz, 1992). HPVs are classified on the basis of difference in genomic sequence, and over 60 HPV types have been described so far, with evidence that more remain to be identified (De Villiers, 1989). HPV infection is common and usually results in the production of benign lesions which eventually spontaneously regress. Despite the frequency with which HPV infection occurs *in vivo*, these viruses are extremely difficult to culture in experimental systems and little is known about their natural life cycle. Research into these viruses has therefore relied heavily on recombinant DNA technology to express viral proteins in suitable cells and has concentrated almost exclusively on a small group of genital HPV types which give rise to lesions with the potential for malignant progression. These so called high risk viral types, such as HPV16 and 18, are found associated with almost all cervical cancers, making them the most common human tumour virus, worldwide (zur Hausen, 1986). The identification of a virus which plays a causative role in the development of a human cancer raises exciting possibilities for prevention or treatment of the disease and an understanding of the mechanisms by which viral infection contributes to malignant development may reveal targets for the design of therapeutic agents. There is now good evidence that these high risk HPV types encode oncoproteins whose functions are consistent with a contribution to oncogenicity.

HPVs AS HUMAN TUMOUR VIRUSES

The epidemiological evidence that certain HPV types are causally linked to the development of human cancers is now extremely convincing, despite the confusion initially generated through the erroneous over-estimation of HPV infection amongst normal women (Schiffman, 1992). As might be predicted from a sexually transmissible agent, genital HPV infection rises rapidly during the late teens and early twenties and then slowly declines in subsequent decades. In the age group in which cervical cancers most frequently develop (over 40 years), the incidence of HPV in women without cervical abnormality drops to less than 5%. The observation that at least 90% of the cancers retain DNA sequences of known HPV types, with evidence of expression of at least some of the viral sequences in the limited numbers analysed in this way, is consistent with a role for the virus in the development, and possibly also the maintenance, of these malignancies. Large-scale prospective studies are now underway to examine directly the risk of developing cancer following HPV infection and initial evidence indicates that infection with these viruses can cause the cervical premalignancies known as cervical intraepithelial neoplasias (Koutsky *et al.*, 1992). One plausible interpretation of the evidence available so far is that infection with genital HPV is common as young women first become sexually active, but that, in most cases, the viral infection is cleared. In rare cases the virus persists and these women are subsequently at risk of developing cervical malignancies. The concept that viral infection results in a premalignant lesion is further supported by studies showing that the presence of high risk viral DNA in a cervical scrape classified as only mildly abnormal by cytology predicts the presence of high grade underlying cervical disease (Cuzick *et al.*, 1992). One of the first clinical benefits of the identification of HPVs as human tumour viruses may be an increase in sensitivity and specificity of cervical screening achieved by the addition of HPV detection to the normal cytological tests.

Although HPV infection is associated with the development of most cervical cancers, two important points should be kept in mind. First, a small minority of these cancers show no evidence of infection with a known HPV type. It is possible that these tumours harbour an unidentified viral type which shows insufficient similarity to the known types to allow detection, or that they originally arose as HPV positive tumours but subsequently lost the HPV sequences. Alternatively, some cervical tumours may arise, albeit rarely, through HPV independent mechanisms. The second point is that HPV infection, by itself, does not appear to be sufficient for cancer development (zur Hausen, 1989). Consistent with the proposed multistage nature of carcinogenesis, infection with the high risk HPV types represents only one event during malignant progression. Additional changes, both genetic and epigenetic, would be predicted to play a role in the development of virus positive and negative cancers, although these are very poorly defined. There is evidence implicating environmental carcinogens in the development of these malignancies, through mechanisms such as the activation of cellular oncogenes, loss of tumour suppressor gene function, induction of cell proliferation and immune suppression (Jackson & Campo, 1993).

Despite the ability of certain HPV types to contribute to cancer development, this is not part of the normal viral life cycle. Although the productive lesion induced by the high risk HPV types has yet to be definitively identified, carcinomas harbouring HPV DNA do not support viral particle production and in many of these tumours only part of the viral genome is conserved and expressed. Any potential oncogenic activities of these viruses should therefore be viewed in the context of an unfortunate byproduct of functions which normally play a role in replication and it seems extremely unlikely that any viral activity exists, *per se*, to participate in malignant progression.

HPV ENCODED ONCOPROTEINS

The epidemiological evidence that infection with certain HPV types participates in the development of cervical cancers is strongly supported by the ability of these viral DNAs to transform cells in culture (Vousden, 1991). Two viral genes, E6 and E7, from high risk HPV types show transforming and immortalizing activities in several rodent cell systems and co-expression of both genes immortalizes primary human genital epithelial cells. These activities are assumed to reflect *in vivo* functions which might contribute to tumour development, although the deregulation of the normal control of cell growth, which these assays often measure, is also consistent with the need of these viruses to induce replication in a normal productive lesion. Drawing distinctions between activities which simply play a role in the viral life cycle and those which also contribute to the oncogenic activity of the virus is therefore not straightforward.

Consideration of the importance of high risk HPV activities in oncogenesis is greatly aided by comparative analysis with a group of genital HPV types which, although successful as infectious agents, give rise to lesions which rarely progress to malignancy. These low risk HPV types encode similar proteins to the high risk group of viruses, including E6 and E7. The E6 and E7 proteins encoded by these viruses function very inefficiently in the cell transformation or immortalization assays (Schlegel *et al.*, 1988; Storey *et al.*, 1988; Barbosa *et al.*, 1991), adding credence to the hypothesis that these assays reflect activities important in oncogenesis. Since these studies have utilized cloned viral genes expressed from the same efficient promoters, they also indicate that at least some of the difference in oncogenic potential between these two groups of viruses stems from biochemical differences in the activities of the viral gene products. This has been extremely valuable in studying the mechanisms of viral oncogenesis and has resulted in some understanding of what is almost certainly only part of this process. Differences between high and low risk viruses resulting from variations in viral entry and expression, target cell type or immune response to infected cells have been largely left unexplored, principally due to the difficulty in growing these viruses.

There has been much enthusiasm to identify activities of high risk HPV types which are lacking in the low risk types and therefore potentially contribute to malignant development. Careful analyses have revealed, however, that most of the activities displayed by the high risk viruses can also be detected, albeit at lower levels, in the low risk viruses (Storey, Osborn & Crawford, 1990; Halbert, Demers & Galloway, 1992). It seems possible that all the genital HPV types use the same mechanisms to perturb the normal controls on proliferation of the epithelial cell in order to allow viral replication and the differences between the high and low risk viruses may simply be in the efficiency with which this loss of cell growth regulation is induced. The malignant potential displayed by the high risk HPV types probably occurs only incidentally, as the result of more aggressive inter-ference with normal cell growth control, since it is difficult to imagine selection for a high risk viral activity which contributes only to oncogenesis. Despite these caveats, however, there are some extremely interesting differences in the activities of E6 and E7 from the high risk HPV types compared to the same proteins encoded by the low risk types and, whilst it is not clear what the consequence of these differences is in the normal viral life cycle, there is persuasive evidence that some of them are important in determining oncogenic activity.

Interestingly, expression of high risk HPV types in cultured cells results in only partial transformation, an observation consistent with the inability of these virus types to cause complete malignant progression *in vivo*. This is particularly well illustrated in the human cells, where expression of E6 and E7 leads to the establishment of immortal cell lines which show differen-tiation patterns characteristic of cervical intraepithelial neoplasia (McCance *et al.*, 1988) but do not form tumours in nude mice (Pirisi *et al.*, 1988; Kaur & McDougall, 1989). Malignant transformation of HPV immortalized cells can be achieved, however, either spontaneously following extensive passage in culture (Hurlin *et al.*, 1991), by treatment with mutagens (Klingelhutz *et al.*, 1993), additional transfection with other cellular oncogenes such as *ras* (DiPaolo *et al.*, 1989) or sequences from other genital viruses such as HSV-2 (DiPaolo *et al.*, 1990). Careful analysis has revealed several steps during this progression, as the cells become increasingly differentiation resistant (Hawley-Nelson *et al.*, 1989), lose responsiveness to negative growth signals such as TGFβ (Woodworth *et al.*, 1992), acquire anchorage independence and, finally, form tumours in animals (Hurlin *et al.*, 1991). It is possible that each of these steps occurs independently and that this therefore presents an excellent model for the identification of genetic alterations which contribute to multistep tumourigenic conversion.

HPV E7 association with cell proteins

The genital HPV E7 proteins show both structural and functional simi-larities to transforming proteins from two other DNA tumour viruses, adenovirus E1A and SV40 LT (Vousden, 1991). These viral oncoproteins share the ability to form complexes with the same cell proteins through a region showing amino acid sequence similarities, and the identification of these proteins as regulators of cell growth prompted speculation that these interactions might be important for the transforming activities of the viral proteins. Cell proteins found in association with E7 include the product of the retinoblastoma tumour suppressor gene, pRb (Dyson *et al.*, 1989) and the related p107 and p130 proteins (Davies *et al.*, 1993), as well as a cyclin dependent kinase activity which may result from a direct interaction between E7 and cyclin A (Tommasino *et al.*, 1993). *In vitro* studies indicating that the low risk E7 proteins complex less efficiently with Rb (Münger *et al.*, 1989; Gage, Meyers & Wettstein, 1990) provided some evidence that these activities might participate in the oncogenic activity of the high risk HPV types. The implication of the observation that low risk HPV E7 proteins bind pRb poorly is that pRb binding is not a key feature of the HPV life cycle. This has been demonstrated using rabbit papillomavirus (Defeo-Jones *et al.*, 1993). In this model system, viruses that produced no E7 protein were unable to induce warts showing that, as expected, the E7 protein plays an important role in normal viral replication. Viral DNA containing a point mutation resulting in the production of a pRb non-binding E7 protein, however, successfully induced warts. pRb binding is therefore dispensable for wart formation in this animal model, although whether the lesions induced by this mutant virus are consequently less likely to undergo malignant conversion than those arising following infection with the wild type virus remains to be determined. It therefore remains possible that pRb binding is important for the oncogenic activity of these viruses, although the significance of this interaction to the normal viral life cycle remains obscure.

Functions of the E7/cell protein interactions

Mutations in E7 which prevent binding to pRb or p107 have also allowed an analysis of the contribution of these activities to cell transformation or immortalization by this viral protein. Initial studies in rodent cells showed a strict requirement for pRb binding by E7 in any of the assays (Edmonds & Vousden, 1989; Barbosa *et al.*, 1990; Phelps *et al.*, 1992) and amino acid substitutions which enhance the ability of the low risk HPV6 E7 protein to bind pRb also greatly increased transforming activity in these systems (Heck *et al.*, 1992; Sang & Barbosa, 1992). Whatever the importance of this activity in the normal viral life cycle, these studies provide reassuring evidence that the E7/pRb interaction is important for oncogenesis. This simple model has been undermined somewhat by analysis of the consequences of these E7 mutations on the ability to immortalize primary human epithelial cells. In the context of the full length viral genome, mutations which prevent E7 interacting with pRb do not prevent immortalizing activity in human cells, although other E7 functions are necessary for the immortalization to occur

(Jewers *et al.*, 1992). Strictly speaking, the human and rodent assays for E7 function are not directly comparable, since use of the whole viral genome in the human cell experiments may allow the expression of one of the other viral genes to substitute for the inability of E7 to complex pRb. It is possible that the pRb binding activity of E7 may be more critical in the immortalization of human cells when the E6 and E7 genes are used in isolation of the rest of the HPV genome. Nevertheless, the apparent dispensability of the pRb binding function in human cell immortalization is somewhat perplexing. It is possible that a closer analysis of the phenotype of the immortalized cells will shed some light on this matter. It would be very informative if, for example, cells immortalized with the pRb binding deficient virus showed a significant difference in the rate of progression to tumourigenicity or a difference in differentiation potential. E6 and E7 immortalized cells show substantial defects in normal differentiation when measured in an organotypic culture system. In these assays, primary human keratinocytes differentiate like normal epithelium whilst HPV immortalized cells become almost indistinguishable from intraepithelial neoplasias, the premalignancies thought to be induced following HPV expression *in vivo* (McCance *et al.*, 1988; Pecoraro *et al.*, 1991). A role for pRb in allowing normal differentiation of muscle cells, in addition to the more widely recognized ability to regulate proliferation, has recently been described (Gu *et al.*, 1993) and the interaction between E7 and pRb may be related to the prevention of normal differentiation, rather than the extension of proliferative capacity measured in the immortalization assay. It should be noted, however, that the rabbit warts formed by a virus expressing E7 defective in pRb binding were pathologically and histologically identical to normal warts (Defeo-Jones *et al.*, 1993), suggesting that these viruses retained the ability to perturb normal differentiation.

The contributions of other interactions between E7 and cell proteins to malignant development are not well established. Since some transformation defective E7 mutants retain the ability to form a complex with p107 (despite losing pRb binding activity) (Davies *et al.*, 1993) and low risk HPV E7 proteins show a stronger affinity for p107 than pRb in *in vitro* assays (Davies & Vousden, unpublished observation), it seems likely that the E7/p107 interaction is more important in normal viral replication. Interestingly, the two pRb binding defective HPV16E7 mutants which function in the human cell immortalization assay both retain at least some ability to complex p107 *in vitro* (Davies *et al.*, 1993). Whether this activity contributes to the immortalizing activity of E7 remains to be determined. The significance of association between E7 and the cyclin dependent kinase is also unknown.

B-myb, a cell gene regulated by E7
Although the significance of the E7/cell protein interactions in viral replication remains unclear, some of the consequences to the cell are now being

Uninfected cell

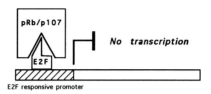

E7 expressing cell

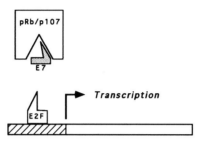

Fig. 1. Effect of E7 expression on E2F activity. In an uninfected cell, transcriptional activity of E2F is negatively regulated during the G0 and G1 stages of the cell cycle by complex formation with pRb or p107. The interaction of E7 with pRb or p107 dissociates the complex with E2F, resulting in transcriptional activity at an inappropriate stage of the cell cycle.

realized. pRb, and probably also p107 and other related proteins, play an important role in negatively regulating cell cycle progression (Goodrich *et al.*, 1991). This activity appears to result from the ability of pRb and p107 to complex with, and control, the action of transcription factors such as E2F (Zamanian & La Thangue, 1992; Dalton, 1992; Hamel *et al.*, 1992; Hiebert *et al.*, 1992; Schwartz *et al.*, 1993), which are involved in controlling the expression of other genes important for cell cycle progression. One example of an E2F regulated gene whose expression is necessary for entry into DNA synthesis is B-myb (Lam & Watson, 1993). Negative control of B-myb expression during quiescence and early stages of the cell cycle is mediated through sequences in the B-myb promoter which are recognized by E2F and there is evidence for the participation of p107 in this regulation. During normal replication, this block is released late in the G1 stage of the cell cycle and the subsequent expression of B-myb is predicted to be at least partially responsible for entry into DNA synthesis. Rodent cells expressing E7, by contrast, are unable to negatively regulate B-myb expression in this way, presumably because the E7 protein can interfere with the formation of the regulatory complexes between E2F and p107 (Lam *et al.*, 1994) (Fig. 1). These cells therefore constitutively express B-myb and escape from at least this facet of normal regulation of cell growth. Several other E2F regulated genes, which include DNA polymerase α (Pearson, Nasheuer & Wang, 1991), DHFR (Slansky *et al.*, 1993), TK (Ogris *et al.*, 1993) and cdc2 (Dalton, 1992), are potentially also subject to a similar E7 mediated deregulation of gene expression. The products of all of these genes are important in regulating orderly and controlled cell proliferation and the possible ability of E7 to interfere with these processes can easily be envisaged as playing a role in oncogenesis. These functions might also be important to the virus by stimulating DNA synthesis in infected cells and so allowing viral replication to proceed, using the host replicative machinery. In this context it is of interest to note that at least some of the low risk E7 proteins retain the ability to form a complex with p107 (Davies & Vousden, unpublished observations). This is supported by the observation that both high and low risk E7 proteins can *trans*-activate transcription from E2F containing promoters (Münger *et al.*, 1991), an activity which, in some cases, may be more closely related to the interaction with p107 rather than pRb.

Cell cycle control of E2F activity by pRb, p107 and related proteins is highly complex and the effects of E7 interactions with these regulatory pathways are likely to be manifold, particularly since at least five proteins have been identified which can heterodimerize to form E2F (Girling *et al.*, 1993; Huber *et al.*, 1993). The potential pleiotropy of E7 function is even further extended by the ability of pRb, and presumably also the related proteins, to interact with a growing number of different cell proteins. Direct interactions with transcription factors such as Myc (Rustgi, Dyson & Bernards, 1991), Elf-1 (Wang *et al.*, 1993), ATF-2 (Kim *et al.*, 1992) and PU.1 (Hageemeier *et al.*, 1993), or indirect effects on the activity of factors such as Sp-1 (Udvadia *et al.*, 1993), may allow pRb to play a role in the transcriptional regulation of a vast array of genes in many different cell types. Other interactions of pRb, for example, with cyclin D1 (Dowdy *et al.*, 1993), could affect cell growth through mechanisms other than direct control of transcription. The effect of E7 on these interactions has not been analysed and E7 sequences involved in the potential disruption of these complexes may differ. The ability of E7 to dissociate the pRb/E2F interaction, for example, requires both the pRb binding region and the C-terminal domain of E7 (Huang *et al.*, 1993; Wu *et al.*, 1993), which plays a role in zinc binding and dimerisation (McIntyre *et al.*, 1993). In contrast, the interaction of pRb with *myc* can be successfully completed using an N-terminal peptide of E7 containing only the pRb binding domain (Rustgi *et al.*, 1991). The high degree of complexity in the cell interactions involving pRb and the related proteins will undoubtedly be reflected in many and varied effects of E7 expression on the regulation of cell proliferation. There are, however, still further activities of E7 which contribute to rodent cell transformation, wart formation and immortalization of human keratinocytes which are not related to pRb binding (Vousden, 1993) and remain to be discovered.

HPV E6 interacts with p53

The function of E6, like E7, is related to its ability to associate with cell proteins. Although there is now evidence that E6 forms several associations, very few of the host partners in these complexes have been identified and studies of E6 interactions have concentrated almost exclusively on the cell encoded tumour suppressor protein, p53 (Werness, Levine & Howley, 1990). Interaction with high risk HPV E6 proteins leads to the rapid degradation of p53 (Scheffner *et al.*, 1990) and although no such degradation can be detected following complex formation between low risk HPV E6 proteins and p53 *in vitro* (Scheffner *et al.*, 1990; Crook, Tidy & Vousden, 1991), there is evidence that the expression of these viral proteins in cells also leads to the reduction of the half-life of the endogenous p53 (Band *et al.*, 1993). Degradation of the p53 protein proceeds by ubiquitin directed proteolysis (Scheffner *et al.*, 1990) and although the half-life of p53 in an uninfected cell is also controlled by this process (Ciechanover *et al.*, 1991), it is not yet clear whether E6 enhances the normal degradative process or whether a novel mechanism is utilized. Although the ability of E6 to induce degradation of p53 is very striking in *in vitro* assays, where limited amounts of each protein are mixed in a cell-free system, the effect on endogenous p53 protein levels in HPV infected cells is less clear. In most E6 expressing cells, either cervical carcinoma cell lines or transfected human epithelial cells, the half-life of newly synthesized p53 is reduced compared to normal cells (Lechner *et al.*, 1992; Hubbert, Sedman & Schiller, 1992). Rather confusingly, this is not consistently accompanied by a decrease in the steady state levels of p53 within the E6 expressing cell (Scheffner *et al.*, 1991; Lechner *et al.*, 1992; Hubbert *et al.*, 1992). This does not seem to be the result of an increase in the rate of p53 synthesis to compensate for the enhanced degradation. This has lead to the development of rather poorly defined hypotheses invoking two pools of p53; one comprising the newly synthesized protein, which is the target for E6 degradation and presumably the active form in the control of cell growth, the other a stable pool which accounts for most of the total p53 protein in the cell. It is worth keeping in mind, however, that the levels of p53 in a normally dividing cell are generally very low and the function of p53 may not be apparent in these cell populations.

One of the normal functions of p53, like pRb, is to negatively control cell proliferation and loss of this activity is frequently seen in many types of human cancer (Levine, Momand & Finlay, 1991). There is accumulating evidence that p53 function is not required during normal division of most cells but becomes important as a checkpoint in the event of DNA damage following exposure to agents such as UV light or γ radiation (Lane, 1992*a,b*; Kuerbitz *et al.*, 1992). A rapid increase in levels of p53 protein, resulting from an increase in protein half-life, prevents the progress of cell division, enabling the cell either to carry out DNA repair or, if the damage is too

Uninfected cell

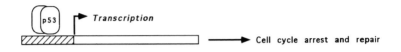

E6 expressing cell

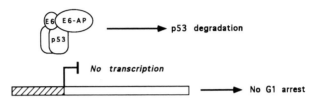

Fig. 2. Effect of E6 expression on p53 activity. In an uninfected cell, induction of p53 activity following DNA damage results in transcriptional activation of p53 responsive genes which might contribute to p53 mediated arrest at the G1 stage of the cell cycle. The interaction of E6 with p53 abrogates p53 mediated transcriptional control and results in the loss of the G1 cell cycle arrest.

extensive, die. The difficulty in showing a clear effect of E6 expression on p53 levels in proliferating cells may therefore be because the E6/p53 interaction only becomes important under circumstances where p53 function is induced. Cells expressing E6 do not show the accumulation of high levels of p53 following irradiation or treatment with actinomycin D (Kessis *et al.*, 1993; Fritsche, Haessler & Brandner, 1993) (Farthing & Vousden, unpublished observations) and consequently do not arrest cell cycle progression in response to these DNA damaging agents (Fig. 2). The mechanism by which p53 carries out the checkpoint function are not understood, although its ability to control transcription of many cell genes, either positively or negatively, is very likely to contribute. It is therefore significant that expression of E6 can abrogate the ability of p53 to both *trans*-activate and *trans*-repress target promoters (Lechner *et al.*, 1992; Mietz *et al.*, 1992).

Consequences of the E6/p53 interaction
The strong association of loss of wild-type p53 activity with the development of many different types of cancer (Hollstein *et al.*, 1991) leads to the prediction that expression of E6, in abrogating p53 function, can also contribute to malignant progression. Loss of p53 function, in the current models, is proposed to lead to the accumulation of mutations as a result of the inability to repair DNA damage, thus increasing the likelihood of developing an oncogenic alteration which could contribute to malignant

progression (Lane, 1992*a*). HPV infection clearly represents an early stage in the development of cervical cancers and the loss of the policing function of p53 at this stage would be consistent with this model. The ability of the high risk HPV E6 proteins to efficiently target p53 for degradation correlates with a strong activity in abrogating the transcriptional control mediated by p53 (Lechner *et al.*, 1992; Mietz *et al.*, 1992) (Crook *et al.*, 1994). The low risk HPV E6 proteins, which show a reduced ability to interact with p53 (Crook *et al.*, 1991; Band *et al.*, 1993), can also interfere with transcriptional regulation by p53, albeit much less efficiently than the high risk E6 proteins. It is therefore possible that this activity plays an important role in normal viral replication, possibly by preventing the cessation of cell proliferation as a p53 mediated stress response to HPV infection. The greater oncogenic activity of the high risk viruses may reflect the greater relative efficiency of the E6/p53 interaction, although it is difficult to understand selection for this activity unless, in addition to the contribution to malignant progression, it also enhances viral replication.

Other cell components involved in the E6/p53 interaction

Another E6 associated cell protein, E6-AP, has been shown to be necessary for the E6/p53 interaction (Huibregtse, Scheffner & Howley, 1991). Although the function of this protein has not yet been identified, it is necessary for E6 induced p53 degradation and may contribute to the proteolytic process (Huibregtse, Scheffner & Howley, 1993). The importance of additional cell proteins for E6 directed p53 proteolysis raises the possibility that different cell types will display varying abilities to support this E6 function. This might be reflected in the observation that, although the half-life of endogenous p53 is shortened in human cells expressing E6, a similar effect has not been detected in E6 expressing rodent cells (Sedman *et al.*, 1992) and it is E7, rather than E6 expression, which can overcome a p53 mediated growth arrest in rat cells (Vousden *et al.*, 1993). Identification of different cell components which participate in E6 mediated p53 degradation may also shed some light on how p53 stability is normally regulated in uninfected cells. The inability of E6 expressing cells to accumulate p53 following DNA damage, however, suggests that the mechanism by which p53 is normally stabilized in these cells is overridden by the action of E6.

Contribution of the E6/p53 interaction to oncogenesis

The importance of the E6/p53 interaction in tumour development has been supported by studies analysing the state of the p53 gene in HPV associated cervical cancers. A large number of HPV positive cancers have now been examined and the incidence of p53 mutations in these tumours is consistently extremely low (Crook *et al.*, 1992; Fujita *et al.*, 1992; Busby-Earle *et al.*, 1992; Borresen *et al.*, 1992). Somatic mutations in p53 are detected more frequently, however, in HPV negative cancers, although in some populations even these cancers only rarely show evidence of a genetic alteration within the p53 sequences (Choo & Chong, 1993). Taken together, these studies support a role for loss of wild-type p53 function in the development of these cancers and indirectly suggest that this can be achieved by expressing E6 in the HPV positive tumours, or following somatic p53 mutation in some of the HPV negative cancers. Alternative mechanisms of inactivating p53 function, such as amplification of the mdm-2 gene, are now being identified in other tumours with a low incidence of somatic p53 mutation (Oliner *et al.*, 1992; Leach *et al.*, 1993). The mdm-2 protein interacts directly with p53, inhibiting p53 mediate *trans*-activation (Momand *et al.*, 1992) and growth suppression (Finlay, 1993). Preliminary analyses of HPV positive cancers have detected no evidence for mdm-2 amplification (Farthing & Vousden, unpublished observations), consistent with the notion that expression of E6 is sufficient to inactivate p53 and allow malignant progression.

Although much emphasis has been placed on the E6/p53 interaction, there is evidence that other functions of E6, for example transcriptional *trans*-activating activity and transformation of established rodent cells, are not dependent on this interaction (Crook *et al.*, 1991; Sedman *et al.*, 1992) and other E6 functions which participate in oncogenesis and viral replication may remain to be identified.

OTHER FACTORS CONTRIBUTING TO THE DEVELOPMENT OF CERVICAL CANCER

Despite the importance of HPV infection to the development of most cervical cancers, additional events are required for full malignant conversion and the identification of these steps may prove to be of broader importance to many other types of cancer. Some of the steps important for oncogenesis may be related to the virus itself, for example integration of the viral genome into the host cell DNA during progression (Cullen *et al.*, 1991; Das *et al.*, 1992) may result in the deregulated expression of E6 and E7 (Choo, Pan & Han, 1987). There is also evidence for an intracellular surveillance mechanism, localized to chromosome 11, which normally regulates HPV E6 and E7 expression and is lost in HPV associated carcinoma cells (zur Hausen, 1986; Smits *et al.*, 1990; Bosch *et al.*, 1990).

Expression of the virus cannot, however, account for every step in cancer development and mutations or alterations in expression of other cell genes are likely to be involved. Amplification of proto-oncogenes such as *myc* have been described in some cancers (Riou, 1988) and several chromosomal abnormalities have been detected (Teyssier, 1989), any of which could contribute to the malignant process. Detection of consistent abnormalities is difficult using primary tumour material and there is hope that cooperating events in oncogenesis will be more easily identified using the experimental model of progression of HPV immortalized human epithelial cells. Preliminary studies have identified several chromosomal abnormalities in these experimental systems (Smith *et al.*, 1989; Hurlin *et al.*, 1991) and a recent study has indicated the possible involvement of the tumour suppressor DCC in carcinogen induced progression (Klingelhutz *et al.*, 1993). Infection with other sexually transmitted agents, such as HSV, may also contribute to this process (DiPaolo *et al.*, 1990; Hildersheim *et al.*, 1991). Although extremely rare in primary tumours, p53 mutations have been detected in some lymph node deposits from HPV positive primary cancers (Crook & Vousden, 1992), suggesting a possible role for mutations of p53 in metastatic progression. At least some mutant p53 proteins are unable to complex with E6, thus escaping the E6 directed degradation (Crook & Vousden, 1992; Scheffner *et al.*, 1992), and the selection for mutant p53 in metastatic disease may reflect the acquisition of a transforming activity unrelated to any potential inactivation of wild-type p53 function through a dominant negative mechanism (Shaulian *et al.*, 1992). The identification of additional events which cooperate with HPV infection in the development of cervical cancers is bound to become one of the major areas of future research activity in this field of study.

CLINICAL APPLICATIONS

The identification of high risk genital HPV types as human tumour viruses and our growing understanding of the molecular mechanisms by which the viral oncoproteins function is leading to a consideration of how this information might be used in the clinic. HPV identification and typing during screening might be used as a diagnostic or prognostic indicator and recent studies have suggested that HPV analysis as an adjunct to cytological screening of cervical smears could enhance the specificity and sensitivity of these tests (Cuzick *et al.*, 1992). The increased expense involved in adding HPV typing to the routine screen might eventually be compensated by a reduction in numbers of patients referred to colposcopy or an increase in the safe time interval between tests. Sensitive methods of HPV detection might also prove useful for monitoring recurrent or residual disease after treatment.

The possibility that interactions between viral oncoproteins and cell proteins might be used as a target for therapeutic intervention has also been the subject of some attention. Several studies indicating that the malignant phenotype of cervical carcinoma cells depends on the continued expression of E6 and E7 (von Knebel Doeberitz *et al.*, 1988; Steele *et al.*, 1992) have raised the hope that inhibiting the function of the viral proteins will prove to be of some therapeutic value, even in cases of invasive cancer. Most attention has been focussed on the E7/pRb interaction, which can be blocked by peptides containing the sequences of E7 necessary for pRb binding. Despite the obvious ability of these peptides to prevent associations between E7 and pRb, p107 and the kinase *in vitro* (Davies *et al.*, 1993) as yet there is no evidence that they inhibit biological activity of the E7 protein in cells. Resolution of the three-dimensional structure of E7 and an understanding of the nature of the interaction with the cell proteins may also be of use in the design of small molecules to inhibit the interaction, but a successful structural analysis of the E7 protein has yet to be achieved. Random screening of compounds for the ability to prevent or dissociate interactions between E7 and pRb may yet provide the quickest route to the identification of potentially therapeutically useful compounds. Unfortunately, the observation that these interactions may not be essential in the normal viral replicative cycle throws doubt on the efficacy of any such compound in the treatment of warts and therefore greatly reduces the economic attraction of their development. The ability of some E7 mutants to bind Rb without disrupting the complex with E2F (Huang *et al.*, 1993; Wu *et al.*, 1993) suggests that compounds may be developed which can act as antagonists, rather than agonists of E7 function. This may not, however, be true for all the consequences of the E7/pRb interaction. The cyclin D family, for example, share sequence similarity with the pRb binding domain of E7 (Dowdy *et al.*, 1993) and it seems likely that compounds which prevent the association of E7 with pRb will also block the cyclin D/pRb interactions. The consequences of this are, at present, unknown.

An alternative strategy is to target the E6/p53 interaction, although there is no evidence for the efficacy of this approach to the treatment of cancers. Other activities of E6 and E7 which have yet to be identified may also act as useful targets for the design of therapeutic drugs and the obvious approach of eliciting an immune response to cells expressing these proteins has been discussed in the previous chapter. The identification of virally encoded oncoproteins whose expression participates in the development of most cervical carcinomas presents an excellent opportunity to develop anti-cancer drugs which need not also interfere with the normal and essential cellular processes.

ACKNOWLEDGEMENTS

I would like to thank Rachel Davies, Tim Crook and Alan Farthing for their helpful comments. I also apologize to those authors whose excellent papers I have been unable to cite.

REFERENCES

Band, V., Dalal, S., Delmolino, L. & Androphy, E. J. (1993). Enhanced degradation of p53 protein in HPV-6 and BPV-1 E6 immortalized human mammary epithelial cells. *EMBO Journal*, **12**, 1847–52.
Barbosa, M. S., Edmonds, C., Fisher, C., Schiller, J. T., Lowy, D. R. & Vousden,

K. H. (1990). The region of the HPV E7 oncoprotein homologous to adenovirus E1a and SV40 large T antigen contains separate domains for Rb binding and casein kinase II phosphorylation. *EMBO Journal,* **9**, 153–60.

Barbosa, M. S., Vass, W. C., Lowy, D. R. & Schiller, J. T. (1991). *In vitro* biological activities of the E6 and E7 genes vary among human papillomaviruses of different oncogenic potential. *Journal of Virology,* **65**, 292–8.

Borresen, A. L., Helland, A., Nesland, J., Holm, R., Trope, C. & Kaern, J. (1992). Papillomaviruses, p53, and cervical cancer. *Lancet,* **339**, 1350–1.

Bosch, F. X., Schwarz, E., Boukamp, P., Fusenig, N. E., Bartsch, D. & zur Hausen, H. (1990). Suppression *in vivo* of human papillomavirus type 18 E6–E7 gene expression in nontumorigenic HeLa × fibroblast hybrid cells. *Journal of Virology,* **64**, 4743–54.

Busby-Earle, R. M. C., Steel, C. M., Williams, A. R. W., Cohen, B. & Bird, C. C. (1992). Papillomaviruses, p53, and cervical cancer. *Lancet,* **339**, 1350.

Choo, K.-B. & Chong, K. Y. (1993). Absence of mutation in the p53 and the retinoblastoma susceptibility genes in primary cervical carcinomas. *Virology,* **193**, 1042–6.

Choo, K.-B., Pan, C.-C. & Han, S.-H. (1987). Integration of human papillomavirus type 16 into cellular DNA of cervical carcinoma: preferential deletion of the E2 gene and invariable retention of the long control region and the E6/E7 open reading frames. *Virology,* **161**, 259–61.

Ciechanover, A., DiGiuseppe, J. A., Bercovich, B. *et al.* (1991). Degradation of nuclear oncoproteins by the ubiquitin system *in vitro*. *Proceedings of the National Academy of Sciences, USA,* **88**, 139–43.

Crook, T., Tidy, J. A. & Vousden, K. H. (1991). Degradation of p53 can be targeted by HPV E6 sequences distinct from those required for p53 binding and *trans*-activation. *Cell,* **67**, 547–56.

Crook, T. & Vousden, K. H. (1992). Properties of p53 mutations detected in primary and secondary cervical cancers suggests mechanisms of metastasis and involvement of environmental carcinogens. *EMBO Journal,* **11**, 3935–40.

Crook, T., Wrede, D., Tidy, J. A., Mason, W. P., Evans, D. J. & Vousden, K. H. (1992). Clonal p53 mutation in primary cervical cancer: association with human-papillomavirus-negative tumours. *Lancet,* **339**, 1070–3.

Crook, T., Fisher, C., Masterson, P. J. & Vousden, K. H. (1994). Modulation of transcriptional regulatory properties of p53 by HPV E6. *Oncogene*, in press.

Cullen, A. P., Reid, R., Campion, M. & Lörincz, A. T. (1991). Analysis of the physical state of different human papillomavirus DNAs in intraepithelial and invasive cervical neoplasm. *Journal of Virology,* **65**, 606–12.

Cuzick, J., Terry, G., Ho, L., Hollingsworth, T. & Anderson, M. (1992). Human papillomavirus type 16 DNA in cervical smears as predictor of high-grade cervical cancer. *Lancet,* **339**, 959–60.

Dalton, S. (1992). Cell cycle regulation of the human *cdc2* gene. *EMBO Journal,* **11**, 1797–804.

Das, B. C., Sharma, J. K., Gopalakrishna, V. & Luthra, U.K. (1992). Analysis by polymerase chain reaction of the physical state of human papillomavirus type 16 in cervical preneoplastic and neoplastic lesions. *Journal of General Virology,* **73**, 2327–36.

Davies, R., Hicks, R., Crook, T., Morris, J. & Vousden, K. H. (1993). HPV16 E7 associates with a histone H1 kinase activity and p107 through sequences necessary for transformation. *Journal of Virology,* **67**, 2521–8.

De Villiers, E. M. (1989). Heterogeneity of the human papillomavirus group. *Journal of Virology,* **63**, 4898–903.

Defeo-Jones, D., Vuocolo, G. A., Haskell, K. M. *et al.* (1993). Papillomavirus E7 protein binding to the retinoblastoma protein is not required for viral induction of warts. *Journal of Virology,* **67**, 716–25.

DiPaolo, J. A., Woodworth, C. D., Popescu, N. C., Koval, D. L., Lopez, J. V. & Doniger, J. (1990). HSV-2-induced tumorigenicity in HPV16-immortalised human genital keratinocytes. *Virology,* **177**, 777–9.

DiPaolo, J. A., Woodworth, C. D., Popescu, N. C., Notario, V. & Doniger, J. (1989). Induction of human cervical squamous cell carcinoma by sequential transfection with human papillomavirus 16 DNA and viral Harvey *ras. Oncogene,* **4**, 395–9.

Dowdy, S. F., Hinds, P. W., Louie, K., Reed, S. I., Arnold, A. & Weinberg, R. A. (1993). Physical interaction of the retinoblastoma protein with human D cyclins. *Cell,* **73**, 499–511.

Dyson, N., Howley, P. M., Münger, K. & Harlow, E. (1989). The human papilloma virus-16 E7 oncoprotein is able to bind to the retinoblastoma gene product. *Science,* **243**, 934–7.

Edmonds, C. & Vousden, K. H. (1989). A point mutational analysis of human papillomavirus type 16E7 protein. *Journal of Virology,* **63**, 2650–6.

Finlay, C. A. (1993). The *mdm-2* oncogene can overcome wild-type suppression of transformed cell growth. *Molecular and Cellular Biology,* **13**, 301–6.

Fritsche, M., Haessler, C. & Brandner, G. (1993). Induction of nuclear accumulation of the tumor-suppressor protein p53 by DNA damaging agents. *Oncogene,* **8**, 307–18.

Fujita, M., Inoue, M., Tanizawa, O., Iwamoto, S. & Enomoto, T. (1992). Alterations of the p53 gene in human primary cervical carcinoma with and without human papillomavirus infection. *Cancer Research,* **52**, 5323–8.

Gage, J. R., Meyers, C. & Wettstein, F. O. (1990). The E7 proteins of the nononcogenic human papillomavirus type 6b (HPV-6b) and of the oncogenic HPV-16 differ in retinoblastoma protein binding and other properties. *Journal of Virology,* **64**, 723–30.

Girling, R., Partridge, J. F., Bandara, L. R. *et al.* (1993). A new component of the transcription factor DRTF1/E2F. *Nature,* **362**, 83–7.

Goodrich, D. W., Wang, N. P., Qian, Y.-W., Lee, E. Y.-H. P. & Lee, W.-H. (1991). The retinoblastoma gene product regulates progression through the G1 Phase of the cell cycle. *Cell,* **67**, 293–302.

Gu, W., Schneider, J. W., Condorelli, G., Kaushal, S., Mahdavi, V. & Nadal-Ginard, B. (1993). Interaction of myogenic factors and the retinoblastoma protein mediates muscle cell commitment and differentiation. *Cell,* **72**, 309–24.

Hageemeier, C., Bannister, A. J., Cook, A. & Kouzarides, T. (1993). The activation domain of transcription factor PU.1 binds the retinoblastoma (RB) protein and the transcription factor TFIID *in vitro*: RB shows sequence similarity to TFIID and TFIIB. *Proceedings of the National Academy of Sciences, USA,* **90**, 1580–4.

Halbert, C. L., Demers, G. W. & Galloway, D. A. (1992). The E6 and E7 genes of human papillomavirus type 6 have weak immortalising activity in human epithelial cells. *Journal of Virology,* **66**, 2125–34.

Hamel, P. A., Gill, R. M., Phillips, R. A. & Gallie, B. A. (1992). Transcriptional repression of the E2-containing promoters EIIaE, *c-myc*, and *RB1* by the product of the RB gene. *Molecular and Cellular Biology,* **12**, 3431–8.

Hawley-Nelson, P., Vousden, K. H., Hubbert, N. L., Lowy, D. R. & Schiller, J. T. (1989). HPV16 E6 and E7 proteins cooperate to immortalize human foreskin keratinocytes. *EMBO Journal,* **8**, 3905–910.

Heck, D. V., Yee, C. L., Howley, P. M. & Münger, K. (1992). Efficiency of binding the retinoblastoma protein correlates with the transforming capacity of the E7 oncoproteins of the human papillomaviruses. *Proceedings of the National Academy of Sciences, USA,* **89**, 4442–6.

Hiebert, S. W., Chellappan, S. P., Horowitz, J. M. & Nevins, J. R. (1992). The interaction of RB with E2F coincides with an inhibition of the transcriptional activity of E2F. *Genes and Development,* **6**, 177–85.

Hildersheim, A., Mann, V., Brinton, L. A., Szklo, M., Reeves, W. C. & Rawls, W. E. (1991). Herpes simplex virus type 2: a possible interaction with human papillomavirus types 16/18 in the development of invasive cervical cancer. *International Journal of Cancer,* **49**, 335–40.

Hollstein, M., Sidransky, D., Vogelstein, B. & Harris, C. C. (1991). p53 mutations in human cancers. *Science,* **253**, 49–53.

Huang, P. S., Patrick, D. R., Edwards, G. *et al.* (1993). Protein domains governing interactions between E2F, the retinoblastoma gene product and human papillomavirus type 16 E7 protein. *Molecular and Cellular Biology,* **13**, 953–60.

Hubbert, N. L., Sedman, S. A. & Schiller, J. T. (1992). Human papillomavirus type 16 E6 increases the degradation rate of p53 in human keratinocytes. *Journal of Virology,* **66**, 6237–41.

Huber, H. E., Edwards, G., Goodhart, P. J. *et al.* (1993). Transcription factor E2F binds DNA as a heterodimer. *Proceedings of the National Academy of Sciences, USA,* **90**, 3525–9.

Huibregtse, J. M., Scheffner, M. & Howley, P. M. (1991). A cellular protein mediates association of p53 with the E6 oncoprotein of human papillomavirus types 16 or 18. *EMBO Journal,* **10**, 4129–35.

Huibregtse, J. M., Scheffner, M. & Howley, P. M. (1993). Cloning and expression of the cDNA for E6-AP, a protein that mediates the interaction of the human papillomavirus E6 oncoprotein with p53. *Molecular and Cellular Biology,* **13**, 775–84.

Hurlin, P. J., Kaur, P., Smith, P. P., Perez-Reyes, N., Blanton, R. A. & McDougall, J. K. (1991). Progression of human papillomavirus type 18-immortalized human keratinocytes to a malignant phenotype. *Proceedings of the National Academy of Sciences, USA,* **88**, 570–4.

Jackson, M. E. & Campo, M. S. (1993). Cooperation between papillomavirus and chemical cofactors in oncogenesis. *Critical Reviews in Oncogenesis,* **4**, 277–91.

Jewers, R. J., Hildebrandt, P., Ludlow, J. W., Kell, B. & McCance, D. J. (1992). Regions of HPV 16 E7 oncoprotein required for immortalization of human keratinocytes. *Journal of Virology,* **66**, 1329–35. (In Press).

Kaur, P. & McDougall, J. K. (1989). HPV-18 immortalization of human keratinocytes. *Virology,* **173**, 302–10.

Kessis, T. D., Slebos, R. J., Nelson, W. G. *et al.* (1993). Human papillomavirus 16 E6 expression disrupts the p53-mediated cellular response to DNA damage. *Proceedings of the National Academy of Sciences, USA,* **90**, 3988–92.

Kim, S.-J., Wagner, S., Liu, F., O'Reilly, M. A., Robbins, P. D. & Green, M. R. (1992). Retinoblastoma gene product activates expression of the human *TGF-b2* gene through transcription factor ATF-2. *Nature,* **358**, 331–4.

Klingelhutz, A. J., Smith, P. P., Garrett, L. R. & McDougall, J. K. (1993). Alterations of the DCC tumor-suppressor gene in tumorigenic HPV-18 immortalised human keratinocytes transformed by nitrosomethylurea. *Oncogene,* **8**, 95–9.

Koutsky, L. A., Holmes, K. K., Critchlow, C. W. (1992). A cohort study of the risk of cervical intraepithelial neoplasia grade 2 or 3 in relation to papillomavirus infection. *New England Journal of Medicine,* **327**, 1272–8.

Kuerbitz, S. J., Plunkett, B. S., Walsh, W. V. & Kastan, M. B. (1992). Wild-type p53 is a cell cycle checkpoint determinant following irradiation. *Proceedings of the National Academy of Sciences, USA,* **89**, 7491–5.

Lam, E. W.-F. & Watson, R. J. (1993). An E2F binding site mediates cell-cycle regulated repression of mouse B-*myb* transcription. *EMBO Journal* (In Press).

Lam, E. W-F., Morris, J. D. H., Davies, R., Crook, T., Watson, R. J. & Vousden, K. H. (1994). HPV16 E7 oncoprotein deregulates B-*myb* expression: correlation with targeting of plO7/E2F complexes. *EMBO Journal.*

Lane, D. P. (1992*a*). p53, guardian of the genome. *Nature,* **358**, 15–16.

Lane, D. P. (1992*b*). Worrying about p53. *Current Biology,* **2**, 581–3.

Leach, F. S., Tokino, T., Meltzer, P. *et al.* (1993). p53 mutation and MDM2 amplification in human soft tissue sarcomas. *Cancer Research,* **53**, 2231–4.

Lechner, M. S., Mack, D. H., Finicle, A. B., Crook, T., Vousden, K. H. & Laimins, L. A. (1992). Human papillomavirus E6 proteins bind p53 *in vivo* and abrogate p53 mediated repression of transcription. *EMBO Journal,* **11**, 3045–52.

Levine, A. J., Momand, J. & Finlay, C. A. (1991). The p53 tumour suppressor gene. *Nature,* **351**, 453–6.

McCance, D. J., Kopan, R., Fuchs, E. & Laimins, L. A. (1988). Human papillomavirus type 16 alters human epithelial cell differentiation *in vitro. Proceedings of the National Academy of Sciences, USA,* **85**, 7169–73.

McIntyre, M. C., Frattini, M. G., Grossman, S. R. & Laimins, L. A. (1993). Human papillomavirus type 18 E7 protein requires intact cys–X–X–cys motifs for zinc binding, dimerization, and transformation but not for Rb binding. *Journal of Virology,* **67**, 3142–50.

Mietz, J. A., Unger, T., Huibregtse, J. M. & Howley, P. M. (1992). The transcriptional transactivation function of wild type p53 is inhibited by SV40 large T-antigen and by HPV-16E6 oncoprotein. *EMBO Journal,* **11**, 5013–20.

Momand, J., Zambetti, G. P., Olson, D. C., George, D. & Levine, A. J. (1992). The *mdm-2* oncogene product forms a complex with the p53 protein and inhibits p53-mediated transactivation. *Cell,* **69**, 1237–45.

Münger, K., Yee, C. L., Phelps, W. C., Pietenpol, J. A., Moses, H. L. & Howley, P. M. (1991). Biochemical and biological differences between E7 oncoproteins of the high- and low-risk human papillomavirus types are determined by amino-terminal sequences. *Journal of Virology,* **65**, 3943–8.

Münger, K., Werness, B. A., Dyson, N., Phelps, W. C., Harlow, E. & Howley, P. M. (1989). Complex formation of human papillomavirus E7 proteins with the retinoblastoma tumor suppressor gene product. *EMBO Journal,* **8**, 4099–105.

Ogris, E., Rotheneder, H., Mudrak, I., Pichler, A. & Wintersberger, E. (1993). A binding site for transcription factor E2F is a target for *trans*-activation of murine thymidine kinase by polyomavirus large T antigen and plays an important role in growth regulation of the gene. *Journal of Virology,* **67**, 1765–71.

Oliner, J. D., Kinzler, K. W., Meltzer, P. S., George, D. L. & Vogelstein, B. (1992). Amplification of a gene encoding a p53-associated protein in human sarcomas. *Nature,* **358**, 80–3.

Pearson, B. E., Nasheuer, H.-P. & Wang, T. S.-F. (1991). Human DNA polymerase a gene: Sequences controlling expression in cycling and serum-stimulated cells. *Molecular and Cellular Biology,* **11**, 2081–95.

Pecoraro, G., Lee, M., Morgan, D. & Defendi, V. (1991). Evolution of *in vitro* transformation and tumorigenesis of HPV16 and HPV18 immortalised primary cervical epithelial cells. *American Journal of Pathology,* **138**, 1–8.

Phelps, W. C., Münger, K., Yee, C. L., Barnes, J. A. & Howley, P. M. (1992). Structure-function analysis of the human papillomavirus type 16 E7 oncoprotein. *Journal of Virology,* **66**, 2418–27.

Pirisi, L., Creek, K. E., Doniger, J. & DiPaolo, J. A. (1988). Continuous cell lines with altered growth and differentiation properties originate after transfection of human keratinocytes with human papillomavirus type 16 DNA. *Carcinogenesis*, **9**, 1573–9.

Riou, G. F. (1988). Proto-oncogenes and prognosis in early carcinoma of the cervix. *Cancer Surveys*, **7**, 441–56.

Rustgi, A. K., Dyson, N. & Bernards, R. (1991). Amino-terminal domains of c-*myc* and N-*myc* proteins mediate binding to the retinoblastoma gene product. *Nature*, **352**, 541–4.

Sang, B.-C. & Barbosa, M. S. (1992). Single amino acid substitutions in 'low-risk' human papillomavirus (HPV) type 6 E7 protein enhance features characteristic of the 'high-risk' HPV E7 oncoproteins. *Proceedings of the National Academy of Sciences, USA*, **89**, 8063–7.

Scheffner, M., Münger, K., Byrne, J. C. & Howley, P. M. (1991). The state of the p53 and retinoblastoma genes in human cervical carcinoma cell lines. *Proceedings of the National Academy of Sciences, USA*, **88**, 5523–7.

Scheffner, M., Takahashi, T., Huibregtse, J. M., Minna, J. D. & Howley, P. M. (1992). Interaction of the human papillomavirus type 16 E6 oncoprotein with wild-type and mutant human p53 proteins. *Journal of Virology*, **66**, 5100–5.

Scheffner, M., Werness, B. A., Huibregtse, J. M., Levine, A. J. & Howley, P. M. (1990). The E6 oncoprotein encoded by human papillomavirus types 16 and 18 promotes the degradation of p53. *Cell*, **63**, 1129–36.

Schiffman, M. H. (1992). Recent progress in defining the epidemiology of human papillomavirus infection and cervical neoplasia. *Journal of the National Cancer Institute*, **84**, 394–7.

Schlegel, R., Phelps, W. C., Zhang, Y.-L. & Barbosa, M. (1988). Quantitive keratinocyte assay detects two biological activities of human papillomavirus DNA and identifies viral types associated with cervical carcinoma. *EMBO Journal*, **7**, 3181–7.

Schwartz, J. K., Devoto, S. H., Smith, E. J., Chellappan, S. P., Jakoi, L. & Nevins, J. R. (1993). Interactions of the p107 and Rb proteins with E2F during the cell proliferation response. *EMBO Journal*, **12**, 1013–20.

Sedman, S., Hubbert, N. L., Vass, W. C., Lowy, D. R. & Schiller, J. T. (1992). Mutant p53 can substitute for human papillomavirus type 16 E6 in immortalization of human keratinocytes but does not have E6-associated *trans*-activation or transforming activity. *Journal of Virology*, **66**, 4201–8.

Shaulian, E., Zauberman, A., Ginsberg, D. & Oren, M. (1992). Identification of a minimal transforming domain of p53: negative dominance through abrogation of sequence specific DNA binding. *Molecular and Cellular Biology*, **12**, 5581–92.

Slansky, J. E., Li, Y., Kaelin, W. G. & Farnham, P. J. (1993). A protein synthesis-dependent increase in E2F1 mRNA correlates with growth regulation of the dihydrofolate reductase promoter. *Molecular and Cellular Biology*, **13**, 1610–18.

Smith, P. P., Bryant, E. M., Kaur, P. & McDougall, J. K. (1989). Cytogenetic analysis of eight human papillomavirus immortalized keratinocyte cell lines. *International Journal of Cancer*, **44**, 1124–31.

Smits, P. H. M., Smits, H. L., Jebbink, M. F. & ter Schegget, J. (1990). The short arm of chromosome 11 likely is involved in the regulation of the human papillomavirus type 16 early enhancer-promoter and in the suppression of the transforming activity of the viral DNA. *Virology*, **176**, 158–65.

Steele, C., Sacks, P. G., Adler-Storthz, K. & Shillitoe, E. J. (1992). Effect on cancer cells of plasmids that express antisense RNA of human papillomavirus type 18. *Cancer Research*, **52**, 4706–11.

Storey, A., Osborn, K. & Crawford, L. (1990). Co-transfection by human papillomavirus types 6 and 11. *Journal of General Virology*, **71**, 165–71.

Storey, A., Pim, D., Murray, A., Osborn, K., Banks, L. & Crawford, L. (1988). Comparison of the *in vitro* transforming activities of human papillomavirus types. *EMBO Journal*, **7**, 1815–20.

Teyssier, J. R. (1989). The chromosomal analysis of human solid tumors: a triple challenge. *Cancer Genetics and Cytogenetics*, **37**, 103–25.

Tommasino, M., Adamczewski, J. P., Carlotti, F. *et al.* (1993). HPV16 E7 protein associates with the protein kinase p33^{CDK2} and cyclin A. *Oncogene*, **8**, 195–202.

Udvadia, A. J., Rogers, K. T., Higgins, P. D. R. *et al.* (1993). Sp-1 binds promoter elements regulated by the RB protein and Sp-1 mediated transcription is stimulated by RB coexpression. *Proceedings of the National Academy of Sciences, USA*, **90**, 3265–9.

von Knebel Doeberitz, M. (1992). Papillomaviruses in human disease: part 1. Pathogenesis and epidemiology of human papillomavirus infections. *European Journal of Medicine*, **1**, 415–23.

von Knebel Doeberitz, M., Oltersdorf, T., Schwarz, E. & Gissmann, L. (1988). Correlation of modified human papilloma virus early gene expression with altered growth properties in C4-1 cervical carcinoma cells. *Cancer Research*, **48**, 3780–6.

Vousden, K. H. (1991). Human papillomavirus transforming genes. *Seminars in Virology*, **2**, 307–17.

Vousden, K. H. (1993). Interactions of human papillomavirus transforming proteins with the products of tumor suppressor genes. *FASEB Journal*, **7**, in press.

Vousden, K. H., Vojtesek, B., Fisher, C. & Lane, D. (1993). HPV-16 E7 or adenovirus E1A can overcome the growth arrest of cells immortalized with a temperature-sensitive p53. *Oncogene*, **8**, 1697–702.

Wang, C.-Y., Petryniak, B., Thompson, C. B., Kaelin, W. G. & Leiden, J. M. (1993). Regulation of the Ets-related transcription factor Elf-1 by binding to the retinoblastoma protein. *Science*, **260**, 1330–5.

Werness, B. A., Levine, A. J. & Howley, P. M. (1990). Association of human papillomavirus types 16 and 18 E6 proteins with p53. *Science*, **248**, 76–9.

Woodworth, C. D., Cheng, S., Simpson, S. *et al.* (1992). Recombinant retroviruses encoding human papillomavirus type 18 E6 and E7 genes stimulate proliferation and delay differentiation of human keratinocytes early after infection. *Oncogene*, **7**, 619–26.

Wu, E. W., Clemens, K. E., Heck, D. V. & Münger, K. (1993). The human papillomavirus E7 oncoprotein and the cellular transcription factor E2F bind to separate sites in the retinoblastoma tumour suppressor protein. *Journal of Virology*, **67**, 2402–7.

Zamanian, M. & La Thangue, N. B. (1992). Adenovirus E1a prevents the retinoblastoma gene product from repressing the activity of a cellular transcription factor. *EMBO Journal*, **11**, 2603–10.

zur Hausen, H. (1986). Intracellular surveillance of persisting viral infections. *Lancet*, **ii**, 489–90.

zur Hausen, H. (1989). Papillomaviruses as carcinomaviruses. In *Advances in Viral Oncology*, ed. G. Klein, p. 1–26, New York: Raven Press.

BOVINE PAPILLOMAVIRUS TYPE 4: FROM TRANSCRIPTIONAL CONTROL TO CONTROL OF DISEASE

M. S. CAMPO, R. A. ANDERSON, M. CAIRNEY and M. E. JACKSON

The Beatson Institute for Cancer Research, CRC Beatson Laboratories, Garscube Estate, Glasgow G61 1BD, UK.

MULTIPLICITY OF BOVINE PAPILLOMAVIRUS

Cattle are infected by several different types of bovine papillomavirus (BPV), each one associated with a specific disease. Six distinct viral types have been identified, BPV-1 to 6, which fall into two subgroups: subgroup A comprises BPV-1, 2 and 5 and subgroup B comprises BPV-3, 4 and 6. The viruses of subgroup A induce fibropapillomas, tumours with both a fibroblastic and an epithelial component, whereas the viruses of subgroup B induce wholly epithelial papillomas with no dermal involvement. In the UK, BPV-1 is the causative agent of fibropapillomas of the penis of bulls and of the teats and udders of cows and of the adjacent skin; BPV-2 induces classical skin warts and fibropapillomas of the alimentary canal; BPV-5 causes the so-called 'rice-grain' fibropapillomas of the teats of cows; BPV-4 induces papillomas of the alimentary canal and BPV-6 papillomas of the teats. BPV-3, first isolated in Australia, causes epithelial papillomas of the skin (Campo & Jarrett, 1987). The different histopathologies are mirrored by the molecular and immunological characteristics of the viruses themselves. The fibropapillomaviruses have a larger genome (approximately 7.8 kb) than the epithelial papillomaviruses, whose genome is approximately 7.2 kb and there is little similarity in DNA sequence and little or no immune cross reactivity between the viruses of the two subgroups (Campo & Jarrett, 1987).

GENOMIC ORGANIZATION AND GENE FUNCTIONS

BPV-1 has long been the paradigm of all papillomaviruses. The DNAs of BPV-1 and human papillomavirus type 1 (HPV-1) were the first to be completely sequenced (Chen *et al.*, 1982; Danos, Katinka & Yaniv, 1982), and, like all the other papillomavirus genomes subsequently sequenced, the coding regions are present in only one DNA strand (Fig. 1). The relative ease with which BPV-1 could be isolated from warts, and, most importantly, its ability to transform cells in culture (Dvoretzky *et al.*, 1980; Law *et al.*,

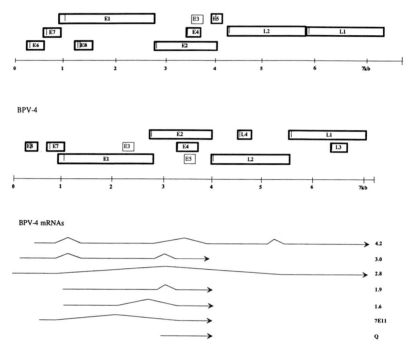

Fig. 1. Genomic organization of BPV-1 and BPV-4 and transcriptional organization of BPV-4. The viral genome is represented as linear and the open boxes represent ORFs. The early ORFs are indicated E and the late ORFs L. The first ATG codon in an ORF is indicated by a vertical line; ORFs without ATG codon are represented as thin-lined boxes. The mRNAs of BPV-4 are below the genome map; horizontal lines indicate exons, diagonal lines indicate splices. The sizes in kb or the designations of the mRNAs are indicated at the right.

1981), have led to the assigning of functions to the individual open reading frames (ORFs).

Thus, the 'early' (E) region, which comprises approximately two-thirds of the genome, encodes proteins that control the replication and transcription of viral DNA, and are instrumental in cell transformation; the remaining third of the genome comprises the 'late' (L) region which encodes the viral capsid proteins and a region of approximately 1000 bp, defined as the long control region (LCR), which contains the *cis*-control elements governing viral functions. This genetic plan, on the whole, is common to all the other papillomaviruses (Iftner, 1990).

The E1 protein is involved in viral DNA replication (Lusky & Botchan, 1986); it is a phosphoprotein found in the nucleus of infected cells (Sun *et al.*, 1990) which recognizes and binds to the viral origin of replication (Ustav *et al.*, 1991). It has helicase and DNA unwinding activities and interacts with DNA polymerase alpha (Lentz *et al.*, 1993), displaying significant functional homology with the large T-antigen of SV40. E1 is aided in its replicative functions by E2, which favours the binding of E1 to DNA through both protein–protein and protein–DNA interactions (Yang *et al.*, 1991, 1992) thus coordinating the assembly of an initiation complex at the replication origin.

In addition to its role in DNA replication, E2 is a transcriptional regulator, both of viral (Spalholz, Yang & Howley, 1985; Haugen *et al.*, 1987; Prakash *et al.*, 1988) and cellular genes (Heike *et al.*, 1989). It binds to the cognate binding sites present in multiple copies in the viral LCR (Androphy, Lowy & Schiller, 1987) and modulates transcription both positively and negatively (Lambert, Spalholz & Howley, 1987). E2 has transcriptional control activity also in HPVs (Sousa, Dostatni & Yaniv, 1990). Indeed, in cervical cancers, integration of the HPV genome often leads to either the deletion or interruption of the E2 ORF, with consequent deregulation of the expression of the transforming E6 and E7 genes (Schwarz *et al.*, 1985; Matsukura *et al.*, 1986). This is considered a critical step in the progression from low grade to high grade lesions.

No function has been ascribed to the E3 ORF, and other papillomaviruses do not possess it.

The E4 protein was first identified in HPV-1 induced warts (Doorbar *et al.*, 1986), where it exists in several forms (Rogel-Gaillard, Breitburd & Orth, 1992). The predominant form is a fusion between the first five residues of E1 and the main body of E4 (Doorbar *et al.*, 1988). E1–E4 interferes with cytokeratin assembly (Doorbar *et al.*, 1991), possibly upsetting the differentiation programme of the cell, and favouring production of virion progeny (Doorbar, 1991).

E5, E6 and E7 are the transforming proteins of papillomavirus but their contribution to cell transformation varies in the different viruses (Campo, 1992). BPV-1 E5 is a dimeric transmembrane protein (Schlegel & Wade-Glass, 1987). It is the major transforming protein of BPV-1 (Schiller *et al.*, 1986), and targets a number of cellular proteins. It has been shown to activate the receptors for epidermal growth factor and colony stimulating factor-1 (Martin *et al.*, 1989), to bind and activate the receptor for platelet-derived growth factor (Petti, Nilson & DiMaio, 1991; Petti & DiMaio, 1992), and to bind the 16k protein of the ductin family (Goldstein *et al.*, 1991; Holzenburg *et al.*, 1993), which is a constituent of both the vacuolar H^+-ATPase and of gap junctions (Finbow *et al.*, 1991). The identification of BPV-1 E5 as an oncoprotein has been instrumental in the analysis of HPV-16 E5: although not as powerful a transforming agent as BPV-1 E5, HPV-16 E5 can transform established keratinocytes (Leptak *et al.*, 1991) and appears to contribute to the early stages of cell transformation by releasing the cell from the need of high levels of growth factors for proliferation (Pim, Collins & Banks 1992; Leechanachai *et al.*, 1992) and by increased recycling of growth factor receptors (Straight *et al.*, 1993).

BPV-1 E6 is the second oncoprotein of the virus (Yang, Okayama & Howley, 1985) but contrary to E5 it is only weakly transforming; it is capable of binding zinc through cys–x–x–cys motifs (Barbosa, Lowy & Schiller, 1989) and is a transcriptional activator (Lamberti *et al.*, 1990). It is possible that deregulation of cellular genes contributes to transformation by E6. HPV-16 E6 is both a transcriptional activator and a transforming agent (Sedman *et al.*, 1991; Desaintes *et al.*, 1992). Although transcriptional activation may play a role in cell transformation, the main action of HPV-16 E6 appears to be the binding (Werness, Levine & Howley, 1990) and the subsequent degradation (Scheffner *et al.*, 1990) of the tumour suppressor protein p53, thus withdrawing it from the control of cell proliferation. Indeed it is possible that the transcriptional activation function of HPV-16 is due to the abrogation of p53-mediated transcription repression (Lechner *et al.*, 1992).

In BPV-1, E7 has been reported to contribute to the control of viral DNA replication and of copy number of viral genomes (Lusky & Botchan, 1986) and it does not appear to have a major role in cell transformation. In contrast, HPV-16 E7 is the major transforming gene of this virus (Watanabe *et al.*, 1990; Phelps *et al.*, 1992). It binds zinc like E6 (Barbosa *et al.*, 1989), and its properties are similar to those of adenovirus E1A. Like E1A, E7 activates the adenovirus E2 promoter (Phelps *et al.*, 1988) and binds to the cellular tumour suppressor protein p105Rb (Dyson *et al.*, 1989; Munger *et al.*, 1989); it also forms complexes with other proteins bound by E1A, i.e. p107, p300, cyclin A and p33^{CDK2} kinase (Dyson *et al.*, 1992; Tommasino *et al.*, 1993). Therefore, E7 binds and deregulates the activity of cellular proteins involved in the control of cell proliferation, thus leading to transformation. The transforming properties of HPV-16 E6 and E7 are discussed in greater detail in the previous chapter.

The function of BPV-1 E8 has not yet been established. L1 and L2 are the structural proteins of all papillomaviruses. L1 is the major and L2 the minor capsid protein of the virion (Favre *et al.*, 1975).

'HIGH RISK' BOVINE PAPILLOMAVIRUSES

Despite being capable of full oncogenic transformation of both established and primary cells *in vitro* (Law *et al.*, 1981; Morgan & Meinke, 1980), of inducing sarcomas in hamsters and in horses (Lancaster, Theilen & Olson, 1979) and in transgenic mice (Sippola-Thiele, Hanahan & Howley, 1989), BPV-1 does not appear to be carcinogenic in its natural host. In Great Britain, two different BPV types are found associated with cancer in cattle: BPV-2 with urinary bladder cancer (Campo *et al.*, 1992) and BPV-4 with alimentary canal cancer (Jarrett *et al.*, 1978; Campo *et al.*, 1980). In Australia, BPV is involved in ocular carcinoma (Spradbrow & Hoffmann, 1980; Ford *et al.*, 1982), and in Zimbabwe in vulvar cancer (VPMG Rutten, personal communication), but these BPVs have not yet been identified. In all cases, viral infection results in production of benign papillomas, at high risk for neoplastic progression in animals exposed to environmental cofactors. Thus, the cofactors critical for the development of urinary bladder and alimentary canal cancers have been identified in the mutagens and immunosuppressants contained in bracken fern present in the diet (Evans, I. A. *et al.*, 1982; Evans W. C. *et al.*, 1982), and for the development of ocular and vulvar carcinomas in high levels of UV rays (Spradbrow & Hoffmann, 1980; VPMG Rutten, personal communication). Progression of alimentary canal papillomas to squamous carcinomas and development of urinary bladder cancers have been experimentally reproduced in animals infected with virus and fed with a diet of bracken (Campo & Jarrett, 1987; Campo *et al.*, 1992), thereby confirming what had been inferred from epidemiological studies (Jarrett *et al.*, 1978).

BPV-4 GENETIC ORGANIZATION

Our work is mainly focussed on the biology of BPV-4 for two reasons, the first being that the virus is involved in a naturally occurring cancer in its own host, therefore providing the opportunity to study carcinogenesis in a natural system, and the second that the bovine system presents itself as a valuable model for HPV-associated carcinogenesis of mucous epithelia in humans, particularly that of the genital tract.

The genetic plan of BPV-4 is similar to that of other papillomaviruses, the main difference being the lack of the E6 ORF and its replacement with the E8 ORF (Fig. 1). These differences are found also in the other two members of subgroup B, BPV-3 and BPV-6, and therefore seem to be a common characteristic of epitheliotropic BPVs (Jackson *et al.*, 1991). The biological significance of the absence of an E6 gene will be discussed later, but it is interesting to speculate how the rather peculiar organization of the early genes of subgroup B viruses might have evolved. These viruses possess a high degree of nucleotide sequence similarity throughout their genome, but they are most dissimilar from each other where the E6 ORF should be. Remnants of the ORF can be recognized in short stretches of nucleotides with limited homology to the 5' end of the E6 ORF of HPV-8, suggesting that the absence of the E6 ORF in subgroup B BPVs is the result of a deletion (Jackson *et al.*, 1991). As will be discussed in greater detail later, the E8 ORF of the epithelial BPVs encodes a protein which is similar in amino acid sequence not to the putative E8 peptide of the fibropapillomaviruses, as might be expected from the similar genomic location (Fig. 1), but to their E5 protein. This suggests translocation of the E5 ORF from the middle to the 5' end of the early region in a progenitor virus, possibly due to an unequal recombination event between two viral genomes. Whether translocation of the E5 ORF and deletion of the E6 ORF were the result of

only one or of two independent recombination events is matter for speculation.

Surely, not all papillomaviruses have been isolated and characterized. If a
virus is found with characteristics common to both subgroup A and subgroup B it might shed light on how the epithelial BPVs have arrived at their
contemporary genomic organization.

BPV-4 gene expression and function

Numerous BPV-4 transcripts, which show a complex pattern of splicing
between the ORFs, have been identified both in papillomas and in transformed cells (Fig. 1: Smith, Patel & Campo, 1986; Stamps & Campo, 1988).
The viral transcripts fall into two classes; those representing the early ORFs
terminate at the polyadenylation site at nucleotide 4004 and are found both
in productive papillomas and in transformed cells, and those representing
the late ORFs terminate at either of the polyadenylation sites at nucleotide
7155 and 7191 and are only found in papillomas.

The E1 ORF is transcribed into a series of mRNAs (Fig. 1) but it is not
known which one of them, if any, encodes the functional E1 protein. The
functions of E1 are not well characterized. The integrity of the E1 ORF is
not necessary for cell transformation (Campo & Spandidos, 1983; Smith &
Campo, 1988), indeed, higher efficiency of focus formation is obtained when
this ORF is interrupted (Smith & Campo, 1988). It is not clear why this
should be so, but it may be due to E1 functions more conducive to virus
growth than to cell transformation. Interruption of the gene would lead to
loss of such functions.

The E2 ORF is transcribed into a mRNA identified as Q (Fig. 1). Q does
not represent the whole E2 ORF but only its 3' half; no transcripts
corresponding to the full ORF have been identified to date. As in BPV-1, E2
is a transcriptional regulator (Jackson & Campo, 1991), which activates the
viral LCR by binding to its cognate sites (Fig. 2(a)), leading to increased
promoter activity (see below).

The E4 ORF is transcribed into several RNA species (Fig. 1). The 7E11
transcript is most likely to encode the E1-E4 fusion protein described for
HPV-1 (Doorbar et al., 1988); the 1.6 kb transcript also encodes a potential
E1-E4 fusion peptide although a different region of E1 is involved. The
functions of E4 have not been analysed but it is reasonable to assume that
they are similar to those of HPV E4. Like HPV-1 E4 (Breitburd, Croissant
& Orth, 1987), the majority of BPV-4 E4 is localized in the differentiating
layers of papillomas, coincident in time and in location with the vegetative
replication of viral DNA.

The E7 and E8 ORFs are transcribed into a 3.0 kb RNA (Fig. 1); they are
instrumental in cell transformation and will be discussed in greater detail
later. The E3 and E5 ORFs have no ATG initiation codon and may be non-

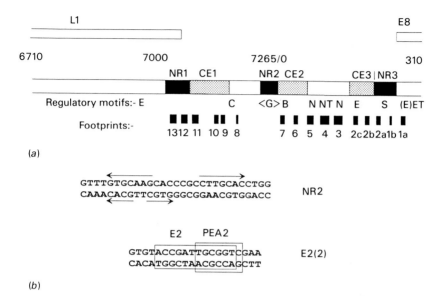

Fig. 2. (*a*), the BPV-4 long control region. The LCR maps between the 3′ end of the L1 ORF and the 5′ end of the E8 ORF. CE1–3 are the positive transcription regulatory elements and NR1–3, the negative elements. B, BPV-1 5′enhancer homology; C, cytokeratin octamer; E, E2-binding sites; (E), degenerate E2-binding site; ⟨G⟩, GC box bounded by inverted repeats; N, NF1 consensus related sequences; S, SV40 promoter homology; T, TATA box. The *in vitro* footprints, identifying binding sites for regulatory factors, are indicated in relation to the positive and negative *cis*-elements (for details, see Jackson & Campo, 1991). (*b*), Structure of the NR2 and E2(2) elements. The 7 bp and the 4 bp inverted repeats of NR2 are underlined and the E2 and PEA2 binding sites in E2(2) are boxed.

functional. The 3.0 kb and the 1.6 kb transcripts are the major viral RNAs present in *in vitro* transformed NIH 3T3 cells, while the 7E11 RNA is the most abundant transcript in papillomas but absent in transformed cells.

The L1 and L2 ORFs encode the structural proteins of the virion; they are transcribed into a 2.8 kb RNA capable of encoding L1, and a 4.2 kb RNA which has the capacity to encode both L1 and L2 (Fig. 1). These RNAs are found only in papillomas, in agreement with the lack of structural proteins in transformed cells. The L3 and L4 ORFs do have ATG start codons but their function, if any, is not known. A similarly sized L3 ORF is found in deer papillomavirus (Groff & Lancaster, 1985).

Control of transcription

The control of viral gene expression is the result of highly complex inter-actions between positive and negative, cellular and viral transcription regulators. Three E2 binding sites [E2(1), (2) and (3)] are found in the

BPV-4 LCR, which mediate the action of the E2 protein (Fig. 2(a)). Mutations that destroy the E2(2) site lead to a dramatic reduction in LCR promoter activity in the presence or absence of E2 protein. In addition to consensus E2 sites, a degenerate E2 binding site (dE2) is found close to E2(3) (Fig. 2(a)). A similar degenerate E2 site is present in an identical position in the LCR of BPV-6, but not in BPV-3 where a consensus site is found instead (Pennie, 1992). A cellular factor binds to the dE2 site in BPV-4 (Jackson & Campo, 1991), suggesting that the balance of nuclear factor binding in this region may have been altered in favour of cellular transcription regulators. Indeed, expression of the viral genes is regulated not only by the viral E2, but also by cellular factors which influence the promoter and enhancer activity of the LCR both in a positive and a negative manner. These cellular regulators bind to positive and negative *cis*-elements which are organized in pairs in the viral LCR (Fig. 2(a)). The pairing of each of the three positive sites (CE1–3) with a negative site (NR1–3) may well have a functional significance and the interplay between viral and cellular transcription regulators will impose an extra level of control on viral expression. For instance, the transcription factor PEA2, first identified as a nuclear factor binding to polyomavirus enhancer A (Piette & Yaniv, 1987), binds to a site overlapping to a large extent with the E2(2) site in CE3 (Fig. 2(b)). Given the overlap between the E2(2) and PEA2 sites, it is likely that the two factors bind to their cognate site in a mutually exclusive manner, and that competition for the binding site contributes to the fine control of viral transcription. Of the negative elements, only NR2 has been studied in detail. NR2 is composed of a seven base pair (bp) inverted repeat whose two halves are separated by a GC box; internal to the left hand 7 bp repeat there is a smaller inverted repeat of 4 bp (Fig. 2(b)). The site binds a cellular factor which may be related to C/EBP and mutation studies have shown that the inverted repeats rather than the GC box are essential for nuclear factor binding and for repression of transcription.

   Given the complexity of the regulation of viral transcription, any change in the balance of the controlling factors would lead to a breakdown in the circuitry, with possible increased expression of the viral transforming genes and thus progression to carcinoma, as already proposed by zur Hausen (1987). It is indeed the case that increased BPV-4 transcription leads to a more aggressive transformed cellular phenotype in *in vitro* systems (Smith & Campo, 1988; Jaggar *et al.*, 1990).

CELL TRANSFORMATION BY BPV-4

BPV-4, although capable of transforming established cells such as NIH 3T3 and C127 (Campo & Spandidos, 1983; Smith & Campo, 1988), is incapable of transforming primary bovine fetal palate (PalF) cells unless these are cotransfected with activated *ras* (Jaggar *et al.*, 1990); the viral genome

Table 1. *Step-wise neoplastic progression of BPV-4 transformed cells*

(+ras)	Morphological transformation	Anchorage independence	Immortality	Tumourigenicity
BPV-4	YES	YES	NO	NO
E7	YES	NO	no	no
E7 + E8	yes	YES	no	no
E7 + E8 + 16E6	yes	yes	YES	no
Q + BPV-4	yes	yes	YES	YES
Q + E7	yes	YES	YES	NO
Q + E7 + E8	yes	NO	n.e	NO

The major steps attributable to individual events are in upper case. n.e., not established. For details, see Jaggar *et al.*, 1990, Pennie & Campo, 1992 and Pennie *et al.*, 1993.

therefore does not contain all the necessary information to fully transform primary cells and needs the co-operation of an additional oncogene, resembling in this respect HPV-16 (Matlashewski *et al.*, 1987). Even in the presence of activated *ras*, the morphologically transformed cells, although non-contact inhibited and anchorage independent, are not immortal or tumourigenic in nude mice (Table 1), pointing to the need for additional events (Jaggar *et al.*, 1990; Pennie *et al.*, 1993).

The transforming functions of BPV-4 have been mapped to the E8 and E7 ORFs (Fig. 1; Smith & Campo, 1988; Jaggar *et al.*, 1990; Pennie *et al.*, 1993). This early ORF region induces gene amplification (Smith & Campo, 1989; Smith *et al.*, 1993), which has been observed also in the case of HPV-16 (Popescu & DiPaolo, 1990) and may be a general feature of papillomavirus transformation. BPV-4 E7 shows amino acid similarity with HPV-16 E7; it contains putative p105Rb and zinc binding domains and it is the major transforming protein of BPV-4, in that, in co-operation with *ras*, it induces morphological transformation of PalF cells in the absence of other viral genes (Table 1). Deletion of the second cys–x–x–cys motif or mutations in the p105Rb binding domain abolish cell transformation, demonstrating the importance of these sites and the pivotal role of E7 in cell transformation. Nevertheless, although morphologically transformed, E7 expressing cells are not capable of growing in semi-solid media (Table 1), thus showing that other viral gene(s) encode function(s) that confer anchorage independence (see below). In papillomas, E7 is expressed in all epithelial layers and in all the developmental stages of the tumour (Fig. 3(*a*) and (*b*)), pointing to its biological significance also *in vivo*.

E8 is the second transforming gene of BPV-4. When introduced into PalF cells, together with *ras* but in the absence of other viral genes, it does not confer any growth advantage and actually appears to accelerate senescence. However, when cotransfected with E7, it imparts anchorage independence (Table 1). It is not clear how E8 achieves this effect, but a comparison with

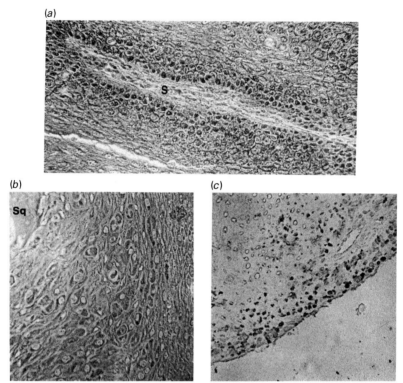

Fig. 3. Immunohistological localization of BPV-4 E7 and E8 proteins. (*a*), Localization of E7 in the basal and suprabasal layers of papillomas; S, stroma. (*b*), Localization of E7 in the superficial layers of papillomas; Sq, keratin squames. (*c*), Localization of E8 in the basal and suprabasal layers of papillomas.

BPV-1 E5 points to some possible mechanisms. Like E5 (Schlegel *et al.*, 1986), the E8 peptide is highly hydrophobic and theoretically capable of forming a transmembrane α-helix (Fig. 4); indeed, like E5 (Burkhardt *et al.*, 1989), E8 is localized in the cell membranes, mostly in the ER and the Golgi apparatus and occasionally in the plasma membrane (Pennie *et al.*, 1993). E8 interacts *in vitro* with 16k ductin (A. M. Faccini, M. E. Finbow and J. D. Pitts, unpublished results) and E8-containing cells have a much diminished gap junctional intercellular communication with surrounding cells (J. D. Pitts, W. D. Pennie and M. Cairney, unpublished results), suggesting that E8 may interact with the gap junctional form of 16k ductin. *In vivo*, E8 is expressed only in the deep layers of papillomas, where little or no vegetative viral DNA replication takes place (Fig. 3(*c*)) and is present only during the early stages of papilloma development; it can therefore be considered a true 'early' protein. This restricted localization in the tumours is not common to BPV-1 E5 which is expressed not only in the deep layers but in the

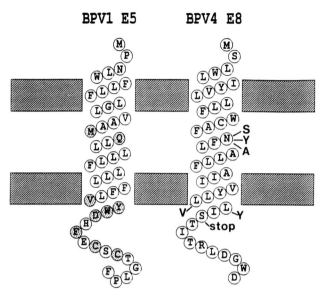

Fig. 4. Putative transmembrane configuration of BPV-1 E5 and BPV-4 E8 proteins. The C-terminus of BPV-1 E5 faces the extracellular space and the lumen of the ER and Golgi vesicles (Burkhardt *et al.*, 1989); BPV-4 E8 has been similarly oriented, although this remains to be proven. The BPV-1 E5 amino acids crucial for cell transformation (Horwitz *et al.*, 1988, 1989) are shaded. The mutations introduced in BPV-4 E8 are indicated; the asparagine residue at position 17 internal to the α-helix has been mutated to serine, tyrosine or alanine; the leucine residues at position 30 and 31 have been mutated to valine and tyrosine respectively, and a stop codon has been introduced at position 33.

differentiated ones as well (Burnett, Jarenborg & Di Maio, 1992). Critical amino acids, which have been shown to be necessary for E5-mediated *in vitro* cell transformation (Horwitz *et al.*, 1988, 1989), are either missing or different in E8. Mutations have been introduced in the E8 gene in an attempt to identify important functional domains. Thus amino acid residues in the C-terminus have been mutated to the corresponding residues in BPV-1 E5 and an asparagine residue in the α-helix portion of the protein, related to the glutamine of BPV-1 E5 crucial for 16k ductin binding, has been changed to a variety of smaller residues (Fig. 4). These mutants will provide an insight into E8 functions.

As mentioned above, BPV-4 transformed cells have an extended life span but are not immortal. This seems to be due, at least in the PalF *in vitro* transformation system, to the absence of the E6 ORF in the viral genome: the addition of the HPV-16 E6 gene to BPV-4 transformed cells leads to immortalization (Table 1). The lack of the E6 gene in BPV-4 has therefore a discernible effect *in vitro*, and the complementation with HPV-16 E6 clearly shows that BPV-4 does not encode E6-like functions. As already discussed, HPV-16 E6 binds and degrades p53, sequestering this protein from its negative control on cell growth; therefore, if p53 dysfunction is important in BPV-4 cell transformation, it may occur by alteration of the cellular gene. Alternatively, cells transformed in vivo by BPV-4 may be capable of evading p53 negative control by other as yet unidentified mechanisms.

Synergism between BPV-4 and chemical cofactors

Even when immortalized by the addition of HPV-16 E6, BPV-4 transformed cells are not tumorigenic in nude mice, indicating that additional events are needed for full transformation. It is worth remembering that in the field neoplastic progression of papillomas to carcinomas takes place only in those animals which are exposed to the mutagenic and immunosuppressive action of bracken fern chemicals (Campo & Jarrett, 1987). Numerous bracken fern mutagens have been identified in several laboratories, but one of the most potent is the 3,5,7,3′,4′-pentahydroxyflavone quercetin. Quercetin binds DNA and induces a variety of genetic lesions in bacteria and cultured mammalian cells (Jackson et al., 1993), including clastogenic damage (Ishidate, 1988). In addition, quercetin activates protein tyrosine kinases (Van Wart-Hood, Linder & Burr, 1989) and inhibits phosphatases and other kinases (Matter, Brown & Vlahos, 1992), therefore interfering with different signal transduction pathways. Quercetin alone does not act as a carcinogen (Morino et al., 1981), but can act as an initiator in a two-stage transformation assay in mammalian cells in vitro, with TPA as a promoter (Sakai et al., 1990). Given the role of the fern in malignant progression and the characteristics of quercetin, the effect of the latter as an initiating agent in the PalF system was investigated. Treatment of PalF cells with quercetin before transfection with BPV-4 DNA and ras induces full transformation and the cells are now oncogenic in nude mice and apparently immortal (Table 1). It has to be noticed that only cells containing the full BPV-4 genome are converted to malignancy by quercetin treatment (Table 1), suggesting that other viral genes may be involved in tumour formation. The observation that quercetin acts as an initiator in PalF cells leads to the formulation of the hypothesis that this flavonoid is a natural initiator of cancer in bracken-eating cattle; BPV-4 provides the initial proliferative stimulus, possibly by E7 binding to p105Rb, which leads to expansion of the initiated cells, and subsequent events, typified by ras mutations (Campo et al., 1990), convert cells to malignancy, in agreement with the requirement for activated ras for in vitro transformation. The rapidity of the conversion of PalF cells to malignancy emphasizes the synergism between the chemical, BPV-4 and ras. We do not yet know whether the target for quercetin is the cell genome, for instance by introducing mutations in critical genes, or whether its action takes place at an epigenetic level, for instance, through its interference with the phosphorylation cascade. An indication of quercetin's mode of action may be provided by the observation that quercetin-treated

Table 2. *Co-operation between BPV-4 and chemical co-factors*

Treatment	Implants	Papillomas (%)	Neoplasia (%)
Virus	9	0	0
Virus + TPA	33	11(33)	4(42)
TPA	25	0	0
Virus + DMBA	20	10(50)	13(65)
DMBA	10	0	0

TPA and DMBA were administered in slow-release pellet form. For details, see Gaukroger *et al.*, 1993.

E7-containing PalF cells are capable of anchorage independent growth in the absence of E8 (Table 1), suggesting that the chemical may be affecting the functions of cell adhesion molecules. This is, however, only one of several possibilities and requires further investigation. Whether the action of quercetin in inducing anchorage independence is similar to that of E8, or follows a different metabolic pathway remains to be seen, particularly in the light of the observation that, paradoxically, quercetin-treated E8 expressing cells are not anchorage-independent (Table 1).

Although BPV-4 needs the co-operation of additional factors to transform cells either *in vivo* or *in vitro*, it can synergize with both tumour initiators and tumour promoters. When chips of fetal bovine palate are infected with BPV-4 and then implanted beneath the renal capsule of nude mice (Kreider *et al.*, 1986), the bovine xenografts develop into virus-producing papillomas (Gaukroger *et al.*, 1989). These remain benign, and spontaneous neoplastic transformation has been observed only once (Gaukroger *et al.*, 1991). In contrast, when the mice are implanted with pellets releasing either the initiator 7-12-dimethylbenz[*a*]anthracene (DMBA) or the promoter 12-*o*-tetradecanoylphorbol-13-acetate (TPA), the rate of conversion of papillomas to carcinomas is dramatically increased (Table 2; Gaukroger *et al.*, 1993). The ability of BPV-4 to synergize with both a classic initiator and a classical promoter is not too surprising in the light of the following observations. DMBA induces activating mutations in *ras* (Quintanilla *et al.*, 1986) and causes the amplification of the epidermal growth factor receptor (EGF-R) gene (Wong, 1987); we do know that *ras* is activated in bovine alimentary canal cancers (Campo *et al.*, 1990) and that there is a greater number of EGF-R in alimentary canal cancer cells (Smith *et al.*, 1987). TPA treatment induces BPV-4 DNA amplification and enhances viral transcription (Smith *et al.*, 1987; Smith & Campo, 1988) and increased transcription results in greater cell transformation (Jaggar *et al.*, 1990). Thus TPA leads to accelerated transformation by BPV-4 probably by increasing the levels of the viral oncoproteins, while DMBA does so probably by rendering the cells more susceptible to their action through mutations in cellular genes.

PROPHYLACTIC AND THERAPEUTIC VACCINATION AGAINST BPV-4

Mucosal papillomavirus infections are often serious lesions which tend to run a prolonged course and may subsequently develop to malignancies, as is the case for HPV-16 infection of the genital tract in humans (zur Hausen, 1991). Prophylactic and therapeutic vaccines which would protect against infection and accelerate tumour rejection respectively would eventually result in a decreased incidence of genital cancer (Campo, 1991). The bovine alimentary canal papilloma-carcinoma system provides an excellent model in which to investigate the development of such vaccines.

Two viral protein have been used in vaccination studies: L2 and E7. L2 is the minor capsid component of the virion and E7 is the major viral oncoprotein. These proteins have been produced in bacteria as glutathione-S-transferase (GST) and β-galactosidase (β-gal) fusion proteins, respectively. Vaccination of calves with L2 induces protection from challenge with BPV-4 (Fig. 5; Campo et al., 1993): while the control non-vaccinated virus-infected animals develop alimentary papillomas as expected, the vaccinated animals are virtually all lesion free; immunity is long-lasting and the animals are refractory to a second virus challenge more than a year after vaccination (Brian O'Neil and others, unpublished results). The vaccinated animals develop high titre L2 antibodies, which are directed primarily against the carboxyl terminus of the protein (Table 3) and are capable of neutralizing virus (J. W. Baird, L. M. Chandrachud and J. M. Gaukroger, unpublished results). These observations imply that the L2 protein encodes non-conformational neutralizing epitopes in its C-terminus which therefore must be exposed on the surface of the virion. Interestingly, vaccination of rabbits with the L2 protein of cottontail rabbit papillomavirus (CRPV) leads to similar results (Christensen et al., 1991; Lin et al., 1992), allowing the optimistic prediction that the same effect will be obtained with L2 of HPV.

Before the outcome of E7 vaccination can be considered, a brief description of the developmental stages of alimentary canal papillomas is needed. Stage 1 consists of round, slightly raised small plaques which appear approximately four weeks after infection; at this stage there is no virus replication. Stage 2 consists of tubular papillomas with obvious cytopathic changes in the granular layer which develop approximately eight weeks after infection; virus replicates and virion progeny is found in large crystalloid arrays. Stage 3 starts approximately twelve weeks after infection and persists until spontaneous regression takes place; it consists of frond papillomas with a constricted base in which virus replication is reduced and then ceases; there is no virus in older tumours. The developmental stages of alimentary

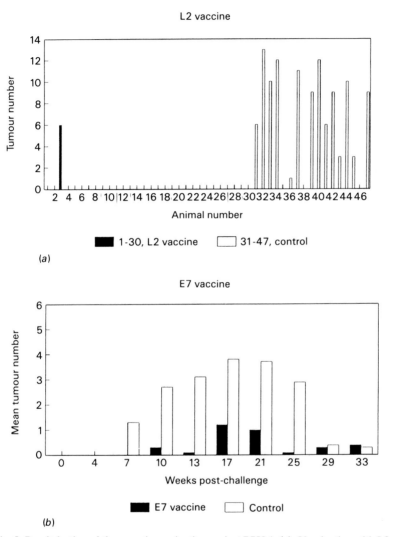

Fig. 5. Prophylactic and therapeutic vaccination against BPV-4. (*a*), Vaccination with L2; papillomas in vaccinated (animal 1–30) and control (animal 31–47) calves. (*b*), Vaccination with E7; number of stage 3 papillomas in vaccinated and control calves. For details, see Campo *et al.*, 1993.

Table 3. *Immunodominant epitopes in BPV-4 proteins*

L2	N-ter (aa 11-200)	middle (aa 201-326)	C-ter (aa 327-524)
positive animals	0/6	0/6	6/6
E7	B1 (aa 1–30)	B2 (aa 79–98)	B3 (aa 50–69)
positive animals	24/27	23/27	6/27
	T1 (aa 31–59)	T2 (aa 70–88)	
positive animals	2/3	1/3	

The L2 epitopes were determined by using bacterially synthesized peptides corresponding to the N-terminus, the middle portion and the C-terminus of the protein. For details, see Campo *et al.*, 1993.
The E7 epitopes were determined by using synthetic overlapping peptides.

papillomas have been described in greater detail by Jarrett (1985). Vacci-nation with E7 does not prevent infection and both non vaccinated and vaccinated animals develop stage 1 and stage 2 lesions. However, while in the control animals stage 2 papillomas convert to stage 3, in the vaccinated animals this conversion takes place rarely (Fig. 5) and the tumours start regressing while still at stage 2 (Campo *et al.*, 1993). Thus vaccination with E7 slows down tumour growth and induces early tumour regression. In contrast to the control animals, the vaccinated animals have high titre antibodies against E7 which are directed to three immunodominant epitopes mapped in the N-terminus (amino acid 1-30), in the middle portion (aa 50-69) and in the C-terminus (aa 79–98) of the protein (Table 3; Chandrachud *et al.*, 1994). B-cell epitopes have been mapped in equivalent areas of the HPV-16 E7 protein both in infected human subjects (Krchnak *et al.*, 1990) and in mice (Comerford *et al.*, 1991; Tindle *et al.*, 1990). E7 vaccination elicits also a very strong cellular immune response, not observed in control animals. T-lymphocytes from vaccinated animals proliferate vigorously *in vitro* when challenged with the antigen, while T-lymphocytes from control animals react much more sluggishly if at all. Two T-cell epitopes have been identified in E7, at aa 31–59 and aa 70–88 (G. M. McGarvie, unpublished results). As in the case of B-cell epitopes, the T-cells epitopes map in equivalent areas of HPV-16 E7 (Comerford *et al.*, 1991; Tindle *et al.*, 1991). The coincidence of epitope sites between BPV-4 and HPV-16 E7 proteins reinforces the contention that the bovine system is a good predictive model for HPV-16 infections. Indeed, vaccination of mice with a live recombinant vaccinia virus containing the E7 gene of HPV-16 resulted in delayed growth and accelerated rejection both of tumours induced by HPV-16 transformed cells (Meneguzzi *et al.*, 1991; Chen *et al.*, 1991) and of grafts of HPV-16 infected human cells (McLean *et al.*, 1993). It is not yet firmly established whether rejection is due to a CD8$^+$ cytotoxic T-cell response (Chen *et al.*, 1991) or a CD4$^+$ response (McLean *et al.*, 1993; Hopfl *et al.*, 1993). This uncertainty notwithstanding, if the results obtained in the bovine and other animal models can be extrapolated to the human disease, in the not so distant future we may have available two immunological ways of containing papillomavirus infection: either prevention of infection by eliciting neutralizing antibodies through a conventionally prophylactic vaccine, or early regression of premalignant lesions by inducing a cell mediated immune response through a therapeutic vaccine.

CONCLUSIONS

Although in recent years research on papillomavirus has focussed, for obvious reasons, on the human viruses, it must not be forgotten that the first steps in the molecular aspects of cell transformation were made with animal papillomaviruses, and the contribution of BPV cannot be overestimated. Indeed, BPV provides a continuous source of information on cellular transformation and immunological mechanisms. For instance, the recent identification of HPV-16 E5 as the third oncogene of this virus followed the discovery of the role of BPV-1 E5 in signal transduction. Also the possibility of vaccination against a mucosal papillomavirus has proven a reality in the BPV-4 system. We can therefore anticipate that further insights into HPV biology and immunology will be provided by research on BPV.

ACKNOWLEDGEMENTS

We are indebted to all our collaborators and colleagues who have made available to us their unpublished results. Our own work is supported by the Cancer Research Campaign, of which M.S.C. is a Fellow.

REFERENCES

Androphy, E. J., Lowy, D. R. & Schiller, J. T. (1987). Bovine papillomavirus E2 *trans*-activating gene product binds to specific sites in papillomavirus DNA. *Nature*, **325**, 70–3.

Barbosa, M. S., Lowy, D. R. & Schiller, J. T. (1989). Papillomavirus peptides E6 and E7 are zinc-binding proteins. *Journal of Virology*, **63**, 1404–7.

Breitburd, F., Croissant, O. & Orth, G. (1987). Expression of human papillomavirus type 1 E4 gene product in warts. *Cancer Cells*, **5**, 115–22.

Burkhardt, A., Willingham, M., Gay, C., Jeang, K.-T. & Schlegel, R. (1989). The E5 oncoprotein of bovine papillomavirus is oriented asymmetrically in Golgi and plasma membranes. *Virology*, **170**, 334–9.

Burnett, S., Jarenborg, N. & DiMaio, D. (1992). Localization of bovine papillomavirus type 1 E5 protein to transformed basal keratinocytes and permissive differentiated cells in fibropapilloma tissue. *Proceedings of the National Academy of Sciences, USA,* **89,** 5665–9.

Campo, M. S. (1991). Vaccination against papillomavirus. *Cancer Cells,* **3,** 421–6.

Campo, M. S. (1992). Cell transformation by animal papillomaviruses. *Journal of General Virology,* **73,** 217–22.

Campo, M. S. & Jarrett, W. F. H. (1987). Papillomaviruses and disease. In *Molecular Basis of Virus Disease,* ed. W. C. Russell and J. W. Almond. pp. 215–243, No. 40, Society of General Microbiology, Cambridge University Press.

Campo, M. S. & Spandidos, D. A. (1983). Molecularly cloned bovine papillomavirus DNA transforms mouse fibroblasts *in vitro. Journal of General Virology,* **64,** 549–57.

Campo, M. S., Grindlay, G. J., O'Neil, B. W., Chandrachud, L. M., McGarvie, G. M. & Jarrett, W. F. H. (1993). Prophylactic and therapeutic vaccination against a mucosal papillomavirus. *Journal of General Virology,* **74,** 945–53.

Campo, M. S., Jarrett, W. F. H., Barron, R., O'Neil, B. W. & Smith, K. T. (1992). Association of bovine papillomavirus type 2 and bracken fern with bladder cancer in cattle. *Cancer Research,* **53,** 1–7.

Campo, M. S., McCaffery, R. E., Doherty, I., Kennedy, I. M. & Jarrett, W. F. H. (1990). The Harvey *ras* 1 gene is activated in papillomavirus-associated carcinomas of the alimentary canal in cattle. *Oncogene,* **5,** 303–8.

Campo, M. S., Moar, M. H., Jarrett, W. F. H. & Laird, H. M. (1980). A new papillomavirus associated with alimentary tract cancer in cattle. *Nature,* **286,** 180–2.

Chandrachud, L. M., O'Neil, B. W., Jarrett, W. F. H., Grindlay, G. J., McGarvie, G. M. & Campo, M. S. (1994). Humoral immune response to the E7 protein of bovine papillomavirus type 4 and identification of B-cell epitopes. *Virology,* in press.

Chen, E. Y., Howley, P. M., Levinson, A. D. & Seeburg, P. H. (1982). The primary structure and genetic organization of bovine papillomavirus type 1. *Nature,* **299,** 529–34.

Chen, L., Thomas, E. K., Hu, S.-L., Hellstrom, I. & Hellstrom, K. E. (1991). Human papillomavirus type 16 nucleoprotein E7 is a tumour rejection antigen. *Proceedings of the National Academy of Sciences, USA,* **88,** 110–14.

Christensen, N. D., Kreider, J. W., Kan, N. C. & DiAngelo, S. L. (1991). The open reading frame L2 of cottontail rabbit papillomavirus contains antibody-inducing neutralizing epitopes. *Virology,* **181,** 572–9.

Comerford, S. A., McCance, D. J., Dougan, G. & Tite, J. P. (1991). Identification of T- and B-cell epitopes of the E7 protein of human papillomavirus type 16. *Journal of Virology,* **65,** 4681–90.

Danos, O., Katinka, M. & Yaniv, M. (1982). Human papillomavirus 1a complete DNA sequence: a novel type of genome organization among Papovaviridae. *EMBO Journal* **1,** 231–6.

Desaintes, C., Hallez, S., van Alphen, P. & Burny, A. (1992). Transcriptional activation of several heterologous promoters by the E6 protein of human papillomavirus type 16. *Journal of Virology,* **66,** 325–33.

Doorbar, J. (1991). An emerging function for E4. *Papillomavirus Report,* **2,** 145–7.

Doorbar, J., Campbell, D., Grand, R. J. A. & Gallimore, P. H. (1986). Identification of the human papillomavirus 1a E4 gene product. *EMBO Journal,* **5,** 355–62.

Doorbar, J., Evans, H. S., Coneron, I., Crawford, L. V. & Gallimore, P. H. (1988). Analysis of HPV-1 E4 gene expression using epitope-defined antibodies. *EMBO Journal,* **7,** 825–33.

Doorbar, J., Ely, S., Sterling, J., McLean, C. & Crawford, L. (1991). Specific interaction between HPV-16 E1-E4 and cytokeratins results in collapse of the epithelial cell intermediate filament network. *Nature,* **352**, 824–7.

Dvoretzky, I., Shober, R., Chattopadhyay, S. K. & Lowy, D. R. (1980). A quantitative *in vitro* focus assays for bovine papillomavirus. *Virology,* **103**, 369–75.

Dyson, N., Howley, P. M., Munger, K. & Harlow, E. (1989). The human papillomavirus 16 E7 oncoprotein is able to bind to the retinoblastoma gene product. *Science,* **243**, 934–6.

Dyson, N., Guida, P., Munger, K. & Harlow, E. (1992). Homologous sequences in adenovirus E1A and human papillomavirus E7 proteins mediate interaction with the same set of cellular proteins. *Journal of Virology,* **66**, 6893–902.

Evans, I. A., Prorok, J. H., Cole, R. C. *et al.* (1982). The carcinogenic, mutagenic and teratogenic toxicity of bracken. *Proceedings of the Royal Society of Edinburgh,* **81**, 65–77.

Evans, W. C., Patel, M. C. & Koohy, Y. (1982). Acute bracken poisoning in homogastric and ruminant animals. *Proceedings of the Royal Society of Edinburgh,* **81**, 29–64.

Favre, M., Breitburd, F., Croissant, O. & Orth, G. (1975). Structural polypeptides of rabbit, bovine and human papillomaviruses. *Journal of Virology,* **15**, 1239–47.

Finbow, M. E., Pitts, J. D., Goldstein, D. J., Schlegel, R. & Findlay, J. B. C. (1991). The E5 oncoprotein target: a 16 kd channel-forming protein with diverse functions. *Molecular Carcinogenesis,* **4**, 441–4.

Ford, J. N., Jennings, P. A., Spradbrow, P. B. & Francis, J. (1982). Evidence for papillomaviruses in ocular lesions in cattle. *Research in Veterinary Science,* **32**, 258–9.

Gaukroger, J., Bradley, A., O'Neil, B. W., Smith, K. T., Campo, M. S. & Jarrett, W. F. H. (1989). Induction of virus-producing tumours in athymic nude mice by bovine papillomavirus type 4. *Veterinary Record,* **125**, 391–2.

Gaukroger, J., Chandrachud, L., Jarrett, W. F. H. *et al.* (1991). Malignant transformation of a papilloma induced by bovine papillomavirus type 4 in the nude mouse renal capsule. *Journal of General Virology,* **72**, 1165–68.

Gaukroger, J. M., Bradley, A., Chandrachud, L., Jarrett, W. F. H. & Campo, M. S. (1993). Interaction between bovine papillomavirus type 4 and cocarcinogens in the production of malignant tumours. *Journal of General Virology,* **74**, 2275–80.

Goldstein, D. J., Finbow, M. E., Andresson, T. *et al.* (1991). Bovine papillomavirus E5 binds to the 16k component of vacuolar H^+-ATPases. *Nature,* **352**, 347–9.

Groff, D. E. & Lancaster, W. D. (1985). Molecular cloning and nucleotide sequence of deer papillomavirus. *Journal of Virology,* **56**, 85–91.

Haugen, T. H., Cripe, T. P., Ginder, G. D., Karin, M. & Turek, L. P. (1987). *Trans*-activation of an upstream early promoter of bovine papillomavirus-1 by a product of the viral E2 gene. *EMBO Journal,* **6**, 145–52.

Heike, T., Miyatake, S., Yoshida, M., Arai, K. & Arai, N. (1989). Bovine papillomavirus encoded E2 protein activates lymphokine genes through DNA elements, distinct from the consensus motif, in the long control region of its own genome. *EMBO Journal,* **8**, 1411–17.

Holzenburg, A., Jones, P. C., Franklin, T. *et al.* (1993). Evidence for a common structure for a class of membrane channels. *European Journal of Biochemistry,* **213**, 21–30.

Hopfl, R., Christensen, N. D., Angell, M. G. & Kreider, J. W. (1993). Skin test to assess immunity against cottontail rabbit papillomavirus antigens in rabbits with progressing papillomas or after papilloma regression. *Journal of Investigative Dermatology,* **101**, 227–31.

Horwitz, B. H., Burkhardt, A. L., Schlegel, R. & DiMaio, D. (1988). 44-amino acid E5 transforming protein of bovine papillomavirus requires a hydrophobic core and specific carboxyl-terminal amino acids. *Molecular and Cell Biology*, **8**, 4071–8.

Horwitz, B. H., Weinstat, D. L. & DiMaio, D. (1989). Transforming activity of a 16-amino acid segment of the bovine papillomavirus E5 protein linked to random sequences of hydrophobic amino acids. *Journal of Virology*, **63**, 4515–19.

Iftner, T. (1990). Papillomavirus genomes: sequence analysis related to functional aspects. In *Papillomaviruses and Human Cancer*, ed. H. Pfister, pp. 181–200. CRC Press Inc.

Ishidate, M. (1988). *Data Book on Chromosomal Aberration Tests in Vitro*. Elsevier.

Jackson, M. E. & Campo, M. S. (1991). Positive and negative E2-independent regulatory elements in the long control region of bovine papillomavirus type 4. *Journal of General Virology*, **72**, 877–83.

Jackson, M. E., Pennie, W. D., McCaffery, R. E., Smith, K. T., Grindlay, G. J. & Campo, M. S. (1991). The B subgroup bovine papillomaviruses lack an identifiable E6 open reading frame. *Molecular Carcinogenesis*, **4**, 382–7.

Jackson, M., Campo, M. S. & Gaukroger, J. M. (1993). Cooperation between papillomavirus and chemical cofactors in oncogenesis. *Critical Reviews in Oncogenesis*, **4**, 277–91.

Jaggar, R. T., Pennie, W. D., Smith, K. T., Jackson, M. E. & Campo, M. S. (1990). Cooperation between bovine papillomavirus type 4 and *ras* in the morphological transformation of primary bovine fibroblasts. *Journal of General Virology*, **71**, 3041–6.

Jarrett, W. F. H. (1985). The natural history of bovine papillomavirus infection. *Advances in Viral Oncology*, **5**, 83–102.

Jarrett, W. F. H., McNeal, P. E., Grimshaw, T. R., Selman, I. E. & McIntyre, W. I. M. (1978). High incidence area of cattle cancer with a possible interaction between an environmental carcinogen and a papillomavirus. *Nature*, **274**, 215–17.

Krchnak, V., Vagner, J., Suchankova, A., Krcmar, M., Ritterova, L. & Vonka, V. (1990). Synthetic peptides derived from E7 region of human papillomavirus type 16 used as antigens in ELISA. *Journal of General Virology*, **71**, 2719–24.

Kreider, J. W., Howett, M. K, Lill, N. L. *et al.* (1986). *In vivo* transformation of human skin with human papillomavirus type 11 from *Condylomata acuminata*. *Journal of Virology*, **59**, 369–76.

Lambert, P. F., Spalholz, B. A. & Howley, P. M. (1987). A transcriptional repressor encoded by BPV-1 shares a common carboxy-terminal domain with the E2 transactivator. *Cell*, **50**, 69–78.

Lamberti, C., Morrissey, L. C., Grossman, S. R. & Androphy, E. J. (1990). Transcriptional activation by the papillomavirus E6 zinc finger oncoprotein. *EMBO Journal*, **9**, 1907–13.

Lancaster, W. D., Theilen, G. H. & Olson, C. (1979). Hybridization of bovine papillomavirus type 1 and type 2 DNA to DNA from virus-induced hamster and naturally occurring equine tumours. *Intervirology*, **11**, 227–33.

Law, M.-F., Lowy, D. R., Dvoretzky, I. & Howley, P. M. (1981). Mouse cells transformed by bovine papillomavirus contain only extrachromosomal viral DNA sequences. *Proceedings of the National Academy of Sciences, USA*, **78**, 2727–31.

Lechner, M. S., Mack, D. H., Finicle, A. B., Crook, T., Vousden, K. H. & Laimins, L. A. (1992). Human papillomavirus E6 proteins bind p53 in vivo and abrogate p53-mediated repression of transcription. *EMBO Journal*, **11**, 3045–52.

Leechanachai, P., Banks, L., Moreau, F. & Matlashewski, G. (1992). The E5 gene from human papillomavirus type 16 is an oncogene which enhances growth factor-mediated signal transduction to the nucleus. *Oncogene*, **7**, 19–25.

Lentz, M. R., Pak, D., Mohr, I. & Botchan, M. R. (1993). The E1 replication protein of bovine papillomavirus type 1 contains an extended nuclear localization signal that includes a p34^{cdc2} phosphorylation site. *Journal of Virology, 67*, 1414-23.

Leptak, C., Ramon y Cajal, S., Kulke, R. *et al.* (1991). Tumorogenic transformation of mouse keratinocytes by the E5 genes of bovine papillomavirus type 1 and human papillomavirus type 16. *Journal of Virology, 65*, 7078–83.

Lin, Y.-L., Borenstein, L. A., Selvakumar, R., Ahmed, R. & Wettstein, F. O. (1992). Effective vaccination against papilloma development by immunization with L1 or L2 structural proteins of cottontail rabbit papillomavirus. *Virology, 187*, 612–19.

Lusky, M. & Botchan, M. R. (1986). Transient replication of BPV-1 plasmids: *cis* and *trans* requirements. *Proceedings of the National Academy of Sciences, USA, 83*, 3609–13.

McLean, C. S., Sterling, J. S., Mowat, J., Nash, A. A. & Stanley, M. A. (1993). Delayed type hypersensitivity response to the human papillomavirus type 16 protein in a mouse model. *Journal of General Virology, 74*, 239–45.

Martin, P., Vass, W. C., Schiller, J. T., Lowy, D. R. & Velu, T. J. (1989). The bovine papillomavirus E5 transforming protein can stimulate the transforming activity of EGF and CSF-1 receptors. *Cell, 59*, 21–32.

Matlashewski, G., Schneiser, J., Banks, L., Jones, N., Murray, A. & Crawford, L. (1987). Human papillomavirus type 16 DNA cooperates with activated *ras* in transforming primary cells. *EMBO Journal, 6*, 1741–6.

Matsukura, T., Kanda, T., Furuno, A., Yoshikawa, H., Kawana, T. & Yoshiike, K. (1986). Cloning of monomeric human papillomavirus type 16 DNA integrated within cell DNA from a cervical carcinoma. *Journal of Virology, 58*, 979–82.

Matter, W. F., Brown, R. F. & Vlahos, C. J. (1992). The inhibition of phosphatidyl-inositol 3-kinase by quercetin and analogs. *Biochemical and Biophysical Research Communications, 186*, 624–31.

Meneguzzi, G., Cerni, C., Kieny, M. P. & Lathe, R. (1991). Immunization against human papillomavirus type 16 tumour cells with recombinant vaccinia viruses expressing E6 and E7. *Virology, 181*, 62–9.

Morgan, D. M. & Meinke, W. (1980). Isolation of clones of hamster embryo cells transformed by the bovine papillomavirus. *Current Microbiology, 3*, 247–51.

Morino, K., Matsukura, N., Kawachi, T., Ohgaki, H., Sugimura, T. & Hirono, I. (1981). Carcinogenicity test of quercetin and rutin in golden hamsters by oral administration. *Carcinogenesis, 3*, 93–7.

Munger, K., Werness, B. A., Dyson, N., Phelps, W. C., Harlow, E. & Howley, P. M. (1989). Complex formation of human papillomavirus E7 proteins with the retinoblastoma tumour suppressor gene product. *EMBO Journal, 8*, 4099–105.

Quintanilla, M., Brown, K., Ramsden, M. & Balmain, A. (1986). Carcinogen-specific mutation and amplification of Ha-ras during mouse skin carcinogenesis. *Nature, 322*, 78–80.

Pennie, W. D. (1992). Analysis of the transforming functions of bovine papillomavirus type 4 in an *in vitro* assay system. PhD Thesis, University of Glasgow.

Pennie, W. D. & Campo, M. S. (1992). Synergism between bovine papillomavirus type 4 and the flavonoid quercetin in cell transformation *in vitro. Virology, 190*, 861–5.

Pennie, W. D., Grindlay, G. J., Cairney, M. & Campo, M. S. (1993). Analysis of the transforming functions of bovine papillomavirus type 4. *Virology, 193*, 614–20.

Petti, L. & DiMaio, D. (1992). Stable association between the bovine papillomavirus E5 transforming protein and activated platelet-derived growth factor receptor in transformed mouse cells. *Proceedings of the National Academy of Sciences, USA,* **89**, 6736–40.

Petti, L., Nilson, L. A. & DiMaio, D. (1991). Activation of the platelet-derived growth factor receptor by the bovine papillomavirus E5 transforming gene. *EMBO Journal,* **10**, 845–55.

Phelps, W. C., Yee, C. L., Munger, K. & Howley, P. M. (1988). The human papillomavirus type 16 E7 gene encodes transactivation and transformation functions similar to those of adenovirus E1A. *Cell,* **53**, 539–47.

Phelps, W. C., Munger, K., Yee, C. L., Barnes, J. A. & Howley, P. M. (1992). Structure–function analysis of the human papillomavirus type 16 E7 protein. *Journal of Virology,* **66**, 2418–27.

Piette, J. & Yaniv, M. (1987). Two different factors bind to the α-domain of the polyomavirus enhancer, one of which also interacts with the SV40 and *c-fos* enhancers. *EMBO Journal,* **6**, 1331–7.

Pim, D., Collins, M. & Banks, L. (1992). Human papillomavirus type 16 E5 gene stimulates the transforming activity of the epidermal growth factor receptor. *Oncogene,* **7**, 27–32.

Popescu, N. C. & DiPaolo, J. A. (1990). Integration of human papillomavirus 16 DNA and genomic rearrangements in immortalized human keratinocyte lines. *Cancer Research,* **50**, 1316–23.

Prakash, S. S., Horwitz, B. H., Zibello, T., Settleman, J. & DiMaio, D. (1988). Bovine papillomavirus E2 gene regulates expression of the viral E5 transforming gene. *Journal of Virology,* **62**, 3608–13.

Rogel-Gaillard, C., Breitburd, F. & Orth, G. (1992). Human papillomavirus type 1 E4 proteins differing by their N-terminal ends have different cellular localizations when transiently expressed *in vitro. Journal of Virology,* **66**, 816–23.

Sakai, A., Sasaski, K., Mizusawa, H. & Ishidate, M. (1990). Effects of quercetin, a plant flavonol on two-stage transformation *in vitro. Teratogenesis, Carcinogenesis, and Mutagenesis,* **10**, 333–40.

Scheffner, M., Werness, B. A., Huibregtse, J. M., Lavine, A. J. & Howley, P. M. (1990). The E6 oncoprotein encoded by human papillomavirus type 16 and 18 promotes the degradation of p53. *Cell,* **63**, 1129–36.

Schiller, J. T., Vass, W. C., Vousden, K. H. & Lowy, D. R. (1986). E5 open reading frame of bovine papillomavirus type 1 encodes a transforming gene. *Journal of Virology,* **57**, 1–6.

Schlegel, R. & Wade-Glass, M. (1987). E5 transforming polypeptide of bovine papillomavirus. *Cancer Cells,* **5**, 87–91.

Schlegel, R., Wade-Glass, M., Rabson, M. S. & Yang, Y.-C. (1986). The E5 transforming gene of bovine papillomavirus encodes a small hydrophobic peptide. *Science,* **233**, 464–67.

Schwarz, E., Freese, U. K., Gissmann, L. *et al.* (1985). Structure and transcription of human papillomavirus sequences in cervical carcinoma cells. *Nature,* **314**, 111–14.

Sedman, S. A., Barbosa, M. S., Vass, W. C. (1991). The full-length E6 protein of human papillomavirus type 16 has transforming and *trans*-activating activities and cooperates with E7 to immortalise keratinocytes in culture. *Journal of Virology,* **65**, 4860–6.

Sippola-Thiele, M., Hanahan D. & Howley, P. M. (1989). Cell-heritable stages of tumour progression in transgenic mice harbouring the bovine papillomavirus type 1 genome. *Molecular and Cell Biology,* **9**, 925–34.

Smith, K. T. & Campo, M. S. (1988). 'Hit and Run' transformation of mouse C127 cells by bovine papillomavirus type 4: the viral DNA is required for the initiation but not the maintenance of the transformed phenotype. *Virology,* **164**, 39–47.

Smith, K. T. & Campo, M. S. (1989). Amplification of specific DNA sequences in C127 mouse4 cells transformed by bovine papillomavirus type 4. *Oncogene,* **4**, 409–13.

Smith, K. T., Campo, M. S., Bradley, J., Gaukroger, J. & Jarrett, W. F. H. (1987). Cell transformation by bovine papillomavirus: cofactors and cellular responses. *Cancer Cells,* **5**, 267–74.

Smith, K. T., Coggins, L. W., Doherty, I., Pennie, W. D., Cairney, M. & Campo, M. S. (1993). BPV-4 induces amplification and activation of 'silent' BPV-1 in a subline of C127 cells. *Oncogene,* **8**, 151–6.

Smith, K. T., Patel, K. R. & Campo, M. S. (1986). Transcriptional organization of bovine papillomavirus type 4. *Journal of General Virology,* **67**, 2381–93.

Sousa, R., Dostatni, N. & Yaniv, M. (1990). Control of papillomavirus gene expression. *Biochimica Biophysica Acta,* **1032**, 19–37.

Spalholz, B. A., Yang, Y.-C. & Howley, P. M. (1985). Transactivation of a bovine papillomavirus transcriptional regulatory element by the E2 gene product. *Cell,* **42**, 183–91.

Spradbrow, P. B. & Hoffmann, D. (1980). Bovine ocular squamous cell carcinoma. *Veterinary Bulletin,* **50**, 449–59.

Stamps, A. C. & Campo, M. S. (1988). Mapping of two novel transcripts of bovine papillomavirus type 4. *Journal of General Virology,* **69**, 3033–45.

Straight, S. W., Hinkle, P. M., Jewers, R. J. & McCance, D. J. (1993). The E5 oncoprotein of human papillomavirus type 16 transforms fibroblasts and affects the downregulation of the epidermal growth factor receptor in keratinocytes. *Journal of Virology,* **67**, 4521–32.

Sun, S., Thorner, L., Lentz, M., MacPherson, P. & Botchan, M. R. (1990). Identification of a 68-kilodalton nuclear ATP-binding phosphoprotein encoded by the bovine papillomavirus type 1. *Journal of Virology,* **64**, 5093–105.

Tindle, R. W., Smith, J. A., Geysen, H. M., Selvey, L. A. & Frazer, I. H. (1990). Identification of B epitopes in human papillomavirus type 16 E7 open reading frame protein. *Journal of General Virology,* **71**, 1347–54.

Tindle, R. W., Fernando, G. J. P., Sterling, J. & Frazer, I. H. (1991). A public T-helper epitope of the E7 transforming protein of human papillomavirus 16 provides cognate help for several B-cell epitopes from cervical cancer-associated human papillomavirus genotypes. *Proceedings of the National Academy of Sciences, USA,* **88**, 5887–91.

Tommasino, M., Adamczewski, J. P., Carlotti, F. *et al.* (1993). HPV-16 E7 protein associates with the protein kinase p33^{CDK2} and cyclin A. *Oncogene,* **8**, 195–202.

Ustav, M., Ustav, E., Szymanski, P. & Stenlund, A. (1991). Identification of the origin of replication of bovine papillomavirus and characterization of the viral origin recognition factor E1. *EMBO Journal,* **10**, 4321–9.

Van Wart-Hood, J., Linder, M. E. & Burr, J. G. (1989). TPCK and quercetin act synergistically with vandate to increase protein–tyrosine phosphorylation in avian cells. *Oncogene,* **4**, 1267–71.

Watanabe, S., Kanda, T., Sato, H., Furuno, A. & Yoshiike, K. (1990). Mutational analysis of human papillomavirus type 16 E7 functions. *Journal of Virology,* **64**, 207–14.

Werness, B. A., Levine, A. J. & Howley, P. M. (1990). Association of human papillomavirus types 16 and 18 E6 proteins with p53. *Science,* **248**, 76–9.

Wong, D. T. W. (1987). Amplification of the c-erb-B1 oncogene in chemically induced oral carcinomas. *Carcinogenesis*, **8**, 1963–5.

Yang, L., Li, R., Mohr, I. J., Clark, R. & Botchan, M. R. (1991). Activation of BPV-1 replication *in vitro* by the transcription factor E2. *Nature*, **353**, 628–32.

Yang, L., Mohr, I. J., Li, R., Mottoli, T., Sun, S. & Botchan, M. R. (1992). The transcription factor E2 regulates BPV-1 DNA replication *in vitro* by direct protein-protein interaction. *Cold Spring Harbor Symposia on Quantitative Biology*, **56**, 335–46.

Yang, Y.-C., Okayama, H. & Howley, P. M. (1985). Bovine papillomavirus contains multiple transforming genes. *Proceeding of the National Academy of Sciences, USA*, **82**, 1030–4.

zur Hausen, H. (1987). Papillomaviruses in human cancer. *Cancer*, **59**, 1692–6.

zur Hausen, H. (1991). Human papillomaviruses in the pathogenesis of anogenital cancer. *Virology*, **184**, 9–13.

TOWARDS HPV VACCINATION

A. ALTMANN[1], L. GISSMANN[2], C. SOLER[3] AND I. JOCHMUS[1]

[1]*Deutsches Krebsforschungszentrum, Angewandte Tumorvirologie, Forschungsschwerpunkt Genomveränderungen und Carcinogenese, 69121 Heidelberg, Germany,* [2]*Loyola University Medical Center, Department of Obstetrics and Gynecology, Maywood, Illinois 60153, USA,* [3]*Inserm U346, Lyon, France.*

INTRODUCTION

Papillomaviruses are a group of heterogeneous viruses, which induce epithelial proliferations in a wide range of vertebrate hosts. In humans, more than 60 different types have been described so far many of which primarily infect the anogenital tract (De Villiers, 1989). The 'low risk' types like HPV 6 or 11, cause benign cell proliferation (Condolymata accuminata, flat warts) which rarely progress to malignancy. In contrast, some mucosotropic HPV types, including HPV 16, 18, 31 and 33 are associated with anogenital cancer, particularly cancer of the uterine cervix. Viral DNA of these 'high risk' types is found in about 90% of tumour biopsies, most frequently DNA of HPV 16. Moreover, precursor lesions of cervical cancer (cervical dysplasias) positive for HPV16 are more likely to progress to high grade malignancy compared with those containing HPV 6 or 11 as indicated by follow-up studies.

The HPV genome encodes two capsid proteins (L1 and L2) and seven or eight early (E) proteins most of which are involved in regulation of transcription, replication and viral persistence. The oncogenic potential of the 'high risk' HPV types depends on the expression of the early genes E6 and E7, which are able to immortalize human foreskin or cervical keratinocytes after transfection (Münger et al., 1989; Hawley-Nelson et al., 1989).

From the long delay between the primary infection and the development of invasive cancer, it is evident that HPV 16 or 18 infection is not sufficient for transformation of epithelial cells. The cofactors that are additionally necessary to promote progression of HPV associated benign lesions into premalignant or malignant tumours still have to be defined. Regardless of these cofactors, there is considerable evidence that the immune system plays an important part in determining the outcome of HPV infection.

CELLULAR IMMUNITY AGAINST PAPILLOMAVIRUS INFECTION

Immunodeficiency and wart regression

Suppression of cellular immunity increases the susceptibility of patients to the development of HPV induced anogenital or skin lesions as a consequence of either primary HPV infection or reactivation of latent HPV (for review see Benton, Shahidullah & Hunter, 1992). Moreover, cellular immunodeficiency seems to be associated with malignant conversion of pre-existing benign HPV-positive lesions, which further underlines the importance of cellular host defence mechanisms directed against the virus. For instance, in patients suffering from the inherited disease Epidermodysplasia verruciformis (EV) skin warts and skin carcinomas occur more frequently than in patients with normal immune status (Orth, 1987). The cancerous lesions usually contain HPV 5 or 8 DNA. Similar results were obtained for transplant recipients after immunosuppressive medication (Barr *et al.*, 1989; Blessing *et al.*, 1990). The tendency towards development of HPV induced lesions increases with the duration of graft life and immunosuppression (Barr *et al.*, 1989). In addition, in female patients with allografts the rate of cervical intraepithelial neoplasia (CIN) was reported to be significantly higher compared to non-immunosuppressed controls (Alloub *et al.*, 1989).

Cell-mediated immune reactions are also involved in the spontaneous regression of papillomavirus associated warts. The regression process is characterized by massive macrophage and lymphocyte infiltration of the infected epithelium and shows similar features for cottontail papillomavirus (CRPV) induced lesions (Kreider, 1980) as well as HPV induced anogenital (Tay *et al.*, 1987) and skin (Tagami *et al.*, 1974; Berman & Winkelmann, 1977) warts. In rabbits, wart regression can be inhibited by treating the animals with agents that suppress cellular immunity (Kreider & Bartlett, 1885). The regression of tumours observed in calves vaccinated with the bovine papillomavirus type 2 (BPV 2) L2 protein resembles the natural regression process (Jarrett, 1985). In humans, where different types of skin warts are caused by different HPV types (De Villiers, 1989), regression is wart type-specific indicating that rejection is mediated by T cells recognizing HPV type-specific antigens.

HPV-specific T cell responses

Peripheral blood lymphocytes obtained from asymptomatic individuals proliferate after *in vitro* stimulation with the proteins HPV 16 L1, E6 (Strang *et al.*, 1990) and E7 (Altmann *et al.*, 1992) and HPV 1 E4 (Steele, Stancovic & Gallimore, 1993). In patients with CIN lesions skin tests using a recombinant HPV 16 L1 fusion protein as antigen elicit delayed type hypersensitivity (Höpfl *et al.*, 1991). Further evidence for the existence of anti-HPV specific

T cell reactivity was obtained from experimental vaccination systems. As demonstrated by McLean *et al.* (1993), CD4$^+$ delayed type hypersensitivity T cells confer rejection of HPV 16 immortalized nontumorigenic keratinocyte grafts in syngeneic mice after challenge with vaccinia virus recombinants expressing the HPV 16 E7 gene. Immunization of rodents with nonmalignant cells (Chen *et al.*, 1991) or recombinant vaccinia viruses (Meneguzzi *et al.*, 1991) expressing either HPV 16 E6 or E7 resulted in protection against transplanted syngeneic tumour cells that were transfected with the HPV 16 E7 gene or with the whole viral genome. CD8$^+$ T lymphocytes are responsible for this anti-tumour immunity. Taken together, these data clearly show that HPV gene products contain T cell-specific epitopes which are able to induce an effective cellular immune response in animals and to primarily stimulate T cells in humans.

NEUTRALIZING ANTIBODIES

While calves immunized with a BPV 2 L1 fusion protein produce serum neutralizing antibodies against this antigen (Jarrett *et al.*, 1991), antibodies obtained against BPV 1 L1 are not neutralizing (Pilacinski *et al.*, 1986; Jin *et al.*, 1990; Ghim *et al.*, 1991). Vaccination of rabbits with the CRPV L1 protein protects animals from CRPV infection and induces high titres of neutralizing anti-L1 antibodies (Lin *et al.*, 1992). The CRPV L2 protein also contains neutralizing epitopes (Christensen *et al.*, 1991; Lin *et al.*, 1992). The neutralizing antibodies are usually type-specific and recognize intact virus particles *in vitro*. IgG type serum-antibodies directed against HPV capsid proteins like HPV 6 and 16 L1 or L2 (Galloway & Jenison, 1990) or intact HPV 11 particles (Bonnez *et al.*, 1992) can be demonstrated in human sera. Some of these antibodies neutralized infection in an athymic mouse xenograft system (Christensen & Kreider, 1990; Christensen *et al.*, 1992). Their biological role, however, is unclear, since so far no correlation has been established between the circulating antibodies and regression or recurrence of papillomavirus induced lesions (Bonnez *et al.*, 1992).

Using different methods (Western blot, ELISA, RIPA) and different reagents several investigators have found an association between the presence of anti-HPV 16 E7 antibodies and invasive cervical cancer (Jochmus-Kudielka *et al.*, 1989; Mann *et al.*, 1990; Köchel *et al.*, 1991; Müller *et al.*, 1992). Similar associations were described for HPV 16 E4 (Kanda *et al.*, 1992), HPV 16 E6 (Müller *et al.*, 1992; Kanda *et al.*, 1992) and HPV 18 E7 (Bleul *et al.*, 1991). In malignant lesions, the proteins E6 and E7 are present in all epithelial layers (Smotkin & Wettstein, 1986) which might explain their increased immunogenicity in cervical cancer patients. Although antibodies to early and late HPV proteins are produced in humans it seems to be unlikely that they significantly influence HPV induced lesions. Nevertheless, they might be important indicators of the disease.

IMMUNE STATUS OF HPV ASSOCIATED LESIONS

Downregulation of MHC class I or β2-microglobulin cell surface expression, has been observed in HPV induced benign and malignant lesions (Connor & Stern, 1990; Viac *et al.*, 1990, 1993). In some of these lesions, dense inflammatory reactions are correlated with focal or homogeneous expression of HLA class II antigens by epithelial cells (Viac *et al.*, 1990, 1993; Glew *et al.*, 1992; Jochmus *et al.*, 1993). Both events occur independently of the presence or absence of HPV DNA or transcripts in the examined specimens (Viac *et al.*, 1990, 1993; Glew *et al.*, 1992; Jochmus *et al.*, 1993). Epithelial HLA class II antigens are often coexpressed with the intercellular adhesion molecule I (ICAM I) (Glew *et al.*, 1992), which might facilitate the recognition of the infected keratinocytes by specific T lymphocytes. On the other hand, reduced expression of HLA class I molecules could be a means for the virus infected cells to escape immune surveillance.

Certain MHC haplotypes may be associated with increased risk for the development of cervical cancer; a strong correlation between the incidence of cervical carcinoma and the HLA class II antigen DQw3 has been reported by Wank and Thomssen (1991), but this finding was not confirmed by others (Glew *et al.*, 1992; Helland *et al.*, 1992) and is therefore still a point of discussion.

Compared to low grade CIN lesions and genital warts the epithelium of high grade CIN lesions (CIN II/III) exhibits low numbers of Langerhans cells which naturally function as antigen presenting cells for CD4$^+$ T lymphocytes in the skin and mucosal immune system (Hughes, Norval & Howie, 1988; Viac *et al.*, 1990). The largest proportion of immunocompetent cells invading the infected epithelium are CD8$^+$ T lymphocytes (Viac *et al.*, 1990). Nevertheless, natural killer (NK) cells could also participate in the anti-HPV immune response, since the persistence of HPV induced lesions correlates with a decrease of NK cell activity against HPV 16 positive target cells (Malejczyk *et al.*, 1989).

TOWARDS HPV VACCINATION

HPV infections are persistent in an immunocompetent host suggesting that either an antiviral immune response is inefficiently or not induced or the virus and infected cells escape immune surveillance. This has been explained by location of the lesions (skin, mucosa) and low levels of viral antigen expressed in the infected tissue. In order to increase HPV directed immune responses two different types of vaccines can be considered: a prophylactic vaccine which would elicit antibody response and prevent infection, and a therapeutic vaccine primarily concentrating on the induction of a specific cell-mediated immune response which would lead to regression of pre-existing lesions or even malignant tumours.

Since only low numbers of virus particles are produced by infected tissues or by *in vitro* culture systems (Meyers *et al.*, 1992) the development of anti-HPV vaccines will depend on the use of synthetic peptides corresponding to individual viral gene products or on the expression of recombinant viral antigens. A prophylactic vaccine will consist either of isolated structural proteins or of viral particles assembled *in vitro*. As recently reported, such particles can be obtained after the expression of the HPV structural proteins in recombinant vector systems (Zhou *et al.*, 1991; Rose *et al.*, 1993; Hagensee, Yaegashi & Galloway, 1993). Prophylactic vaccines are supposed to induce virus-specific neutralizing antibodies (IgG, sIgA) that are able to prevent initial infections. For the immune surveillance of pre-existing infections, however, T cell-mediated immune responses seem to be more relevant than HPV-specific antibodies. Capsid proteins cannot be detected in proliferating basal cells of HPV induced lesions which are expected to be the targets for cytotoxic T lymphocytes. In contrast, products of the early open reading frames are expressed in all epithelial layers and consequently might play a role in induction of cellular immunity. The ability of HPV 16 E7 to elicit a cytotoxic T cell response in experimental vaccination systems (Meneguzzi *et al.*, 1991; Chen *et al.*, 1991) supports this assumption and indicates the potential of this protein as a component of a therapeutic vaccine.

The identification of specific T cell epitopes that are presented in association with particular MHC molecules is a critical step for the development of vaccines composed of synthetic peptides. Viral peptides that complex with common MHC alleles will be more effective than others with regard to the immunization of a population. Recently, immunogenic epitopes of HPV16 E7 recognized by human CD4[+] T lymphocytes have been identified (Altmann *et al.*, 1992). In order to increase the cytotoxic T cell response directed against HPV-bearing keratinocytes, both the HLA class I- and class II-restricted epitopes of HPV 16 E7 or any other immunogenic HPV protein have to be defined and simultaneously delivered as proposed by Widmann *et al.* (1992) for other systems. In animal models peptide vaccinations that induce cytotoxic T lymphocyte-mediated responses have been demonstrated to be protective against different viruses (Deres *et al.*, 1989; Kast *et al.*, 1991; Schulz, Zinkernagel & Hengartner, 1991; Reinholdsson-Ljunggren *et al.*, 1992). In some of these cases, coupling of the immunogenic peptides to lipid components before immunization was necessary to activate virus-specific T lymphocytes (Deres *et al.*, 1989; Schulz *et al.*, 1991).

In order to optimize the vaccination efficiency immunoregulatory events in the environment of papillomavirus associated lesions have to be characterized. Binding of the T cell receptor to specific peptide/MHC complexes is sufficient to trigger effector functions in already activated T cells. Primary stimulation of resting T lymphocytes, however, requires costimulatory signals that are independent of specific antigens and are mediated mainly by cellular adhesion molecules. For instance, binding of the TCR to specific peptide/MHC complexes in the absence of interaction of CD28 and B7 leads to inactivation of the particular T cells (clonal anergy). Recently, Chen *et al.* (1992) demonstrated the capacity of melanoma cells expressing HPV 16 E7 and the murine B7 antigen to induce E7-specific anti-tumour reactivity, while the rejection of B7-negative tumours requires prior immunization with the E7 protein. As observed in earlier studies, antigen presentation by keratinocytes can result in tolerance rather than clonal expansion of specific T lymphocytes (Bal *et al.*, 1990). This is probably due to reduced expression of accessory molecules at the keratinocyte surface and/or an inefficient production of lymphokines and might explain the inadequate immune response to HPV infection. Once the interactions between keratinocytes and T cells have been characterized substitution of missing immunoregulatory signals could initiate or enhance a specific immune response in patients suffering from HPV-positive lesions. For example, the lack of costimulatory signals provided by CD28/B7 interaction can be counteracted by exogenous delivery of interleukin 2 (Lanzavecchia, 1993).

Finally, the target population for vaccination has to be defined. Clearly, as soon as an effective prophylactic vaccine without serious side effects has been developed, people at young age could be addressed, particularily in developing countries with high prevalence of cervical cancer. On the other hand, therapeutic vaccines preventing the conversion of benign HPV induced lesions into malignant tumours would be useful worldwide.

REFERENCES

Alloub, M. I., Barr, B. B. B., McLaren, K. M., Smith, I. W., Bunney, M. H. & Smart, G. E. (1989). Human papillomavirus infection and cervical intraepithelial neoplasia in women with renal allografts. *British Medical Journal*, **298**, 153–6.

Altmann, A., Jochmus-Kudielka, I., Frank, R. *et al.* (1992). Definition of immunogenic determinants of the human papillomavirus type 16 nucleoprotein E7. *European Journal of Cancer*, **28**, 326–33.

Bal, V., McIndoe, A., Denton, G. *et al.* (1990). Antigen presentation by keratinocytes induces tolerance in human T cells. *European Journal of Immunology*, **20**, 1893–7.

Barr, B. B. B., McLaren, K., Smith, I. W. *et al.* (1989). Human papilloma virus infection and skin cancer in renal allograft recipients. *The Lancet*, **i**, 124–9.

Benton, C., Shahidullah, H. & Hunter, J. A. A. (1992). Human papillomavirus in the immunosuppressed. *Papillomavirus Report*, **3**, 23–6.

Berman, A. & Winkelmann, R. K. (1977). Flat warts undergoing involution: histopathological findings. *Archives of Dermatology*, **113**, 1219–21.

Blessing, K., McLaren, K., Morris, R. *et al.* (1990). Detection of human papillomavirus in skin and genital lesions of renal allograft recipients by *in situ* hybridization. *Histopathology*, **16**, 181–5.

Bleul, C., Müller, M., Frank, R. *et al.*, (1991). Human papillomavirus type 18 E6 and E7 antibodies in human sera: increased anti-E7 prevalence in cervical cancer patients. *Journal of Clinical Microbiology*, **29**, 1579–88.

Bonnez, W., Kashima, H. K., Leventhal, B. *et al.* (1992). Antibody response to human papillomavirus (HPV) type 11 in children with juvenile-onset recurrent respiratory papillomatosis (RRP). *Virology*, **188**, 384–7.

Chen, L., Thomas, E. K., Hu, S.-L., Hellström, I. & Hellström, K. E. (1991). Human papillomavirus type 16 nucleoprotein E7 is a tumour rejection antigen. *Proceedings of the National Academy of Sciences, USA*, **88**, 110–14.

Chen, L., Ashe, S., Brady, W. A. *et al.* (1992). Costimulation of antitumour immunity by the B7 counterreceptor for the T lymphocyte molecules CD28 and CTLA-4. *Cell*, **71**, 1093–102.

Christensen, N. D. & Kreider, J. W. (1990). Antibody-mediated neutralization *in vivo* of infectious papillomaviruses. *Journal of Virology*, **64**, 3151–6.

Christensen, N. D., Kreider, J. W., Kan, N. C. & DiAngelo, S. L. (1991). The open reading frame L2 of cottontail rabbit papillomavirus contains antibody-inducing neutralizing epitopes. *Virology*, **181**, 572–9.

Christensen, N. D., Kreider, J. W., Shah, K. V. & Rando, R. F. (1992). Detection of human serum antibodies that neutralize infectious human papillomavirus type 11 virions. *Journal of General Virology*, **73**, 1261–7.

Connor, M. E. & Stern, P. (1990). Loss of MHC class-I expression in cervical carcinomas. *International Journal of Cancer*, **46**, 1029–34.

Deres, K., Schild, H., Wiesmüller, K. H., Jung, G. & Rammensee, H.-G. (1989). *In vivo* priming of virus-specific cytotoxic T lymphocytes with synthetic lipopeptide vaccine. *Nature*, **342**, 561–4.

De Villiers, E.-M. (1989). Minireview: Heterogeneity of the human papillomavirus group. *Journal of Virology*, **63**, 4898–903.

Galloway, D. A. & Jenison, S. A. (1990). Characterization of the humoral immune-response to genital papillomaviruses. *Molecular Biology and Medicine*, **7**, 59–72.

Ghim, S., Christensen, N. D., Kreider, J. W. & Jenson, A. B. (1991). Comparison of neutralization of BPV-1 infection of C127 cells and bovine fetal skin xenografts. *International Journal of Cancer*, **49**, 285–9.

Glew, S. S., Duggan-Keen, M., Cabrera, T. & Stern, P. L. (1992). HLA class II antigen expression in human papillomavirus-associated cervical cancer. *Cancer Research*, **52**, 4009–16.

Glew, S. S., Stern, P. L., Davidson, J. A. & Dyer, P. A. (1992). HLA antigens and cervical carcinoma. *Nature*, **356**, 22.

Hagensee, M. E., Yaegashi, N. & Galloway, D. A. (1993). Self-assembly of human papillomavirus type 1 capsids by expression of the L1 protein alone or by coexpression of the L1 and L2 capsid proteins. *Journal of Virology*, **67**, 315–22.

Hawley-Nelson, P., Vousden, K. H., Hubbert, N. L., Lowy, D. R. & Schiller, J. T. (1989). HPV16 E6 and E7 proteins cooperate to immortalize human foreskin keratinocytes. *EMBO Journal*, **8**, 3905–10.

Helland, A., Borresen, A. L., Ronningen, K. S. & Thorsby, E. (1992). HLA antigens and cervical carcinoma. *Nature*, **356**, 23.

Höpfl, R., Sandbichler, M., Sepp, N. *et al.* (1991). Skin test for HPV type 16 proteins in cervical intraepithelial neoplasia. *Lancet*, **1**, 373–4.

Hughes, R. G., Norval, M. & Howie, S. E. M. (1988). Expression of major histocompatibility class II antigens by Langerhans' cells in cervical intraepithelial neoplasia. *Journal of Clinical Pathology*, **41**, 253–9.

Jarrett, W. F. H. (1985). The natural history of bovine papillomavirus infection. *Advances in Viral Oncology*, **5**, 83–102.

Jarrett, W. F. H., Smith, K. T., O'Neil, B. W. *et al.* (1991). Studies on vaccination against papillomaviruses: prophylactic and therapeutic vaccination with recombinant structural proteins. *Virology*, **184**, 33–42.

Jin, X. W., Cowsert, L., Marshall, D. *et al.* (1990). Bovine serological response to a recombinant BPV-1 major capsid protein vaccine. *Intervirology*, **31**, 345–54.

Jochmus-Kudielka, I., Schneider, A., Braun, R. *et al.* (1989). Antibodies against the human papillomavirus type 16 early proteins in human sera: correlation of anti-E7 reactivity and cervical cancer. *Journal of the National Cancer Institute*, **81**, 1698–704.

Jochmus, I., Dürst, M., Reid, R. *et al.* (1993). Major histocompatibility complex and human papillomavirus type 16 E7 expression in high-grade vulvar lesions. *Human Pathology*, **24**, 519–24.

Kanda, T., Onda, T., Zanma, S. *et al.* (1992). Independent association of antibodies against human papillomavirus type 16 E1/E4 and E7 proteins with cervical cancer. *Virology*, **190**, 724–32.

Kast, W. M., Roux, L., Curren, J. *et al.* (1991). Protection against lethal Sendai virus infection by *in vivo* priming of virus-specific cytotoxic T lymphocytes with free synthetic peptide. *Proceedings of the National Academy of Sciences, USA*, **88**, 2283–87.

Köchel, H. G., Monazahian, M., Sievert, K. *et al.* (1991). Occurrence of antibodies to L1, L2, E4f and E7 gene products of human papillomavirus types 6b, 16 and 18 among cervical cancer patients and controls. *International Journal of Cancer*, **48**, 682–8.

Kreider, J. W. (1980). Neoplastic progression of the Shope rabbit papilloma. In *Viruses in Naturally Occurring Cancers*. (7th Conference on Cell Proliferation), pp. 283–300. New York: Cold Spring Harbor Laboratories.

Kreider, J. W. & Bartlett, G. L. (1985). Shope rabbit papilloma-carcinoma complex: a model system of human papillomavirus infections. *Clinical Dermatology*, **3**, 20–6.

Lanzavecchia, A., (1993). Identifying strategies for immune intervention. *Science*, **260**, 937–44.

Lin, Y.-L., Borenstein, L. A., Selvakumar, R., Ahmed, R. & Wettstein, F. O. (1992). Effective vaccination against papilloma development by immunization with L1 or L2 structural protein of cottontail rabbit papillomavirus. *Virology*, **187**, 612–19.

McLean, C. S., Sterling, J. S., Mowat, J., Nash, A. A. & Stanley, M. A. (1993). Delayed-type hypersensitivity response to the human papillomavirus type 16 E7 protein in a mouse model. *Journal of General Virology*, **74**, 239–45.

Mann, V. M., Loo de Lao, S., Brenes, M. *et al.* (1990). Occurrence of IgA and IgG antibodies to select peptides representing human papillomavirus type 16 among cervical cancer cases and controls. *Cancer Research*, **50**, 7815–19.

Malejczyk, J., Majewski, S., Jablonska, S., Rogozinski, T. T. & Orth, G. (1989). Abrogated NK-cell lysis of human papillomavirus (HPV)-16-bearing keratinocytes in patients with precancerous and cancerous HPV-induced anogenital lesions. *International Journal of Cancer*, **43**, 209–14.

Meneguzzi, G., Cerni, C., Kieny, M. P. & Lathe, R. (1991). Immunization against human papillomavirus type 16 tumour cells with recombinant vaccinia viruses expressing E6 and E7. *Virology*, **181**, 62–9.

Meyers, C., Frattini, M. G., Hudson, J. B. & Laimins, L. L. (1992). Biosynthesis of human papillomavirus from a continuous cell line upon epithelial differentiation. *Science*, **257**, 971–3.

Müller, M., Viscidi, R. P., Sun, Y. *et al.* (1992). Antibodies to HPV 16 E6 and E7 proteins as markers for HPV 16-associated invasive cervical cancer. *Virology*, **187**, 508–14.

Münger, K., Phelps, W. C., Bubb, V., Howley, P. M. & Schlegel, R. (1989). The E6 and E7 genes of human papillomavirus type 16 together are necessary and sufficient for transformation of primary human keratinocytes. *Journal of Virology*, **63**, 4417–21.

Orth, G. (1987). *Epidermodysplasia verruciformis*. In *The Papovaviridae. Vol. 2: The Papillomaviruses*, ed. N. P. Salzman & P. Howley, pp. 199–243. New York: Plenum Press.

Pilacinski, W. P., Glassman, D. L., Glassman, K. F. *et al.* (1986). Immunization against bovine papillomavirus infection. *Ciba Foundation Symposium*, **120**, 136–56.

Reinholdsson-Ljunggren, G., Ramqvist, T., Ährlund-Richter, L. & Dalianis, T. (1992). Immunization against polyoma tumours with synthetic peptides derived from the sequences of middle- and large-T antigens. *International Journal of Cancer*, **50**, 142–6.

Rose, C. R., Bonnez, W., Reichman, R. C. & Garcea, R. L. (1993). Expression of human papillomavirus type 11 L1 protein in insect cells: *in vivo* and *in vitro* assembly of viruslike particles. *Journal of Virology*, **67**, 1936–44.

Schulz, M., Zinkernagel, R. M. & Hengartner, H. (1991). Peptide-induced antiviral protection by cytotoxic T cells. *Proceedings of the National Academy of Sciences, USA*, **88**, 991–3.

Smotkin, D. & Wettstein, F. O. (1986). Transcription of human papillomavirus type 16 early genes in a cervical cancer and cancer derived cell line and identification of the E7 protein. *Proceedings of the National Academy of Sciences, USA*, **83**, 4680–4.

Steele, J. C., Stancovic, T. & Gallimore, P. H. (1993). Production and characterization of human proliferative T-cell clones specific for human papillomavirus type 1 E4 protein. *Journal of Virology*, **67**, 2799–806.

Strang, G., Hickling, J. K., McIndoe, G. A. J. *et al.* (1990). Human T cell responses to human papillomavirus type 16 L1 and E6 synthetic peptides: identification of T cell determinants, HLA-DR restriction and virus type specificity. *Journal of General Virology*, **71**, 423–31.

Tagami, H., Ogino, A., Takigawa, M., Imamura, S. & Ofuji, S. (1974). Regression of plane warts following spontaneous inflammation: a histopathological study. *British Journal of Dermatology*, **90**, 147–54.

Tay, S. K., Jenkins, D., Maddox, P., Hogg, N. & Singer, A. (1987). Tissue macrophage response in human papillomavirus infection and cervical intraepithelial neoplasia. *British Journal of Obstetrics and Gynaecology*, **94**, 1094–7.

Viac, J., Guerin-Reverchon, Y., Chardonnet, Y. & Bremond, A. (1990). Langerhans cells and epithelial cell modifications in cervical intraepithelial neoplasia: correlation with human papillomavirus infection. *Immunobiology*, **180**, 328–38.

Viac, J., Soler, C., Chardonnet, Y., Euvrard, S. & Schmitt, D. (1993). Expression of immune associated surface antigens of keratinocytes in human papillomavirus-derived lesions. *Immunobiology*, in press.

Wank, R. & Thomssen, C. (1991). High risk of squamous cell carcinoma of the cervix for women with HLA-DQw3. *Nature*, **352**, 723–5.

Widmann, C., Romero, P., Marjanski, J. L., Corradin, G. & Valmori, D. (1992). T helper epitopes enhance the cytotoxic response of mice immunized with MHC class I-restricted malaria peptides. *Journal of Immunological Methods*, **155**, 95–9.

Zhou, J., Sun, X. Y., Stenzel, D. J. & Frazer, I. H. (1991). Expression of vaccinia recombinant HPV 16 L1 and L2 ORF proteins in epithelial cells is sufficient for assembly of HPV virion-like particles. *Virology*, **185**, 251–7.

EBV INFECTION AND EBV-ASSOCIATED TUMOURS

A. B. RICKINSON

CRC Laboratories, Department of Cancer Studies, University of Birmingham, The Medical School, Birmingham B15 2TJ, UK.

INTRODUCTION

One of the central themes which emerges from the study of candidate human tumour viruses is that such agents are frequently widespread in host populations, can be carried as persistent and largely asymptomatic infections in the immunocompetent host, and are only associated with malignant change in specific cell types and under specific circumstances. The Epstein–Barr virus (EBV), a lymphotropic herpesvirus first discovered by virtue of its association with a rare B cell malignancy, Burkitt's lymphoma (BL), illustrates these lessons particularly well. Thus EBV is endemic in all human populations, is transmitted to naive individuals early in life and is carried thereafter as a life-long latent infection of lymphoid tissues with only limited reactivation to infectivity. In the great majority of EBV carriers, this life-long infection is completely apathogenic. Yet the virus is potentially oncogenic because a subset of its genes, associated with latent rather than lytic infection, have the capacity to deregulate cell growth (see accompanying chapter by Kieff *et al.*). Indeed, these latent genes, identified by virtue of their constitutive expression in latently infected growth-transformed B cell lines *in vitro*, are found conserved in all natural EBV isolates. This strongly suggests that they are crucially important for the virus' normal strategy for persistence as an asymptomatic infection in the lymphoid system. Nevertheless, when expressed inappropriately, either in an unusual cell type or in the usual cell type but in an immunocompromised host, they may contribute towards malignant change.

PRIMARY VIRUS INFECTION

EBV is transmitted orally either by virus particles or via lytically infected cells present in buccal fluid. Sero-epidemiological studies suggest that, in most human communities, primary infection occurs during the first few years of life, where it appears to be almost always asymptomatic. When delayed until adolescence or later, however, as happens increasingly in affluent societies, primary infection is often manifest clinically as infectious mononucleosis (IM). Indeed, most of our ideas about the usual events of primary infection are extrapolations from the study of acute IM patients.

It is not known whether the initial target cell for orally transmitted virus is a B lymphocyte, made accessible via damage to the oropharyngeal mucosa, or an epithelial cell of the mucosa itself. However, by the time the clinical symptoms of IM appear some weeks later, there is clear evidence that the virus has established foci of productive infection in the oropharynx, involving mucosal epithelium (Sixbey et al., 1984) and perhaps also local mucosa-associated lymphocytes as sources of permissive cells. Entry into B lymphocytes is known to occur via a ligand receptor interaction between the major viral envelope glycoprotein gp340 and the cell surface complement receptor molecule CR2 (Nemerow et al., 1987), whereas the means of viral access into pharyngeal epithelium has yet to be determined. Squamous epithelial cells express a 200 kD surface protein which shares some antigenic epitopes with CR2 (Young et al., 1989a) but which does not appear to have virus receptor function. Very low expression of CR2 itself on epithelial cells, as has been reported for certain epithelial cell lines in culture (Birkenbach et al., 1992), remains a possible route for viral entry; alternatively, the virus may be delivered directly from a lytically infected B cell either through a postulated membrane fusion event or through some other form of close intercellular contact naturally occurring in lympho-epithelium. Infectious virus shed from productively infected epithelial and/or lymphoid foci of replication is easily detectable as cord blood B cell transforming activity in the buccal fluid of IM patients not only during the acute phase of the disease but also from the very few prospective cases studied, well before the onset of disease symptoms (Svedmyr et al., 1984).

Primary infection also leads to the establishment of a large pool of latently infected B cells in vivo. Their presence both in the blood and in the lymphoid tissues of IM patients has been directly demonstrated by immunostaining for virus latent antigens and indirectly assayed by explantation into cell culture, where the latently infected cells reactivate into lytic cycle and release transforming virus. The results from immunostaining of lymphoid tissues (Pallesen, G. and Hamilton-Dutoit, S., personal communication) strongly suggest that virus-transformed LCL-like cells, expressing the complete spectrum of EBV latent proteins (the nuclear antigens EBNAs 1, 2, 3A, 3B, 3C, -LP and the latent membrane proteins LMPs 1 and 2) are generated in large numbers during primary infection. This is an important observation since it documents the activity of the virus directly in its natural site of persistence, in lymphoid tissues, rather than by inference from the minor population of virus-positive cells in peripheral blood. The inference is that following primary infection and before development of the host cell-mediated response, there is a period in which the virus-driven proliferation of infected B cells can serve to generalize the infection throughout the lymphoid system. Indeed, the main evolutionary advantage to the virus of many of the latent genes may be their ability rapidly to amplify the pool of latently infected cells in the naive host.

T cell response to primary virus infection

The T cell response seen in IM involves activated cells of the CD4+ and particularly of the CD8+ subset which appear in large numbers not just in peripheral blood but also generalized throughout the body. As such the response is functionally complex, and contains not only virus-specific components but also what appear to be many coincidentally activated T cell clones which have no obvious role in combating the infection. Indeed, the fact that large-scale T cell activation is coincident with the onset of clinical symptoms in IM patients suggests that the response contains immunopathological elements. Some indication of possible pathogenetic mechanisms comes from the finding that IM T cells have a wide spectrum of suppressive activities over mitogen and antigen-induced T cell responses *in vitro* (Reinhertz *et al.*, 1980), and this may underlie the depression of cell-mediated immune responses to recall antigens that is seen in IM patients. Cytotoxicity assays with IM T cell effectors are complicated by the presence of activated CD8+ cells, many of which do not express conventional natural killer cell markers such as CD16 but which nevertheless display complex patterns of non-HLA-restricted cytotoxicity against allogeneic target cell panels (Strang & Rickinson, 1987). These various 'non-specific' reactivities within the IM T cell fraction bear witness to an acute polyclonal T cell activation of what appear to be short-term effector cells; interestingly, these cells die rapidly by apoptosis on explantation *in vitro* (Moss *et al.*, 1985) and may well be programmed to a similar fate *in vivo*.

Despite the preponderance of these non-specific reactivities, it is nevertheless clear that EBV-specific cytotoxic T lymphocytes (CTLs) are contained within both the activated CD8+ and activated CD4+ subpopulations in acute IM blood, and that these display respectively HLA class I-restricted and HLA class II-restricted recognition of EBV-transformed LCL target cells (Strang & Rickinson, 1987; Misko, I. S. and Moss, D. J., personal communication). The viral target epitopes which elicit these responses are still largely undefined, although they are presumed to be derived from the spectrum of virus latent proteins that are constitutively expressed in LCL cells. Now that the immunogenic epitopes for memory CTL responses to persistent EBV infection are becoming defined (see later), it will be possible to look directly for evidence of the corresponding reactivities in IM blood. The involvement of CD4+ as well as CD8+ effector CTLs in the primary response to EBV is a very interesting feature of this virus infection and presumably reflects the fact that the major antigenic challenge to the T cell system in IM comes from B lymphocytes and these are both HLA class I and class II-positive.

It is still not known why primary infection in infancy is almost always asymptomatic whereas that in later life can produce disease in up to 50% of cases. However, one interesting possibility is that it relates to size of the initial virus dose rather than to age *per se*, the argument being that high doses of an orally transmitted virus are much more likely to be delivered by salivary exchange between consenting adults than by adult to child or child to child. Infection with a large virus dose makes it much more likely that virus-driven expansion of the infected B cell pool will exceed a critical threshold beyond which the atypical T cell response (complete with immunopathological elements) is activated and disease ensues. In this regard it is interesting to note that reported examples of 'post-perfusion' mononucleosis, a disease induced by the transfusion of seropositive donor blood into a seronegative recipient, all involve cases where the donor was in the incubation period of IM (Gerber *et al.*, 1969; Blacklow *et al.*, 1971); in other words, cases where an unusually large number of EBV-infected B cells were inadvertently introduced.

EBV-positive malignancies in the context of primary infection

The clearest demonstration that EBV is indeed potentially oncogenic comes from rare clinical situations where primary infection occurs in patients who are immunologically compromised. Foremost amongst these are young bone marrow transplant recipients, not infected with EBV prior to transplant, who receive EBV-infected donor B cells within the T cell-depleted marrow graft (Shapiro *et al.*, 1988; Zutter *et al.*, 1988; Gratama, 1993). Several such patients have developed EBV-positive lymphoproliferative disease ('immunoblastic lymphoma') within 45–100 days of transplantation; the lesions are multifocal and are composed of EBV-transformed donor B cell clones expressing the same spectrum of latent proteins as are present in LCLs (Young *et al.*, 1989*b*; Thomas *et al.*, 1990; Gratama *et al.*, 1991). These therefore appear to be classic examples of virus-driven proliferations whose progressive growth remains unchecked in the absence of CTL surveillance.

Acute lymphoproliferative disease with a fatal course is also a frequent consequence of primary EBV infection in boys homozygous at the XLP (X-linked lymphoproliferative) gene locus (Sullivan & Woda, 1989). This is a poorly understood syndrome characterized by an extreme, and apparently quite selective, susceptibility to EBV. Primary infection leads to a severe IM-like illness with evidence both of EBV-driven B cell activation and of massive T cell proliferation, the latter containing multiple CTL reactivities typical of those seen in acute IM. The patients tend to succumb not to EBV-driven immunoblastic lymphoma, however, but to a progressive (and possibly T cell-mediated) destruction of lymphoid and haemopoietic tissues; this is followed by extensive infiltration of the tissues by histiocytes to produce lesions identical to those described in the virus-associated haemophagocytic syndrome (VAHS) (Sullivan & Woda, 1989). Important potential insights into the pathogenesis of XLP have in fact come very recently from studies on rare non-familial cases of young children who suffer symptomatic primary EBV infection but present with clinical and histological criteria resembling VAHS (Chan *et al.*, 1992; Craig *et al.*, 1992; Gaillard *et al.*, 1992). In each case the underlying lesion was identified as a monoclonal EBV-positive lymphoma of T cell origin. This very interesting clinical observation suggests that in some circumstances the virus not only can gain access to the T cell system, perhaps via the CR2 molecule which is reported to be expressed at low levels on certain immature T cells (Watry *et al.*, 1991), but also can contribute to malignant change in that cell type.

PERSISTENT VIRUS INFECTION

EBV shares with all other members of the herpesvirus family the capacity for life-long persistence in the immunocompetent host. Thus all EBV-seropositive individuals, whether primary infection was sub-clinical or manifest as IM, continue to shed infectious virus at low levels into buccal fluid (detectable as transforming activity in *in vitro* assays) and to harbour small numbers of non-productively infected B cells in the circulation (detectable through their reactivation into lytic cycle *in vitro* and subsequent release of transforming virus) (Yao, Rickinson & Epstein, 1985). It is widely believed that EBV persistence requires the establishment of a long-term latent infection within a target tissue, and arguments have been made on the one hand for the self-renewing basal cells of mucosal epithelium (Allday & Crawford, 1988), and on the other hand for long-lived B lymphocytes as the essential target cell. Though this question is by no means completely resolved, the current balance of evidence favours the concept of a B lymphoid reservoir which, once established, is not dependent upon continued replication of the virus at permissive sites, i.e. not dependent upon the continued recruitment of newly infected B cells. Thus, complete destruction of lymphoid tissues in EBV seropositive patients prior to bone marrow transplantation (a procedure which leaves epithelial tissues relatively intact) has been found to eradicate the resident EBV infection (Gratama *et al.*, 1988). Secondly, patients in whom chronic virus replication (including that at epithelial sites) is completely blocked by sustained acyclovir therapy show no consequent reduction in the numbers of latently infected B cells in the circulating pool (Yao *et al.*, 1989). Thirdly, careful screening of oral hairy leukoplakia lesions, where there is obvious differentiation dependent replication of EBV in the outer layers of stratified lingual epithelium, has found no evidence of latent infection markers in the underlying basal epithelial cells (Niedobitek *et al.*, 1991).

The mechanism of virus persistence within the B lymphoid system, and the virus-cell interactions which underlie it, are still only poorly understood (Qu & Rowe, 1992). In this context it is worth remembering that the LCL-like cells which arise *in vivo* during primary virus infection and which display the full spectrum of virus latent proteins are extremely immunogenic to the

CTL response. It is therefore difficult to conceive of virus persistence being achieved via a reservoir of long-lived LCL-like cells unless this reservoir were situated in some immunologically privileged site. Another possibility is that long-lived B cells harbour the virus as a less immunogenic form of latent infection (Rickinson, 1988), perhaps like that seen in Burkitt's lymphoma (BL) cells where virus latent protein expression is limited to the genome maintenance protein, EBNA1 (see later).

Whatever the precise nature of the virus reservoir in the lymphoid system, it is more likely to be a dispersed pool of cells than concentrated at a single site, since virus-positive cells are regularly present within all lymphoid tissues tested as well as in the blood (Nilsson *et al.*, 1971; Yao *et al.*, 1985). Furthermore, it is clear that immunogenic LCL-like cells must be continually emerging from this reservoir and priming the T cell system, whilst other signals (perhaps local factors in mucosal lympho-epithelium) may be able to trigger a latently infected B cell from the reservoir back into lytic cycle, thereby establishing new transient foci of virus production.

T cell response to persistent infection

It is something of a paradox that EBV, a virus with potent cell growth transforming potential, is carried by the vast majority of immunocompetent individuals as a completely asymptomatic infection. A key factor in maintaining this virus–host balance appears to be the cell-mediated response, since all healthy EBV-carrying individuals possess relatively high numbers of virus-specific memory CTLs in the circulating T cell pool. Their functional activity is detectable following *in vitro* re-stimulation with autologous virus-infected B cells, and the effectors thus produced are largely CD8+ HLA class I antigen-restricted and operationally virus-specific in that they recognized LCL cells but not mitogen-activated B lymphoblasts (Moss *et al.*, 1992).

As in all viral systems, these CTLs recognize degraded forms of viral proteins generated by cytoplasmic processing and presented on the cell surface as HLA class I:peptide complexes. All eight EBV latent proteins therefore represent potential sources of target peptides for CTL responses, and one of the major objectives of recent work in this field has been to identify immunodominant EBV epitopes presented by a range of different HLA molecules. Effector CTL preparations reactivated from more than 30 virus-immune donors have now been analysed for antigen specificity by testing on an appropriately HLA-matched target cell panel expressing individual EBV latent proteins from recombinant vaccinia virus vectors (Murray *et al.*, 1992; Khanna *et al.*, 1992). The following points emerged: (i) HLA class I type was a key determinant of target antigen choice, any one particular allele usually focussing the response on a single viral protein, (ii) certain HLA class I alleles, where present in the host genotype, tended to be

Table 1. *Defined CTL epitopes of EBV*

HLA allele	Epitope sequence	Epitope location	Type specificity
A11	IVTDFSVIK	EBNA3B 416–424	Type 1
B8	FLRGRAYGL	EBNA3A 339–347	Type 1
B27.05/02/04	RRIYDLIEL	EBNA3C 258–266	Type 1 + Type 2
B44	EENLLDFVRF	EBNA3C 290–299	Type 1 + Type 2
A2.1	CLGGLLTMV	LMP2 426–434	Type 1 + Type 2

dominant restriction elements within the overall polyclonal CTL response, (iii) the major responses on a range of HLA backgrounds were frequently directed against the EBNA3A, 3B, 3C family of proteins, with less frequent examples of individual reactivities against EBNA2, EBNA-LP, LMP1 and LMP2, (iv) there were no examples, within polyclonal populations or derived CTL clones, of responses to EBNA1.

Several of the reactivities that are consistently strong in donors with the relevant HLA alleles, for instance the HLA-A11-restricted response to EBNA3B (Gavioli *et al.*, 1993), the HLA-B8-restricted response to EBNA3C (Burrows *et al.*, 1990*a*), the HLA-B27 restricted response to EBNA3C (Brooks *et al.*, 1993) and the HLA-B44-restricted response to EBNA3A (Burrows *et al.*, 1990*b*), have now been mapped to a single immunodominant peptide epitope within the primary sequence of these viral proteins (see Table 1). Weaker reactivities, for instance that restricted through the HLA-A2.1 allele and directed against LMP2, are likewise being precisely mapped (Lee *et al.*, 1993). Such detailed analysis allows one to begin to ask questions about epitope conservation between different virus isolates. Thus the above studies involved Caucasian donors whose resident virus was of type 1 (prototype B95.8) rather than type 2 (prototype AG876), and whose CTLs were reactivated *in vitro* by stimulation with a type 1 virus (usually B95.8)-transformed LCL. In some cases, such as the B27-restricted and B44-restricted responses, the CTLs showed broad cross-recognition of all type 1 and all type 2 virus isolates tested, reflecting conservation of the epitope sequence; this was also true of the A2.1-restricted response to LMP2. However, given that the immunodominant EBNA3A, 3B, 3C family of proteins are polymorphic between type 1 and type 2 viruses (Sample *et al.*, 1990), it was not surprising that certain other reactivities against these antigens, such as the major A11-restricted and B8-restricted responses, were type 1 specific. In both these particular instances there was additional evidence of variation even between individual viruses in the type 1 family, certain isolates not being recognized because of critical single amino acid changes in the epitope sequence (Apolloni *et al.*, 1992; de Campos-Lima *et al.*, 1993).

The EBV-induced memory CTL response is therefore focussed upon a relatively small number of viral target epitopes, and precise epitope choice will be dependent not only upon the HLA class I type of the host but also upon the identity of the resident viral strain.

EBV-POSITIVE MALIGNANCIES IN THE CONTEXT OF PERSISTENT INFECTION

Immunoblastic lymphomas

Immunoblastic lymphomas, essentially similar to those seen as multifocal lesions soon after bone marrow transplantation from a sero-positive donor, can also arise as a long-term consequence of immune suppression. Such cases are occasionally seen as late-onset disease in bone marrow recipients who show incomplete engraftment and where the lesions often prove to be of host rather than donor B cell origin (Gratama, 1993). Other organ allograft recipients under long-term immunosuppression are also at risk, as are end stage AIDS patients with profoundly impaired T cell function. The lymphoproliferations seen in these groups again tend to appear at multiple sites, often involving the central nervous system and/or alimentary tract, and each focus consists of either one or a small number of unique B cell clones. Furthermore such lesions are almost always EBV genome-positive (Ballerini et al., 1993; MacMahon et al., 1991; McGrath et al., 1991). In this context, one quite unexpected feature of the lymphomas seen in AIDS patients, as opposed to other immunologically compromised groups, is the frequency with which they carry a type 2 rather than a type 1 EBV isolate (Boyle et al., 1991). This finding is counter-intuitive in that type 2-infected B cells grow less well than their type 1 counterparts both in vitro and in SCID mice (Rickinson, Young & Rowe, 1987; Rowe et al., 1991). It remains to be seen, therefore, whether the above association reflects some special growth advantage conferred upon type 2-infected cells in the particular setting of AIDS, or simply an increased prevalence of type 2 virus infection amongst AIDS patients.

Where analysed, the above immunoblastic lymphomas (like those arising in association with primary EBV infection) display a lympho-blastoid cellular phenotype and have retained the full spectrum of virus latent gene expression (Young et al., 1989b; Thomas et al., 1990); this pattern of infection is now referred to as Latency III. Such observations are consistent with EBV being the primary driving force behind these proliferations. There is no clear evidence for any 'second event' in disease pathogenesis, even in the case of monoclonal foci, and certainly most lesions retain a normal diploid karyotype with none of the specific chromosomal translocation characterizing, for instance, Burkitt's lymphoma (Gratama, 1993). Interestingly, there are anecdotal cases in which allograft recipients with a history of EBV-driven lympho-proliferative disease have presented much later with a histologically confirmed, translocation-positive BL (D. H. Crawford, personal communication), but it is not yet clear whether these truly reflect disease progression or the generation of a quite independent malignant clone.

In the vast majority of cases therefore, immunoblastic lymphomas of the immunosuppressed represent 'opportunistic' proliferations of EBV-transformed B cells growing out in the absence of virus-specific CTL surveillance; in such circumstance the virus may be both necessary and sufficient for tumour growth. By contrast, all other examples of EBV-associated malignancies display a much more complex multi-step pathogenesis, and the virus is but one of several factors involved in the neoplastic process.

Burkitt's lymphoma

This is, for historical reasons, perhaps the best known EBV-associated malignancy and the nature of that association is discussed fully in an accompanying Chapter (see Chapter by Farrell and Sinclair). The salient point to emphasize here is that this tumour is clinically, histologically and cytogenetically quite distinct from the immunoblastic lymphomas described above. Thus BL usually presents in children who are not overtly immunosuppressed; indeed, even the very interesting cases of BL now being seen in HIV-infected patients usually appear as early rather than late manifestations of AIDS. Secondly, the BL cell surface phenotype shows classic germinal centre cell markers and lacks many of the lymphocyte activation antigens (CD23, CD30, CD39, CD70) and cellular adhesion molecules (ICAM1, LFA3) so characteristic of LCL-like cells. Thirdly, BL cells consistently display one of three specific chromosomal translocations, t(8:14), t(2:8), t(8:22), which place the c-myc oncogene under the transcriptional influence of an immunoglobulin gene locus (Magrath, 1990).

There are also major distinctions between BL and the immunoblastic lymphomas in virological terms. Thus, whilst virtually all cases of endemic BL are EBV genome-positive (with both virus types being represented), the virus is found in only 10–15% of the 'sporadic' tumours seen in Europe and North America; even the special category of AIDS-BL is EBV genome-positive in less than 50% cases (Subar et al., 1988; Ballerini et al., 1993). The virus therefore appears to play an important, but not essential, role in BL development. A second distinguishing characteristic of this malignancy is that virus latent protein expression is restricted to the genome maintenance protein, EBNA1 (Rowe et al., 1987). Selective expression of EBNA1 in BL cells is achieved by mRNA transcription from a novel promoter Fp, and the latent cycle promoters which in LCL-like cells are responsible for expression of all six EBNAs and the LMPs are completely silent (Sample et al., 1991; Schaeffer et al., 1991). This form of infection is now called Latency I.

These observations highlight how different is the behaviour of EBV in the context of BL to that seen in the *in vitro*-transformation system. Furthermore, the findings established the concept that alternative forms of EBV latent infection can and do exist. It seems inconceivable that EBV has evolved the novel patterns of promoter usage and transcription characterizing Latency I solely for use in the context of a rare malignancy. A much more likely explanation is that the situation in BL reflects a type of virus:cell interaction which EBV uses as part of its normal strategy for persistence in the healthy virus-carrying host.

Nasopharyngeal carcinoma (NPC)

Nasopharyngeal carcinoma of undifferentiated or poorly differentiated type is another EBV-associated malignancy whose incidence varies widely between different regions of the world. In this case, however, all tumours are consistently EBV genome-positive irrespective of the population in which they occur (Raab-Traub, 1993). It is widely believed that these differences in incidence rate worldwide reflect the combined influences of genetic predisposition in particular racial groups and of local environmental or dietary factors, although the biological basis for any of these effects remains conjectural. Another possibility, which still cannot be ruled out entirely, is that particular strains of EBV which carry a higher risk for NPC happen to be more prevalent in certain parts of the world. In this context, recent work has shown that EBV isolates from Southern China (a very high NPC risk area) are distinguishable from Caucasian and from African isolates at a number of polymorphic loci; furthermore these polymorphisms cut across the conventional division of EBV into types 1 and type 2 (Lung *et al.*, 1991; Abdel-Hamid *et al.*, 1992). Whether there are subtle biological differences between Chinese and Caucasian African isolates, particularly in relation to epithelial cell infection, remains to be tested.

Our understanding of the likely pathogenesis of NPC is fragmentary, not least because the presumed cellular genetic changes which complement the contribution of the virus in tumour development are not yet known (Raab-Traub, 1993). The tumour displays a characteristic histological appearance with considerable infiltration of the malignant epithelium by non-malignant T lymphocytes. The latter do not appear to reflect a specific immune reaction *per se*, but are instead thought to be responding to lymphokine release by the tumour cells themselves (Busson *et al.*, 1987). This feature appears closely linked to, perhaps determined by, the presence of EBV in the malignant epithelial cells; thus carcinomas with an exactly similar histological appearance are occasionally found in the salivary gland, in thymic epithelium and in gastric mucosa, and these particular tumours are also consistently EBV genome-positive (Saemundsen *et al.*, 1982; Leyvraz *et al.*, 1985; Shibata *et al.*, 1991).

One important unresolved question concerns the frequency with which EBV gains access to the normal nasopharyngeal epithelium in healthy virus carriers. A preliminary study of tumour-free nasopharyngeal biopsies from a population at high risk for NPC was unable to detect evidence of latent or lytic cycle viral markers (Sam *et al.*, 1993), but the possibility that EBV does normally colonize specific localized epithelial sites in the nasopharynx cannot yet be excluded. The finding that serum IgA responses to EBV lytic cycle antigens are a risk indicator of subsequent NPC development (Zeng, 1985) suggests that increased virus replication at one or more mucosal sites may be a key step in tumour pathogenesis. One interesting proposal is that such IgA responses may actually facilitate viral access to the relevant epithelial target population from which NPC develops, as a consequence of the natural process of secretory IgA transport across mucosal surfaces. Thus, in an *in vitro* model system, virus particles complexed with anti-gp340-specific polymeric IgA were able to enter and initiate infection in an epithelial cell line via the cells' polymeric IgA receptor (Sixbey & Yao, 1992). If substantiated in an *in vivo* situation, these observations would certainly help to explain an important and currently controversial aspect of NPC pathogenesis.

The pattern of virus latent gene expression in NPC is distinct from that seen either in BL or in the immunoblastic lymphomas, and is designated Latency II. Thus NPC cells use the Fp promoter to express EBNA1 in the absence of all the other EBNAs, but also show expression of the latent membrane proteins LMPs 1 and 2, albeit at levels which can vary considerably between individual tumours. Thus in some tumours LMP1 is easily detectable by immunoblotting in tumour extracts and by immunostaining in every malignant cell (Young *et al.*, 1988; Fahraeus *et al.*, 1988), whereas in other tumours the protein is below the threshold of detectability and LMP1 expression can only be inferred from the results of reverse transcription/ polymerase chain reaction (RT/PCR) assays for the presence of LMP1 mRNA. The work to date on LMP2 expression in NPC is entirely based on such transcriptional analysis and, whilst the relevant transcripts are regularly detectable by such methods (Smith & Griffin, 1991; Brooks *et al.*, 1992; Busson *et al.*, 1992), questions about the levels of LMP2 expression (*vis-à-vis* LCL cells) and about the homogeneity of expression within the tumour cell population have still to be resolved.

We still know very little about the natural behaviour of EBV in epithelial tissue, except that full virus replication is linked to terminal differentiation of the infected cell (Sixbey *et al.*, 1984; Greenspan *et al.*, 1985). There appears to be no demonstrably latent phase in the EBV-associated oral hairy leukoplakia lesions of lingual epithelium (Niedobitek *et al.*, 1991), and so what little information we do have on the establishment of latent infections has to be gleaned from *in vitro* models. It is interesting to note in this context that introduction of episomal viral genomes into an epithelial environment via fusion between an LCL and the HeLa cell line generated epithelial hybrids with the same Latency II form of infection as seen in NPC (Contreras-Brodin *et al.*, 1991). Efficient infection of epithelial cells by a conventional route has, to date, only been possible in targets expressing the B cell receptor molecule CR2 from an introduced vector (Li *et al.*, 1992). In the early latent phase following experimental infection, these cells showed detectable expression only of the EBNA1 protein, although this was accompanied by some apparently low level transcription of LMP1 and LMP2 mRNA detectable by RT/PCR (Q.X. Li and L. S. Young, unpublished data).

In the context of NPC pathogenesis therefore, once EBV does access the relevant target epithelium in the nasopharynx (perhaps cells in which other genetic changes have already occurred), the above observations suggest that it will naturally adopt the EBNA1, LMP1, LMP2 pattern of expression which one sees subsequently in the malignant clone. Of these virus latent proteins, LMP1 clearly has the potential to contribute to malignant change since expression of the protein has been shown to alter the epithelial cell phenotype quite dramatically *in vitro*, including a reduced responsiveness of the cells to terminal differentiation signals (Dawson *et al.*, 1990). Similar *in vitro* experiments have not shown any obvious phenotype for LMP2, but there are early indications that EBNA1 may have unexpected effects on an epithelial cell background since cells expressing this protein are unusually difficult to maintain (C. Dawson and L. S. Young, unpublished data).

Hodgkin's Disease and the EBV-positive T cell lymphomas

These are considered together because they are members of a growing group of lymphoid, but non-B-cell, malignancies which are now being linked to EBV. Whilst several epidemiological features of HD had been suggestive of an EBV association, the key evidence was the finding that a subset of HD biopsies contained monoclonal EBV episomes detectable by Southern blotting, and that, in such biopsies, the viral DNA could be localized by *in situ* hybridization to the malignant Reed–Sternberg cells (Anagnostopoulos *et al.*, 1989; Weiss *et al.*, 1989). Immunostaining then showed that these cells were EBNA2-negative but LMP1-positive (Pallesen *et al.*, 1991; Herbst *et al.*, 1991) and subsequent transcriptional analysis by PCR has confirmed a Latency II pattern of infection, i.e. expression of EBNA1, LMP1 and LMP2 (Deacon *et al.*, 1993). Overall, almost half the total cases of HD seen in Europe and North America are EBV-associated. There is, furthermore, an interesting correlation with histological subtype in that >80% of the more aggressive mixed cellularity/lymphocyte-depleted forms of HD appear to be EBV genome-positive, whilst the figure is closer to 30% for the nodular sclerosing and even lower for the lymphocyte-predominant subtype. Type 1 virus strains are prevalent in the tumours analysed to date, reflecting the

Table 2. *Overview of EBV-associated human tumours*

Tumour	Subtype	% genome positivity	Latent protein expression	Pattern of latency
Burkitt's lymphoma	Endemic	100%	EBNA1	I
	Sporadic	10–15%		
	AIDS-associated	30–40%		
Nasopharyngeal carcinoma	Undifferentiated	100%	EBNA1, LMPs 1–2	II
Hodgkin's disease	Mixed cell/Lymph. depleted	>80%	EBNA1, LMPs 1–2	II
	Nodular sclerosing	30%		
	Lymph. predominant	<10%		
T cell lymphoma	*VAHS/?XLP-associated	?100%	?	II
	Nasal (mid-line granuloma)	100%	EBNA1, LMPs 1, 2	
	AILD-type	?40%	?	
	Others	<10%	?	
Immunoblastic lymphoma	*Transplant (early onset)	100%	EBNAs 1, 2, 3A, 3B, 3C,-LP	III
	Transplant (late onset)	100%	LMPs 1,2	
	AIDS-associated	>90%		

*Tumours associated with primary rather than persistent EBV infection.

predominance of such strains in most Western societies, but again there is a suggestion that HD arising in an immunocompromised setting may more frequently carry type 2 virus (Boyle *et al.*, 1993). The corresponding figures for EBV association with HD in other societies are awaited with interest.

There is continuing debate over the lineage of cells from which HD is derived, not least because the paucity of tumour cells in most HD biopsies makes conventional immunogenetic analysis for immunoglobulin/T cell receptor gene rearrangement difficult. The balance of current evidence from phenotypic studies indicates a lymphoid origin, but there is heterogeneity between tumours in terms of their expression of early lymphoid versus mature B versus mature T cell markers (Herbst *et al.*, 1989). Interestingly, a similar heterogeneity is also seen in another (much rarer) malignancy, anaplastic large cell lymphoma, where the tumour cells bear some histological resemblance to Reed–Sternberg cells and where a proportion of tumours are also EBV genome-positive (Herbst, Stein & Niedobitek, 1993). Perhaps all these various malignancies originate from a common progenitor population of immature lymphoid cells into which EBV occasionally gains entry.

There is also increasing evidence linking EBV to specific lymphomas that are indisputably of T cell origin. Thus, in addition to the T cell tumours seen very occasionally in the context of a primary EBV infection (see earlier), EBV is also being consistently detected in a nasal T cell lymphoma ('midline granuloma') of adults (Harabuchi *et al.*, 1990) and in up to 40% of cases of other peripheral T cell lymphomas of angioimmunoblastic lymphoadenopathy type (Anagnostopoulos *et al.*, 1992). Only for the nasal tumour is there yet clear evidence of monoclonal EBV DNA in all the lymphoma cells, as opposed to what might be an innocent 'passenger' infection of just some cells in the malignant clone. Interestingly, recent work on this nasal lymphoma suggests that the resident pattern of viral gene expression resembles the Latency II seen in NPC and HD (Minarovits *et al.*, 1993).

An overview of these various associations between EBV and human tumours is given in Table 2.

IMMUNE T CELL CONTROL OF EBV-ASSOCIATED MALIGNANCIES

The existence of different forms of latency in the different EBV associated tumours has important implications for the whole question of immune T cell control over these malignancies.

Immunoblastic lymphomas all arise in the context of immune T cell dysfunction, whether iatrogenic ('post-transplant') or virus-induced ('AIDS-associated'). These tumours display a LCL-like cellular phenotype, which is known to confer efficient antigen-presenting function, and also express the full spectrum of virus latent proteins from which the target epitopes for CTL recognition derive. This strongly implies that such lesions will remain sensitive to a recovery of host T cell surveillance, a fact actually borne out by clinical observations (Starzl *et al.*, 1984).

In contrast, both the cellular and viral phenotype in BL strongly favour escape from immune recognition. First, recent work has shown that several components of the HLA class I antigen processing pathway are expressed only at very low levels in BL cells (M. Rowe, unpublished data); this, combined with the very low levels of cellular adhesion molecules such as ICAM1 and LFA3 at the BL cell surface, clearly makes for an inefficient antigen presenting function. Secondly, the only viral protein detectably expressed in BL tumour cells, EBNA1, appears not to be a preferred target for virus-induced CTL responses. Whether the lack of CTL epitopes in EBNA1 is absolute remains an interesting question; one intriguing possibility is that the protein has somehow evolved a means of avoiding the usual HLA class I antigen-processing pathway.

There is as yet relatively little information available on the antigen-presenting function of the other EBV-associated malignancies, namely NPC, HD and the T cell lymphomas. This largely reflects the difficulties of access to pure tumour cell populations and the paucity of *in vitro* models. Certainly for NPC and HD, however, what is known of the tumour cell phenotype is at least consistent with retention of antigen-presenting ability. In virological terms, these tumours all show down-regulation of the EBNA2, 3A, 3B, 3C and -LP proteins from which the immunodominant CTL epitopes are frequently derived. However, there does appear to be continued expression of LMP1 and LMP2, proteins which can provide epitopes for at least some HLA class I-restricted responses. One can therefore begin to consider the long-term prospects for CTL-based immunotherapy. The HLA-A2.1-restricted LMP2-specific response identified in Table 1, for instance, is at best a minor component of EBV-induced CTL preparations from A2.1-positive donors but, because the A2.1 allele is common in virtually all human populations, this particular example could, in the long run, be relevant to large numbers of patients. An important priority for the future will be to understand how best to amplify such minor responses, perhaps by some form of selective reactivation, and to target them against malignant cells.

REFERENCES

Abdel-Hamid, M., Chen, J. J., Constantine, N., Massoud, M. & Raab-Traub, N. (1992). EBV strain variation: Geographical distribution and relation to disease state. *Virology,* **190**, 168–75.

Allday, M. J. & Crawford, D. H. (1988). Role of epithelium in EBV persistence and pathogenesis of B-cell tumours. *Lancet,* **i**, 855–857.

Anagnostopoulos, I., Herbst, H., Niedobitek, G. & Stein, H. (1989). Demonstration of monoclonal EBV genomes in Hodgkin's disease and Ki-1-positive anaplastic large cell lymphoma by combined Southern blot and *in situ* hybridisation. *Blood*, **74**, 810–16.

Anagnostopoulos, I., Hummel, M., Finn, T. *et al.* (1992). Heterogeneous Epstein–Barr virus infection patterns in peripheral T-cell lymphomas of angioimmunoblastic lymphadenopathy type. *Blood*, **80**, 1804–12.

Apolloni, A., Moss, D., Stumm, R. *et al.* (1992). Sequence variation of cytotoxic T cell epitopes in different isolates of Epstein–Barr virus. *European Journal of Immunology*, **22**, 183–9.

Ballerini, P., Gaidano, G., Gong, J. Z. *et al.* (1993). Multiple genetic lesions in acquired immunodeficiency syndrome-related non-Hodgkin's lymphoma. *Blood*, **81**, 166–76.

Birkenbach, M., Tong, X., Bradbury, L. E., Tedder, T. F. & Kieff, E. (1992). Characterisation of an Epstein–Barr virus receptor on human epithelial cells. *Journal of Experimental Medicine*, **176**, 1405–14.

Blacklow, N. R., Watson, B. K., Miller, G. & Jacobson, B. M. (1971). Mononucleosis with heterophile antibodies and EB virus infection. Acquisition by an elderly patient in hospital. *American Journal of Medicine*, **51**, 549–52.

Boyle, M. J., Sewell, W. A., Sculley, T. B. *et al.* (1991). Subtypes of Epstein–Barr virus in human immunodeficiency virus-associated non-Hodgkin's lymphoma. *Blood*, **78**, 3004–11.

Boyle, M. J., Vasak, E., Tschuchnigg, M. *et al.* (1993). Subtypes of Epstein–Barr virus (EBV) in Hodgkin's Disease: association between B-type EBV and immunocompromise. *Blood*, **81**, 468–74.

Brooks, J. M., Murray, R. J., Thomas, W. A., Kurilla, M. G. & Rickinson, A. B. (1993). Different HLA-B27 subtypes present the same immunodominant Epstein–Barr virus peptide. *Journal of Experimental Medicine*, **178**, 879–87.

Brooks, L., Yao, Q. Y., Rickinson, A. B. & Young, L. S. (1992). Epstein–Barr virus latent gene transcription in nasopharyngeal carcinoma cells: coexpression of EBNA1, LMP1, and LMP2 transcripts. *Journal of Virology*, **66**, 2689–97.

Burrows, S. R., Sculley, T. B., Misko, I. S., Schmidt, C. & Moss, D. J. (1990*a*). An Epstein–Barr virus-specific cytotoxic T cell epitope in EBNA 3. *Journal of Experimental Medicine*, **171**, 345–50.

Burrows, S. R., Misko, I. S., Sculley, T. B., Schmidt, C. & Moss, D. J. (1990*b*). An Epstein–Barr virus-specific cytotoxic T cell epitope present on A- and B-type transformants. *Journal of Virology*, **64**, 3974–6.

Busson, P., Braham, K., Ganem, G. *et al.* (1987). Epstein–Barr virus-containing epithelial cells from nasopharyngeal carcinoma produce interleukin 1α. *Proceedings of the National Academy of Sciences, USA*, **84**, 6262–6.

Busson, P., McCoy, R., Sadler, R., Gilligan, K., Tursz, T. & Raab-Traub, N. (1992). Consistent transcription of the Epstein–Barr virus LMP2 gene in nasopharyngeal carcinoma. *Journal of Virology*, **66**, 3257–62.

Chan, L. C., Srivastava, G., Pittaluga, S., Kwong, Y. L., Liu, H. W. & Yuen, H. L. (1992). Detection of clonal Epstein–Barr virus in malignant proliferation of peripheral blood CD3+ CD8+ T cells. *Leukemia*, **6**, 952–6.

Contreras-Brodin, B. A., Anvret, M., Imreh, S., Altiok, E., Klein, G. & Masucci, M. G. (1991). B cell phenotype-dependent expression of the Epstein–Barr virus nuclear antigens EBNA-2 to EBNA-6: Studies with somatic cell hybrids. *Journal of General Virology*, **72**, 3025–33.

Craig, F. E., Clare, C. N., Sklar, J. L. & Banks, P. M. (1992). T cell lymphoma and the virus-associated haemophagocytic syndrome. *American Journal of Clinical Pathology*, **97**, 189–94.

Dawson, C. W., Rickinson, A. B. & Young, L. S. (1990). Epstein–Barr virus latent membrane protein inhibits human epithelial cell differentiation. *Nature*, **344**, 777–80.

Deacon, E. M., Pallesen, G., Niedobitek, G. *et al.* (1993). Epstein–Barr virus and Hodgkin's disease: transcriptional analysis of virus latency in the malignant cells. *Journal of Experimental Medicine*, **177**, 339–49.

de Campos-Lima, P-O., Gavioli, R., Zhang, Q-J. *et al.* (1993). HLA-A11 epitope loss isolates of Epstein–Barr virus from a highly A11+ population. *Science*, **260**, 98–100.

Fahraeus, R., Fu, H. L., Ernberg, I. (1988). Expression of Epstein–Barr virus-encoded proteins in nasopharyngeal carcinoma. *International Journal of Cancer*, **42**, 329–38.

Gaillard, F., Mechinaud-Lacroix, F., Papin, S. *et al.* (1992). Primary Epstein–Barr virus infection with clonal T-cell lymphoproliferation. *American Journal of Clinical Pathology*, **98**, 324–33.

Gavioli, R., Kurilla, M. G., de Campos-Lima, P. O. *et al.* (1993). Multiple HLA-A11-restricted cytotoxic T-lymphocyte epitopes of different immunogenicities in the Epstein–Barr virus-encoded nuclear antigen 4. *Journal of Virology*, **67**, 1572–8.

Gerber, P., Walsh, J. H., Rosenblum, E. N. & Purcell, R. H. (1969). Association of EB virus infection with the post-perfusion syndrome. *Lancet*, **i**, 593–6.

Gratama, J. W., Oosterveer, M. A. P., Zwaan, F. E., Lepoutre, J., Klein, G. & Ernberg, I. (1988). Eradication of Epstein–Barr virus by allogeneic bone marrow transplantation: implications for the site of viral latency. *Proceedings of the National Academy of Sciences, USA*, **85**, 8693–9.

Gratama, J. W., Zutter, M. M., Minarovits, J. *et al.* (1991). Expression of Epstein–Barr virus growth-transformation-associated proteins in lymphoproliferations of bone-marrow transplant recipients. *International Journal of Cancer*, **47**, 188–92.

Gratama, J. W. (1993). Epstein–Barr virus infections in bone marrow transplant recipients. In *Bone Marrow Transplantation*, ed. S. J. Forman, K. G. Blume & E. D. Thomas. New York: (in press). Blackwell Scientific Publications.

Greenspan, J. S., Greenspan, D., Lennette, E. T. *et al.* (1985). Replication of Epstein–Barr virus within the epithelial cells of oral 'hairy' leukoplakia, an AIDS-associated lesion. *New England Journal of Medicine*, **313**, 1564–71.

Harabuchi, Y., Yamanaka, N., Kakaura, A. *et al.* (1990). Epstein–Barr virus in nasal T-cell lymphomas in patients with lethal midline granuloma. *Lancet*, **335**, 128–30.

Herbst, H., Dallenbach, F., Hummel, M. *et al.* (1991). Epstein–Barr virus latent membrane protein expression in Hodgkin and Reed-Sternberg cells. *Proceedings of the National Academy of Sciences, USA*, **88**, 4766–70.

Herbst, H., Stein, H. & Niedobitek, G. (1993). Epstein–Barr virus & CD30+ malignant lymphomas. *Critical Reviews in Oncogenesis*, **4**, 191–239.

Herbst, H., Tippelmann, G., Anagnostopoulos, I. *et al.* (1989). Immunoglobulin and T cell receptor gene rearrangements in Hodgkin's Disease and Ki-1 positive anaplastic large cell lymphoma: dissociation between genotype and phenotype. *Leukaemia Research*, **13**, 103–16.

Khanna, R., Burrows, S. R., Kurilla, M. G. *et al.* (1992). Localisation of Epstein–Barr virus cytotoxic T cell epitopes using recombinant vaccinia: Implications for vaccine development. *Journal of Experimental Medicine*, **176**, 169–78.

Lee, S. P., Thomas, W. A., Murray, R. J. *et al.* (1993). HLA A2.1-restricted cytotoxic T cells recognising a wide range of Epstein–Barr virus isolates through a defined epitope in the latent membrane protein LMP2. *Journal of Virology*, **67**, 7428–35.

Leyvraz, S., Henle, W., Chahinian, A. P. *et al.* (1985). Association of Epstein–Barr virus with thymic carcinoma. *New England Journal of Medicine, 312,* 1296–9.

Li, Q. X., Young, L. S., Niedobitek, G. *et al.* (1992). Epstein–Barr virus infection and replication in a human epithelial cell system. *Nature,* 356, 347–50.

Lung, M. L., Lam, W. P., Sham, J. *et al.* (1991). Detection and prevalence of the 'f' variant of Epstein–Barr virus in Southern China. *Virology,* 185, 67–71.

McGrath, M. S., Shiramizu, B., Meeker, T. C., Kaplan, L. D. & Herndier, B. (1991). AIDS-associated polyclonal lymphoma: identification of a new HIV-associated disease process. *Journal of Acquired Immune Deficiency Syndrome,* 4, 408–15.

MacMahon, E. M., Glass, J. D., Hayward, S. D. *et al.* (1991). Epstein–Barr virus in AIDS-related primary central nervous system lymphoma. *Lancet,* 338, 969–73.

Magrath, I. (1990). The pathogenesis of Burkitt's lymphoma. *Advances in Cancer Research,* 55, 133–269.

Minarovits, J., Hu, L. F., Imai, S. *et al.* (1993). Clonality, expression and methylation patterns of the Epstein–Barr virus genomes in lethal mid-line granulomas (LMGs) classified as peripheral angiocentric T cell lymphomas. *Journal of General Virology,* in press.

Moss, D. J., Bishop, C. J., Burrows, S. R. & Ryan, J. M. (1985). T lymphocytes in infectious mononucleosis. I. T cell death *in vitro. Clinical and Experimental Immunology,* 60, 61.

Moss, D. J., Burrows, S. R., Khanna, R., Misko, I. S. & Sculley, T. B. (1992). Immune surveillance against Epstein–Barr virus. *Seminars in Immunology,* 4, 97–104.

Murray, R. J., Kurilla, M. G., Brooks, J. M. *et al.* (1992). Identification of target antigens for the human cytotoxic T cell response to Epstein–Barr virus (EBV): Implications for the immune control of EBV-positive malignancies. *Journal of Experimental Medicine,* 176, 157–68.

Nemerow, G. R., Mold, C., Schwend, V. K., Tollefson, V. & Cooper, N. R. (1987). Identification of gp350 as the viral glycoprotein mediating attachment of Epstein–Barr virus (EBV) to the EBV/C3d receptor of B cells: sequence homology of gp350 and C3 complement fragment C3d. *Journal of Virology,* 61, 1416–20.

Niedobitek, G., Young, L. S., Lau, R. *et al.* (1991). Epstein–Barr virus infection in oral hairy leukoplakia: virus replication in the absence of a detectable latent phase. *Journal of General Virology,* 72, 3035–46.

Nilsson, K., Klein, G., Henle, W. & Henle, G. (1971). The establishment of lymphoblastoid cell lines from adult and from foetal human lymphoid tissue and its dependence on EBV. *International Journal of Cancer,* 8, 443–50.

Pallesen, G., Hamilton-Dutoit, S. J., Rowe, M. & Young, L. S. (1991). Expression of Epstein–Barr virus latent gene products in tumour cells of Hodgkin's disease. *Lancet,* 337, 320–2.

Qu, L. & Rowe, D. T. (1992). Epstein–Barr virus latent gene expression in uncultured peripheral blood lymphocytes. *Journal of Virology,* 66, 3715–24.

Raab-Traub, N. (1993). Epstein–Barr virus and nasopharyngeal carcinoma. In *Seminars in Cancer Biology,* Vol. 3, ed. A. B. Rickinson, pp. 297–307. London: Saunders Scientific Publications/Academic Press.

Reinhertz, E. L., O'Brien, C., Rosenthal, P. & Schlossman, S. F. (1980). The cellular basis for viral-induced immunodeficiency: analysis by monoclonal antibodies. *Journal of Immunology,* 125, 1269–74.

Rickinson, A. B. (1988). Novel Forms of Epstein–Barr virus persistence. In *Immunobiology and Pathogenesis of Persistent Virus Infections*, ed. C. Lopez, pp. 294–305. Washington. DC: American Society of Microbiology.

Rickinson, A. B., Young, L. S. & Rowe, M. (1987). Influence of the EBV nuclear antigen EBNA 2 on the growth phenotype of virus-transformed B cells. *Journal of Virology*, **61**, 1310–17.

Rowe, M., Rowe, D. T., Gregory, C. D. *et al.* (1987). Differences in B cell growth phenotype reflect novel patterns of Epstein–Barr virus latent gene expression in Burkitt's lymphoma cells. *European Molecular Biology Organisation Journal*, **6**, 2743–51.

Rowe, M., Young, L. S., Crocker, J., Stokes, H., Henderson, S. & Rickinson, A. B. (1991). Epstein–Barr virus (EBV)-associated lymphoproliferative disease in the SCID mouse model: implications for the pathogenesis of EBV-positive lymphomas in man. *Journal of Experimental Medicine*, **173**, 147–58.

Saemundsen, A. K., Albeck, H., Hansen, J. P. H. *et al.* (1982). Epstein–Barr virus nasopharyngeal and salivary gland carcinomas in Greenland Eskimoes. *British Journal of Cancer*, **46**, 721–8.

Sam, C. K., Brooks, L. A., Niedobitek, G., Young, L. S., Prasad, U. & Rickinson, A. B. (1993). Analysis of Epstein–Barr virus infection in nasopharyngeal biopsies from a group at high risk of nasopharyngeal carcinoma. *International Journal of Cancer*, **53**, 957–62.

Sample, J., Young, L., Martin, B. *et al.* (1990). Epstein–Barr virus type 1 (EBV-1) and 2 (EBV-2) differ in their EBNA-3A, EBNA-3B and EBNA-3C genes. *Journal of Virology*, **64**, 4084–92.

Sample, J., Brooks, L., Sample, C. *et al.* (1991). Restricted Epstein–Barr virus protein expression in Burkitt's lymphoma is due to a different Epstein–Barr nuclear antigen 1 transcriptional initiation site. *Proceedings of the National Academy of Sciences, USA*, **88**, 6343–7.

Schaefer, B. C., Woisetschlaeger, M., Strominger, J. L. & Speck, S. H. (1991). Exclusive expression of Epstein–Barr virus nuclear antigen 1 in Burkitt lymphoma arises from a third promoter, distinct from the promoters used in latently infected lymphocytes. *Proceedings of the National Academy of Sciences, USA*, **88**, 6550–4.

Shapiro, R. S., McClain, K., Frizzera, G. *et al.* (1988). Epstein–Barr virus-associated B cell lymphoproliferative disorders following bone marrow transplantation. *Blood*, **71**, 1234–43.

Shibata, D., Tokunaga, M., Uemura, Y., Sato, E., Tanaka, S. & Weiss, L. M. (1991). Association of Epstein–Barr virus with undifferentiated gastric carcinomas with intense lymphoid infiltration. Lymphoepithelioma-like carcinoma. *American Journal of Pathology*, **139**, 469–74.

Sixbey, J. W., Nedrud, J. G., Raab-Traub, N., Hanes, R. A. & Pagano, J. S. (1984). Epstein–Barr virus replication in oropharyngeal epithelial cells. *New England Journal of Medicine*, **310**, 1225–30.

Sixbey, J. W. & Yao, Q. Y. (1992). Immunoglobulin A-induced shift of Epstein–Barr virus tissue tropism. *Science*, **255**, 1578–80.

Smith, P. R. & Griffin, B. E. (1991). Differential expression of Epstein-Barr viral transcripts for two proteins (TP1 & LMP) in lymphocyte and epithelial cells. *Nucleic Acids Research*, **19**, 2435–40.

Starzl, T. E., Nalesnik, M. A., Porter, K. A. *et al.* (1984). Reversibility of lymphomas and lymphoproliferative lesions developing under cyclosporin A-steroid therapy. *Lancet*, **i**, 583–7.

Strang, G. & Rickinson, A. B. (1987). Multiple HLA class I-dependent cytotoxicities constitute the non-HLA-restricted response in infectious mononucleosis. *European Journal of Immunology*, **17**, 1007–13.

Subar, M., Neri, A., Inghirami, G., Knowles, D. M. & Dalla-Favera, R. (1988).

Frequent c-myc oncogene activation and infrequent presence of Epstein–Barr virus genome in AIDS-associated lymphoma. *Blood,* **72**, 667–71.

Sullivan, J. L. & Woda, B. A. (1989). X-linked lymphoproliferative syndrome. *Immunedeficiency Review,* **1**, 325–47.

Svedmyr, E., Ernberg, I., Seeley, J. *et al.* (1984). Virologic, immunologic, and clinical observations on a patient during the incubation, acute, and convalescent phases of infectious mononucleosis. *Clinical Immunology and Immunopathology,* **30**, 437–50.

Thomas, J. A., Hotchin, N., Allday, M. J., Yacoub, M. & Crawford, D. H. (1990). Immunohistology of Epstein–Barr virus associated antigens in B cell disorders from immunocompromised individuals. *Transplantation,* **49**, 944–53.

Watry, D., Hedrick, J. A., Siervo, S. *et al.* (1991). Infection of human thymocytes by Epstein–Barr virus. *Journal of Experimental Medicine,* **173**, 971–80.

Weiss, L. M., Mohared, L. A., Warnke, R. A. & Sklar, J. (1989). Detection of Epstein–Barr viral genomes in Reed–Sternberg cells of Hodgkin's disease. *New England Journal of Medicine,* **320**, 502–6.

Yao, Q. Y., Rickinson, A. B. & Epstein, M. A. (1985). A re-examination of the Epstein–Barr virus carrier state in healthy seropositive individuals. *International Journal of Cancer,* **35**, 35–42.

Yao, Q. Y., Ogan, P., Rowe, M., Wood, M. & Rickinson, A. B. (1989). Epstein–Barr virus-infected B cells persist in the circulation of acyclovir-treated virus carriers. *International Journal of Cancer,* **43**, 67–71.

Young, L. S., Dawson, C. W., Clark, D. *et al.* (1988). Epstein–Barr virus gene expression in nasopharyngeal carcinoma. *Journal of General Virology,* **69**, 1051–65.

Young, L. S., Dawson, C. W., Brown, K. W. & Rickinson, A. B. (1989a). Identification of a human epithelial cell surface protein sharing an epitope with the C3d/Epstein–Barr virus receptor molecule of B lymphocytes. *International Journal of Cancer,* **43**, 786–94.

Young, L., Alfieri, C., Hennessy, K. *et al.* (1989b). Expression of Epstein–Barr virus transformation-associated genes in tissues of patients with EBV lymphoproliferative disease. *New England Journal of Medicine,* **321**, 1080–85.

Zeng, Y. (1985). Seroepidemiological studies on nasopharyngeal carcinoma in China. *Advances in Cancer Research,* **44**, 121–38.

Zutter, M. M., Martin, P. J., Sale, G. E. *et al.* (1988). Epstein–Barr virus lymphoproliferation after bone marrow transplantation. *Blood,* **72**, 520–9.

BURKITT'S LYMPHOMA

P. J. FARRELL AND A. J. SINCLAIR

Ludwig Institute for Cancer Research, St Mary's Hospital Medical School, Norfolk Place, London W2 1PG, UK.

HISTORICAL PERSPECTIVE

Burkitt's lymphoma was first recognized as a distinct disease because of its exceptionally high incidence in children in central Africa. While travelling in Africa in the 1950s, Burkitt described the tumour and collected remarkable incidence data (for review see Burkitt, 1983). At this time, following progress in the analysis of oncogenic retroviruses in animals and the observation that incidence rates of certain tumours showed marked geographic variations, medical research had become interested in the possibility that viruses might contribute to cancer in humans. After reporting his results in London, Burkitt sent tumour biopsy material to Epstein who succeeded in culturing a virus (the Epstein-Barr virus, EBV) from the tumour cells (Epstein, Achong & Barr, 1964). The virus proved to be extremely efficient at transforming human lymphocytes (Pope, Horne & Scott, 1968) and was the first human virus isolated capable of transforming human cells. It thus caused intense interest as a possible human cancer virus.

In due course it became apparent that infection with EBV was not confined to the districts where BL was endemic and, in fact, the vast majority of the world's population are infected with EBV and carry the virus for life. It was also realized that high grade malignant B cell lymphomas of the Burkitt's type also occur at a lower incidence throughout the world. There is nevertheless substantial evidence linking EBV to the development of many cases of BL.

BURKITT'S LYMPHOMA

Three different forms of BL are now recognized (for review see Magrath, 1990). The endemic form is found in Africa and New Guinea and is restricted to areas where infection by *Plasmodium falciparum*, the parasite causing malaria, is frequent. This tumour is the most frequent cancer in African children and is the form originally recognized by Burkitt. All the malignant cells of the great majority (95%) of BL cases occurring in high incidence areas carry EBV DNA.

The second, sporadic, form of BL occurs with a 20 to 100-fold lower incidence worldwide and is not associated with malarial infection. It represents less than 3% of childhood cancer. Viral markers are present in the tumour cells in less than 15% of cases. Thus BL in Europe and the USA is usually EBV negative and the few EBV negative tumours (5%) in areas of endemic BL probably represent the underlying incidence of sporadic EBV negative tumours worldwide. The third form of BL is found in patients suffering from AIDS; the EBV genome is present in the tumours in 30–45% of these cases of BL. These variations in EBV content occur even though the great majority (about 95%) of people worldwide carry EBV.

Type variation of EBV has been recognized for many years (for review see Kieff and Liebowitz, 1990), with isolates usually sorting readily into type A or B (also called 1 and 2). The types are defined by differences in the sequence of EBNA-2 but also differ in the EBER RNAs (Arrand, Young & Tugwood, 1989), the EBNA-3 genes (Rowe *et al.*, 1989; Sample *et al.*, 1990; Sculley *et al.*, 1989) and the 3' end of EBNA-LP (Huen, Grand & Young, 1988). Type A strains are more efficient immortalizing viruses than type B (Rickinson, Young & Rowe, 1987). Type B EBV tends to be somewhat more common in African populations than in European or US people (Zimber *et al.*, 1986, Young *et al.*, 1987), but there is no indication that this is significant for the incidence of BL.

Independent of geographical distribution or association with EBV, all BLs present the same cell phenotype and display a characteristic translocation bringing the *c-myc* proto-oncogene on chromosome 8 into the proximity of one of the immunoglobulin genes on chromosome 14, 22 or 2, leading to the deregulation of *c-myc*. There are, however, differences in clinical presentation of sporadic and endemic BL particularly in organs affected and the younger age distribution of endemic BL.

B CELL PHENOTYPE OF BL CELLS

BL cells synthesize immunoglobulins and are therefore of B cell origin (Klein *et al.*, 1967). The vast majority of tumours express surface IgM, but a few percent of tumours lack IgM and express other heavy chain classes (Gunven *et al.*, 1980). Some tumours have a pre-B phenotype in which μ chains are present in the cytoplasm but are not expressed on the surface (Cohen *et al.*, 1987). All BL cells express the surface antigens CD19 and CD20 (Rowe *et al.*, 1985); the expression of these membrane glycoproteins is restricted to cells in the B-lymphoid lineage from early pre-B to mature B cells, correlating with the presence of immunoglobulin (Ig) gene rearrangements characteristic of B cells.

BL cells consistently express two markers which are characteristic of B cells at specific stages of maturation, CD10 (Rowe *et al.*, 1985) and CD77 (Nudelman *et al.*, 1983). Normal B cells expressing CD10 and CD77 are found mostly in lymph node germinal centres and have the morphology of centroblasts (Gregory *et al.*, 1987). Centroblasts might therefore represent a normal correlate of the BL cell but, unlike BL cells, centroblasts do not express surface immunoglobulin. Generally, endemic BLs express cytoplasmic or surface immunoglobulin but do not secrete immunoglobulin whereas in sporadic BLs IgM is often secreted, indicating that the sporadic tumours may have arisen in a later stage of B cell development (Benjamin *et al.*, 1982; Pelicci *et al.*, 1986). So the best physiological correlate of BL cells might be an intermediate stage between follicular B blasts (activated by antigen and T lymphocytes outside the germinal centre) and centroblasts in the germinal centre (Ling *et al.*, 1989), although it should be noted that the tumour is only rarely found in the lymph node.

HLA class I antigens are usually found on BL cells but defective expression of alleles within the HLA-A and -C loci is a common feature (Masucci *et al.*, 1987). Expression of the cell adhesion molecules, LFA-1, ICAM-1 and LFA-3 is also low or absent on BL cells (Gregory *et al.*, 1988).

Many cell lines have been established from BL tumours. Whilst EBV negative cell lines have a homogeneous phenotype mostly similar to the original tumour cells (Rowe *et al.*, 1986), EBV positive cell lines display an important heterogeneity with regard to surface antigens and growth properties and this is due to a phenotypic drift during the early passage of the cell lines *in vitro* (Rowe *et al.*, 1987; Rowe & Gregory, 1989; Klein, 1989). They have been divided into several groups according to their cell and viral gene expression.

Group I BL lines display the CD10, CD 77 surface phenotype characteristic of freshly isolated BL cells (Rooney *et al.*, 1986) and are the best cell culture equivalent of BL cells available (Rowe *et al.*, 1987). Group III BL lines have acquired surface antigens normally found on activated B cells (CD23, CD30, CD39, CD70), lost the CD10 and CD77 markers characteristic of BL biopsy cells and also show up-regulation of surface adhesion molecules (LFA-1, ICAM-1 and LFA-3; Gregory *et al.*, 1988). Group II cells have an intermediate phenotype. These phenotypic changes are associated with modification of growth properties. Group III lines grow in multicellular clumps, whereas group I lines grow as a carpet of single cells. Many commonly studied EBV positive BL cell lines fall into the group II or III category and therefore differ from authentic BL cells *in vivo*.

EBV GENE EXPRESSION IN BL

For many years it was assumed that EBV gene expression in BL cells was the same as in B lymphocytes immortalized by EBV (LCLs) but careful analysis of Group I BL cell lines that faithfully retain the BL cell phenotype showed that this is not the case.

An LCL immortalized by EBV normally expresses up to 11 EBV genes (Kieff & Liebowitz, 1990) of which six encode nuclear proteins (EBNA-1, 2, 3A, 3B, 3C, LP), three encode cytoplasmic membrane proteins LMP (LMP1), TP1 (LMP2A) and TP2 (LMP2B) and two are small RNAs

Table 1. *Nomenclature of EBV genes*

EBNA-1		
EBNA-2		
EBNA-3A	EBNA-3	
EBNA-3B	EBNA-4	
EBNA-3C	EBNA-6	
EBNA-LP	EBNA-5	EBNA-4
LMP1		LMP
LMP2A		TP1
LMP2B		TP2
EBER		

Names on the same horizontal line refer to the same gene.
EBNA = Epstein–Barr nuclear antigen
LMP   = Latent membrane protein
EBER = Epstein–Barr encoded RNA

(EBER1, EBER2). For a summary of the alternative nomenclatures used for EBV latent genes by different groups see Table 1. Of the genes expressed when EBV immortalizes a B cell, EBNA-2 (Cohen *et al.*, 1989; Hammerschmidt & Sugden, 1989) EBNA-3A, EBNA-3C (Tomkinson, Robertson & Kieff, 1993), EBNA-LP (Mannick *et al.*, 1991) and LMP have been shown by mutagenesis to be required for efficient immortalization of B lymphocytes but these proteins are not expressed in group I BL cell lines. In BL biopsies and in group I BL cell lines EBNA-1 and the EBER RNAs are expressed but EBNA-2, EBNA-3A, 3B, 3C, EBNA-LP, LMP and TP1, 2 (LMP2A, B) are absent (Rowe *et al.*, 1987).

BL cell lines classified as group III have become like LCLs in their surface markers and EBV gene expression. There may be a selective advantage in cell culture for cells expressing the full set of EBV immortalization genes, and this may be the reason for the spontaneous outgrowth of the group III lines. Recent results have indicated that BL cell lines also express a family of RNAs covering part of the BamHI A region of EBV (Brooks *et al.*, 1993; Karran *et al.*, 1992; Smith *et al.*, 1993). These RNAs are found only at low levels in BL cell lines and no function or protein product has yet been demonstrated from them but it remains possible that they contribute to the transformed phenotype or the control of EBV in the cells.

If there is a continued role for EBV in the growth of mature EBV positive BL cells, EBNA-1 seems to be an obvious candidate for *trans*-regulation of cell functions since it is the only EBV protein known to be expressed. EBNA-1 activates an enhancer within the ori-P origin of replication in EBV and thus has the capacity to affect gene expression (Sugden & Warren, 1989). The EBER RNAs are also expressed in group I BL cells (Rowe *et al.*, 1987) but they play no obvious role in EBV immortalization (Swaminathan,

Tomkinson & Kieff, 1991) so it is more difficult to envisage a function for them in BL cell growth. The necessary maintenance replication function of EBNA-1 complicates the isolation of virus EBNA-1 deletion mutants and this has so far precluded testing whether EBNA-1 has a further direct role in the immortalization process. Unfortunately, there are no data on EBV gene expression in the cells in the lymph node germinal centre which may be the precursors of BL cells. In BL cell lines and tumour biopsies we are only observing the end stage of presumably a multistep carcinogenic process; this is of relevance when considering possible roles for EBV in BL.

The failure to express most of the EBNAs and the LMPs in BL cells may be an important part of the mechanism by which BL cells evade cytotoxic T lymphocyte (CTL) mediated immune surveillance. Some of the EBNAs (excluding EBNA-1) and LMPs have been shown to encode peptide epitopes which mediate class I restricted CTL killing of cells infected with EBV (Murray et al., 1992). The reduction in surface LFAs and ICAM-1 (and sometimes class I HLA) doubtless also plays a role in evasion of immune surveillance.

To understand the mechanism of immortalization by EBV, considerable effort has been put into identifying cell genes which might be regulated in trans by EBV, either in immortalized cells or BL cells. Thus it was found that CD21 (Cordier et al., 1990), CD23 (Wang et al., 1990), c-fgr (Cheah et al., 1986; Knutson, 1990), EBI-1, EBI-2 (Birkenbach et al., 1993) and cyclin D2 (Palmero et al., 1993) are all expressed at a much higher level in cells expressing the EBV immortalizing genes (for example, in LCLs and in group III BL cell lines) than in EBV negative BL cells or in group I BL cells. For some of these genes progress has been made in the mechanism of activation. CD23 is the first B lineage restricted surface marker to appear on B lymphocytes following infection with EBV (Thorley-Lawson & Mann, 1985). Transfection assays of the promoter of CD23 linked to a reporter gene revealed that EBNA-2 is able to upregulate the CD23 promoter and that EBNA-2 and LMP co-operate in the induction (Wang et al., 1990, 1991). Since CD23 has been proposed to be part of an autocrine growth factor system for LCLs by giving rise to a soluble proteolytic product that has growth stimulatory activity (Thorley-Lawson, Swendeman & Edson, 1986; Gordon et al., 1984; Swendeman & Thorley-Lawson, 1987), it is likely that this is a significant contribution of EBNA-2 to immortalization. Similarly, transfection of EBNA-2 is sufficient to upregulate c-fgr (Knutson, 1990), although the mechanism of this has not been studied in so much detail as CD23. The two G protein genes EBI-1 and EBI-2 were isolated by virtue of their up regulation after EBV conversion of an EBV negative BL cell line (Birkenbach et al., 1993) and these genes may be important in signal transduction controlling the growth of these EBV converted cells. Presumably, however, CD21, CD23, EBNA-2 and c-fgr are not important in the growth of group I BL cell lines or BL cells in vivo since these cells do not express them. The expression status of EBI-1 and EBI-2 in group I BL cells has not yet been reported.

For cyclin D2 it is unclear whether it is EBV gene products or the differentiation state of the B cell that is determining expression. This important member of the cyclin family acts in the G1 phase of the cell cycle, modulating activity of cdk kinases and regulating progression through the cell cycle (Hunter & Pines, 1991; Honigberg, McCarroll & Esposito, 1993; Motokura & Arnold, 1993). Cyclin D2 is expressed in LCLs and group III BL cell lines but not in group I BL cells (Palmero *et al.*, 1993). It is not expressed in the P3HR1 and Daudi BL lines, which lack EBNA-2 and have a truncated EBNA-LP (Sinclair A. J., Palmero I., Farrell P. J. & Peters G, personal communication). So cyclin D2 is a candidate for transregulation by EBNA-2, though this has not been demonstrated directly and preliminary studies transfecting EBNA-2 to EBV negative BL cell lines failed to activate expression of cyclin D2. Again, this regulation of cyclin D2 seems unlikely to be important in BL tumour cells since it is not expressed in group I BL cell lines, which are the best cell culture model available for the tumour cell *in vivo*. No cell gene has yet been described which is regulated by EBNA-1, the only EBV protein known to be expressed in group I BL cells.

ORGANIZATION OF EBV GENE EXPRESSION IN BLs AND LCLs

Although their coding sequences are distributed over a large part of the EBV genome, in LCLs the EBNA genes are all part of the same transcription unit which extends about 100 kb from the promoter (Cp) to the 3' end of the EBNA-1 gene (Fig. 1). In BL cells the downregulation of all the EBNAs except EBNA-1 is achieved by Cp being off and another promoter (Fp) becoming active (Sample *et al.*, 1991; Schaefer *et al.*, 1991; Smith & Griffin, 1992). RNA from this promoter appears to splice only to EBNA-1 and not to the EBNA-3 family. The Wp promoter which mediates EBNA expression immediately after infection of normal resting B cells and becomes relatively downregulated as immortalization is established (Woisetschlaeger *et al.*, 1990) is also off in BL cells. In group III BL cell lines which have arisen recently in cell culture and where artefactual reactivation of the EBV genome has occurred, the Cp promoter is again active, as in LCLs (Woisetschlaeger *et al.*, 1989; Bodescot, Perricaudent & Farrell, 1987; Altiok *et al.*, 1992). Some of the BL lines such as Daudi, P3HR1, Raji and Namalwa which have been in culture for many years and have been used in many early studies, have deleted or integrated EBV genomes and do not conform readily to the clear group I/III and EBNA promoter usage pattern of newly established lines.

The mechanism of EBNA promoter regulation in BL cells has two components, the inactivation of Cp and Wp and the activation of Fp. Little has yet been reported about the regulation of Fp in latency but some studies

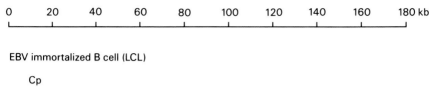

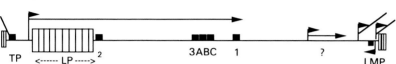

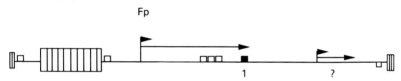

Fig. 1. EBV gene layout in immortalized cells compared to BL. The positions of the Cp and Fp promoters referred to in the text are shown. The coding segments of EBNA-1, 2, 3A, 3B, 3C, LMP and TP1, 2 are illustrated as filled boxes in the upper map. EBNA-LP is composed of many small exons crossing the major internal repeat. The ? shows the region transcribed into the spliced Bam HI A RNAs presently of unknown significance.

have been made of Cp and Wp. DNA methylation has a role in keeping Wp switched off (Allday *et al.*, 1990; Jannson, Masucci & Rymo, 1992). Changes in DNA methylation have also been noted at Cp (Altiok *et al.*, 1992) and treatment of group I BL cell lines with the demethylating agent 5-azacytidine results in some activation of EBNA-2 and LMP expression (Gregory, Rowe & Rickinson, 1990; Masucci *et al.*, 1989), although it is not yet clear whether this is a direct effect of demethylation of promoter elements. Other studies have also suggested a mechanism for down-regulation of Cp in BL cells which involves a reduction in the activity of the glucocorticoid receptor signalling pathway (Sinclair *et al.*, 1994), a transcription factor upon which Cp is dependent (Kupfer & Summers, 1990; Schuster *et al.*, 1991; Sinclair, Brimmell & Farrell, 1992). The promoters expressing LMP and TP1, 2 (LMP2A, B) are also inactive in group I BL cells. This may partly be the result of lack of EBNA-2 expression since EBNA-2 activates or derepresses these promoters (for review see Sinclair & Farrell, 1992).

Generally the EBV DNA remains episomal in BL cells, but in some cell lines integrated EBV genomes occur (Delecluse *et al.*, 1993). It is difficult to evaluate the significance of these at present; there is no evidence for a consistent site of integration in the cell genome between different BL lines.

Transcription has been studied from the integrated EBV in the Namalwa BL line where RNA from the TP promoters is transcribed apparently across the virus/cell junction and could, in principle, activate expression of an adjacent cell gene (Zimber-Strobl *et al.*, 1991), but there is no evidence that this RNA is important, and no cell gene product has been identified in the transcripts.

Proto-oncogenes

The c-myc protein is an important cell growth control gene; c-myc forms a heterodimer with the max protein, the dimer acting as a DNA binding transcription factor affecting the expression of other cell genes (for review see Evan & Littlewood, 1993). The c-myc locus on one of the two copies of chromosome 8 is always reciprocally translocated in BL to one of the immunoglobulin loci (for review see Klein, 1983; Lenoir & Bornkamm, 1987; Magrath, 1990). 8:14 translocations to the heavy chain gene are the most common but 2:8 and 8:22 to the kappa and lambda light chain genes are also found. The translocations do not simply result in the immunoglobulin promoter driving c-myc expression at high level but cause several complex and subtle changes to c-myc expression. The locations of the breakpoints in many cases of sporadic and endemic BLs have been mapped and can occur 5' or 3' of the c-myc gene (Magrath, 1990). Comparing tumours from endemic (Africa) and sporadic areas (USA), there is a tendency for the breakpoints in sporadic EBV negative tumours to be very close to the c-myc gene and to be clustered near the 5' end (Shiramizu *et al.*, 1991). In the EBV positive tumours the breakpoints tend to be more widely dispersed around the c-myc gene and can be hundreds of kb away (Joos *et al.*, 1992). A subsequent smaller study, however, using tumours from Brazil did not show a correlation of breakpoint with EBV status (Gutierrez *et al.*, 1992). The 2:8 and 8:22 translocation breakpoints tend to be 3' of the c-myc gene (telomeric) whereas the 8:14 breakpoints tend to be 5' of the c-myc gene.

In the sporadic BLs the breakpoint in the immunoglobulin gene is normally at the switch region (Gelmann *et al.*, 1983; Showe *et al.*, 1985; Neri *et al.*, 1988) where the switch recombinase mediates isotype switching. As lymph node germinal centres are associated with Ig isotype switching and somatic mutation (MacLennan & Gray, 1986), the type of chromosome translocations characteristic of sporadic BL may be a result of errors in this process within the germinal centre. The breakpoints on the immunoglobulin loci in endemic BL are generally at the J region (Haluska *et al.*, 1986; Neri *et al.*, 1988) but occasionally in the V or D regions, consistent with the translocation occurring as a result of errors in the normal process of VDJ joining during immunoglobulin gene rearrangement, which normally occurs in the bone marrow. The further mutations in exon 1 of c-myc, which have

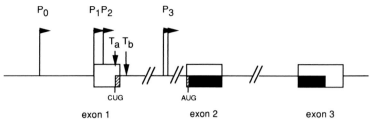

Fig. 2. The c-myc gene. Exons 1–3 and positions of transcription initiation points from promoter P0 to P3 are indicated. The coding part of c-myc is shown by filled boxes. The positions of premature transcription termination (Ta, Tb) are shown by vertical arrows.

been found in some endemic BLs where the break point is distant from the c-myc gene, might then arise in the germinal centre through a misdirection of the somatic mutation mechanism.

Several different mRNAs are normally transcribed through the c-myc gene and the expression of the proto oncogene is tightly regulated (for review see Spencer & Groudine, 1991). Transcription of the gene can start from one of four promoters (P0 to P3, see Fig. 2), resulting in at least four different mRNAs. The major coding sequence, which initiates with an AUG start codon and encodes a 64 kd protein, is located within exons 2 and 3 which are common to all four mRNAs. Exon 1 also contains a CUG initiation codon and further coding sequence which extends the N terminus of the protein giving a 67 kd product. The expression of c-myc is also controlled by premature termination of transcription in exon 1 (Bentley & Groudine, 1986, 1988; Eick & Bornkamm, 1986). This control appears to be specific for transcripts which initiate at promoter P2, the most abundant in normal B cells.

In BL cells, transcription from the normal allele is reduced to a very low level (Spencer and Groudine, 1991) and is mainly from the P3 promoter (Eick *et al.*, 1990). On the translocated allele a switch in promoter usage is observed and cytoplasmic c-myc RNA is now found initiating both at P1 and P2, with P1 RNA predominating. RNAs initiated at P0 are also enhanced from the translocated allele. Whilst it is clear that control of transcription elongation through c-myc is changed on the translocated allele, the details of the mechanism have become controversial. One interpretation argues that avoidance of transcription termination controls at the 3' end of exon 1 specific for P2 initiated transcripts (Spencer *et al.*, 1990) permits enhanced expression. Other experiments have suggested, however, that in the normal c-myc gene RNA, pausing of pol II just after initiation at P2 impedes transcription from P1 and in the translocated c-myc gene, this control is lost (Krumm *et al.*, 1992; Strobl *et al.*, 1993). A detailed analysis of kinetic nuclear run on data has suggested that in some situations the premature termination at the exon 1/intron boundary might be an artefactual property of the systems used to study it (Strobl & Eick, 1992) but doubts have also been raised about the interpretation of these run on data and the detailed mechanism of regulation of c-myc still requires further clarification. In some cell lines, mutations in the first exon have been found (Rabbitts, Hamlyn & Baer, 1983) and these could in those cases relate to the loss of pausing, although this has not been tested directly.

The translocation results in the immunoglobulin enhancer being brought into the vicinity of the myc gene and at first it had seemed that this would account fully for the altered regulation. But direct testing of this hypothesis using transfection of c-myc constructs including the enhancer did not result in the characteristic promoter usage shift (Polack *et al.*, 1991). It is possible that further powerful enhancer sequences in the immunoglobulin locus (Petterson *et al.*, 1990) more distant from the gene will prove to be the important element. These were not included in the constructs of Polack *et al.* (1991). Loss in tumour cells of a negative feedback regulation operating on c-myc may be important in tumour development (Lombardi *et al.*, 1990) and a stabilization of c-myc RNA in response to EBV in some cell lines (Lacy, Summers & Summers, 1989) may also be relevant. Further support for the changed control of c-myc transcription in BL translocations comes from the resistance of transcription from the translocated c-myc gene to inhibition by theophylline (Sandlund *et al.*, 1993). The mechanism by which the trans-location achieves deregulation of c-myc may not be the same in every case and different combinations of enhancer positioning and mutation of exon 1, perhaps complemented by EBV, could achieve equivalent overall effects on c-myc expression.

The importance of the c-myc deregulation to B lymphoma development has been demonstrated by constructing transgenic mice which contain a c-myc gene linked to the $E\mu$ immunoglobulin enhancer (Adams *et al.*, 1985; Harris *et al.*, 1988; Schmidt *et al.*, 1988). These mice develop plasmacytomas (a murine equivalent of BL) at a high rate, strongly implicating c-myc as the key gene on chromosome 8 in BL. The development of lymphomas in the $E\mu$-myc transgenic mice was greatly enhanced relative to normal mice but does not occur in every B lymphocyte, clearly indicating that other changes must accumulate in the lymphocyte for lymphoma development. There are many candidates for these changes but none has been fully defined yet.

One approach to identifying other genes that might be important in BL has used retrovirus mediated lymphomagenesis in mice containing the $E\mu$-myc transgene (Haupt *et al.*, 1991; van Lohuizen *et al.*, 1991). Retroviral infection causes rapid tumour formation, apparently as a result of chance integration of the retrovirus near key growth control genes. Analysis of the proviral integration sites in the tumours revealed frequent integration near the pim-1 gene and identified a new gene called bmi-1 and three new loci called pal-1, bal-1 and emi-1, implicating these in tumour growth (Haupt *et al.*, 1991; van Lohuizen *et al.*, 1991). Elevated levels of pim-1 RNA have been reported in a BL cell line (Meeker *et al.*, 1990). A subsequent survey of group I BL cell lines showed several lines which overexpress pim-1 RNA (Jacquemin *et al.*, 1993). The contributions of the bmi-1 gene and the pal-1, bal-1 and emi-1 loci to BL have not yet been reported.

In a separate approach, study of the expression of the c-fps/c-fes proto-oncogene RNA has revealed a difference between BL cell and normal cells. Normal and leukaemic myeloid cells express a 3.0kb fps/fes RNA but a shorter transcript of 0.9 kb, potentially able to encode a short protein has been found in Hodgkin's lymphoma (Jucker *et al.*, 1990) and 50% of BL cell lines (Jucker *et al.*, 1992). Its function is uncertain at present, but this change in expression makes it potentially important in BL.

Translocation of bcl-2 and consequent overexpression is a regular feature of the low grade, follicular lymphomas but this translocation is not found in BL (Klein, 1991). Transgenic mice containing a bcl-2/IgH construct structurally similar to the translocation in follicular lymphomas eventually develop malignant lymphomas, of which many contain a rearranged c-myc gene (McDonnell & Korsmeyer, 1991). Expression of bcl-2 prevents apoptosis in BL cells (Gregory *et al.*, 1991) and the EBV LMP protein is associated with induction of bcl-2 (Henderson *et al.*, 1991). So if the EBV LMP gene were expressed in an EBV infected cell in an early stage of BL development, this could promote cell survival within the lymph node (where many B cells that fail to bind antigen at high affinity to their surface immunoglobulin normally die). This could in turn expand the B cell pool and prolong the life of precursors to BL cells. Inappropriate expression of c-myc when cell growth is arrested can lead to apoptosis, at least in fibroblasts (Evan *et al.*, 1992), and in the bcl-2/IgH transgenic mice expression of bcl-2 could have a role in maintaining cell viability in lymphocytes which undergo an aberrant Ig rearrangement to c-myc.

Tumour suppressor genes p53 and Rb

Loss of function of the p53 gene is implicated in many types of cancer (Hollstein *et al.*, 1991). About 30% of BL biopsies and a higher proportion of BL cell lines contain a mutated p53 gene and frequently show loss of the other allele of p53 (Gaidano *et al.*, 1991; Farrell *et al.*, 1991; Wiman *et al.*, 1991). Most BL cell lines also show raised levels of p53 mRNA relative to B cell lines immortalized by EBV. In several cases these mutations of p53 have been shown to alter the p53 function, reducing its normal ability to suppress DNA synthesis in BL cells (Vousden *et al.*, 1993; Ramqvist *et al.*, 1993). So it is likely that mutation of p53 is an important step in tumour development in some BL cases.

The RB protein seems normally to bind and inactivate several cell transcription factors at certain stages in the cell cycle (Rustgi, Dyson & Bernards, 1991; Shirodkar *et al.*, 1992). It can also form an inactive complex with the adenovirus Ela or HPV16 E7 transforming proteins (Vousden,

1991; Green, 1989). Careful immunofluorescence studies have shown that RB and the EBNA-LP protein colocalize within the cell nucleus (Jiang *et al.*, 1991). Recently Szekely *et al.* (1993) have reported that EBNA-LP can bind *in vitro* both to RB and p53 using bacterial fusion proteins. The part of RB required for this association was not the pocket required for SV40LT, adenovirus E1A and HPV16 E7 binding since this could be mutated without affecting binding. If these observations are extended to demonstrate functionally important interactions in cells, consistent with the colocalization of EBNA-LP and RB, this will provide an important insight into one of the pathways by which EBV can modulate cell growth control. Since EBNA-LP is not expressed in BL tumours, a role for EBNA–LP interactions with p53 or RB could only be proposed in BL for hypothetical early stages of tumour development in which EBNA-LP might be expressed. BL cells do not have frequent gross deletions of Rb since the protein is present and the mRNA is not obviously changed in size but the gene is very large and careful studies of possible point mutations have not yet been attempted.

ROLE OF EBV IN BL

Arguments in favour of a role for EBV in BL remain circumstantial rather than directly mechanistic and are most readily applied to the endemic disease where the incidence of EBV positive BL is highest. In EBV positive cases of BL every tumour cell retains the virus, perhaps indicating a continued role for EBV in growth of the mature BL cell. Clonality of EBV terminal repeat number (Raab-Traub & Flynn, 1986) in BL cells has been used as an argument in favour of the virus being present from a very early stage (perhaps the beginning) of tumour development (Neri *et al.*, 1991; Brown *et al.*, 1988), again consistent with an aetiological role for EBV. It is clear that the factor that correlates with the much higher incidence of BL in Africa and New Guinea is hyperendemic malaria (for review see de Thé, 1982). The vast majority of people worldwide are infected with EBV but the geometric mean titre of EBV antibodies is significantly raised in African children who go on to develop the disease (de Thé *et al.*, 1978). If one accepts the arguments about monoclonality of BL with respect to the cell (all cells in a tumour show the same immunoglobulin gene rearrangement) and virus (terminal repeat copy number) this indicates that the tumour arose from a single cell that was probably infected with EBV at the time of malignant transformation. Since the number of circulating B-lymphocytes containing EBV is normally very low (less than 0.1% of B cells; Yao, Rickinson & Epstein, 1985) but the proportion of EBV positive tumours is high (90% in Africa) it follows that there was a much higher chance of malignant conversion occurring in a B cell infected with EBV than in an uninfected B cell. A weakness of this numerical argument is that we do not know the proportion of B cells infected with EBV in the lymph nodes of children in Africa with malaria. It is known that the proportion of circulating B cells infected with EBV is raised during malaria (Lam *et al.*, 1991) but it is still a relatively small proportion of the B cells. A potential criticism of the clonality experiments based on EBV TR number lies in the possible artefactual cloning of the virus in cell outgrowth by progressive dominance of the culture by the progeny of a few cells. It could thus, in principle, be possible for one virus TR number to be selected in culture by outgrowth of the host cell that happens to contain it. Such a phenomenon could occur during tumour outgrowth *in vivo* and give a false impression of virus clonality during the original transformation events.

The role of EBV was at first seen as enhancement of the size of the precursor B lymphocyte pool where Ig rearrangements are happening, thus increasing the possibility of an aberrant rearrangement to the myc locus occurring (Klein, 1987). However, the lack of expression of most of the immortalizing genes of EBV in BL cells and failure to detect LCL-like EBV infected cells in lymph nodes makes this type of explanation difficult to sustain. Because of the location of immunoglobulin breakpoints in endemic BL in the VDJ region, Lenoir and Bornkamm (1987) suggested that, in these tumours, the c-myc translocation may have occurred as an early event (in a pre- or pro-B cell) and EBV infection would follow. The link with EBV would then be based on complementation of the c-myc translocation by an EBV function to permit outgrowth. Malaria is the key variable determining the enhanced incidence of BL in Africa and thus seems likely to be the factor that encourages the endemic BL type of translocation. Malaria causes the B cell pool size to be enhanced and it simultaneously causes a reduction in T cell mediated immune surveillance, reflected in altered T4/T8 ratios (Whittle *et al.*, 1984). HIV has a similar effect on polyclonal activation of the B cell pool (Clifford-Lane *et al.*, 1983) and this could also be a basis for the enhanced incidence of BL in AIDS patients. The nature of the hypothetical complementing function of EBV in the virus positive tumours is obscure. It is still possible that the simple and obvious solution that EBNA-1 is the important gene will be correct but no role for EBNA-1 in cell growth control has yet been discovered. The intriguing possibility (Shiramizu *et al.*, 1991) that the location of the breakpoint relative to c-myc sufficient to cause a tumour is much less restricted in EBV positive endemic BL than EBV negative sporadic BL might also point to a role of EBV. If a higher proportion of erroneous translocations became tumorigenic in the presence of EBV this could cause the enhanced tumour incidence but the difference might equally relate to the different Ig enhancer elements controlling c-myc in the two situations.

The contrary view that EBV is simply a passenger present as a result of subsequent infection of BL tumour cells is unlikely because of the clonality, the fact that all tumour cells carry the virus in EBV positive tumours and a tendency of EBV to often follow a different fate when it infects established

EBV negative BL cell lines, leading to integration of the virus DNA (Hurley *et al.*, 1991). It seems unlikely that the role of EBV in BL will be settled by any epidemiological argument but a convincing molecular mechanism for its contribution to tumour cell growth will be needed to solve the problem. Solutions where EBV substitutes for an otherwise obligatory oncogenic change in EBV negative BL or enhances the effects of otherwise marginal oncogene changes would be perhaps the most attractive and consistent with current general views on the role of viruses in multistep carcinogenesis.

ACKNOWLEDGEMENT

We thank Martin Allday for helpful discussion.

REFERENCES

Adams, J. M., Harris, A. W., Pinkert, C. A. *et al.* (1985). The c-myc oncogene driven by immunoglobulin enhancers induce lymphoid malignancy in transgenic mice. *Nature,* **318**, 533–8.

Allday, M. J., Kundu, D., Finerty, S. & Griffin, B. E. (1990). CpG methylation of viral DNA in EBV associated tumours. *International Journal of Cancer,* **45**, 1125–30.

Altiok, E., Minarovits, J., Li-Fu, H., Contreras-Brodin, B., Klein, G. & Ernberg, I. (1992). Host-cell-phenotype-dependent control of the BCR2/BWR1 promoter complex regulates the expression of Epstein–Barr virus nuclear antigens 2–6. *Proceedings of the National Academy of Sciences, USA,* **89**, 905–9.

Arrand, J. R., Young, L. S. & Tugwood, J. D. (1989). Two families of sequences in the small RNA-encoding region of Epstein–Barr virus (EBV) correlate with types A and B. *Journal of Virology,* **63**, 983–6.

Benjamin, D., Magrath, I. T., Maguire, R., Janus, C., Todd, H. D. & Parsons, R. G. (1982). Immunoglobulin secretion by cell lines derived from African and American undifferentiated lymphomas of Burkitt's and non-Burkitt's type. *Journal of Immunology,* **129**, 1336–42.

Bentley, D. L. & Groudine, M. (1986). A block to elongation is largely responsible for decreased transcription of c-myc in differentiated HL60 cells. *Nature,* **321**, 702–6.

Bentley, D. L. & Groudine, M. (1988). Sequence requirements for premature termination of transcription in the human *c-myc* gene. *Cell,* **53**, 245–56.

Birkenbach, M., Josefsen, K., Yalamanchili, R., Lenoir, G. & Kieff, E. (1993). Epstein–Barr virus-induced genes: first lymphocyte-specific G protein-coupled peptide receptors. *Journal of Virology,* **67**, 2209–20.

Bodescot, M., Perricaudent, M. & Farrell, P. J. (1987). A promoter for the highly spliced EBNA family of RNAs of Epstein–Barr virus. *Journal of Virology,* **61**, 3424–30.

Brooks, L. A., Lear, A. L., Young, L. S. & Rickinson, A. B. (1993). Transcripts from the Epstein–Barr virus BamHI A fragment are detectable in all three forms of virus latency. *Journal of Virology,* **67**, 3182–90.

Brown, N. A., Liu, C.-R., Wang, Y.-F. & Garcia, C. R. (1988). B-cell lymphoproliferation and lymphomagenesis are associated with clonotypic intracellular terminal regions of the Epstein–Barr virus. *Journal of Virology,* **62**, 962–9.

Burkitt, D. (1983). The discovery of Burkitt's lymphoma. *Cancer,* **51**, 1777–86.

Cheah, M. S. C., Ley, T. J., Tronick, S. R. & Robbins, K. C. (1986). *fgr* proto-oncogene mRNA induced in B lymphocytes by Epstein–Barr virus infection. *Nature,* **319**, 238–40.

Clifford-Lane, H., Masur, H., Lynn, C. *et al.* (1983). Abnormalities of B-cell activation and immunoregulation in patients with the acquired immunodeficiency syndrome. *New England Journal of Medicine,* **309**, 453–8.

Cohen, J. I., Wang, F., Mannick, J. & Kieff, E. (1989). Epstein–Barr virus nuclear protein 2 is a key determinant of lymphocyte transformation. *Proceedings of the National Academy of Sciences, USA,* **86**, 9558–62.

Cohen, J. H., Revillard, J. P., Magaud, J. P. *et al.* (1987). B-cell maturation stage of Burkitt's lymphoma cell lines according to Epstein–Barr virus status and type of chromosome translocation. *Journal of the National Cancer Institute,* **78**, 235–42.

Cordier, M., Calender, A., Billaud, M. *et al.* (1990). Stable transfection of Epstein–Barr virus (EBV) nuclear antigen 2 in lymphoma cells containing the EBV P3HR1 genome induces expression of B-cell activation molecules CD21 and CD23. *Journal of Virology,* **64**, 1002–13.

de-Thé, G., Geser, A., Day, N. E. *et al.* (1978). Epidemiological evidence for causal relationship between Epstein–Barr virus and Burkitt's lymphoma from Ugandan prospective study. *Nature,* **274**, 756–61.

de-Thé, G. (1982). Epidemiology of Epstein–Barr virus and associated diseases in man. In *The Herpesviruses 1,* ed. B. Roizman, pp. 25–104.

Delecluse, H-J., Bartnizke, S., Hammerschmidt, W., Bullerdiek, J. & Bornkamm, G. W. (1993). Episomal and integrated copies of Epstein–Barr virus coexist in Burkitt Lymphoma cell lines. *Journal of Virology,* **67**, 1292–9.

Eick, D. & Bornkamm, G. W. (1986). Transcriptional arrest within the first exon is a fast control mechanism in c-myc gene expression. *Nucleic Acids Research,* **14**, 8331–46.

Eick, D., Polack, A., Kofler, E., Lenoir, G. M., Rickinson, A. B., Bornkamm, G. W. (1990). Expression of P_0- and P_3-RNA from the normal and translocated c-myc allele in Burkitt's lymphoma cells. *Oncogene,* **5**, 1397–402.

Epstein, M. A., Achong, B. G. & Barr, Y. M. (1964). Virus particles in cultured lymphoblasts from Burkitt's lymphoma. *Lancet,* **i**, 702–3.

Evan, G. I., Wyllie, A. H., Gilbert, C. S. *et al.* (1992). Induction of apoptosis in fibroblasts by c-myc protein. *Cell,* **69**, 119–28.

Evan, G. I. & Littlewood, T. D. (1993). The role of c-myc in cell growth. *Current Opinions in Genetic Development,* **3**, 44–9.

Farrell, P. J., Allan, G. J., Shanahan, F., Vousden, K. H. & Crook, T. (1991). p53 is frequently mutated in Burkitt's lymphoma cell lines. *EMBO Journal,* **10**, 2879–87.

Gaidano, G., Ballerini, P., Gong, J. Z. *et al.* (1991). *p53* mutations in human lymphoid malignancies: association with Burkitt's lymphoma and chronic lymphocytic leukemia. *Proceedings of the National Academy of Sciences, USA,* **88**, 5413–17.

Gelmann, E. P., Psallidopoulos, M. C., Papas, S., Dalla-Favera, R. (1983). Identification of reciprocal translocation sites within the c-myc oncogene and immunoglobulin μ locus in a Burkitt's lymphoma. *Nature,* **306**, 799–803.

Gordon, J., Ley, S. C., Melamed, M. D., English, L. S. & Hughes-Jones, N. C. (1984). Immortalised B-lymphocytes produce B-cell growth factor. *Nature,* **310**, 145–7.

Green, M. R. (1989). When the products of oncogenes and anti-oncogenes meet. *Cell,* **56**, 1–3.

Gregory, C. D., Dive, C., Henderson, S., Smith, C. A., Williams, G. T., Gordon, J.

& Rickinson, A. B. (1991). Activation of Epstein–Barr virus latent genes protects human B cells from death by apoptosis. *Nature,* **349**, 612–14.

Gregory, C. D., Murray, R. J., Edwards, C. F. & Rickinson, A. B. (1988). Down-regulation of cell adhesion molecules LFA-3 and ICAM-1 in Epstein–Barr virus-positive Burkitt's lymphoma underlies tumour cell escape from virus-specific T cell surveillance. *Journal of Experimental Medicine,* **167**, 1811–24.

Gregory, C. D., Tursz, T., Edwards, C. F. *et al.* (1987). Identification of a subset of normal B cells with a Burkitt's lymphoma (BL)-like phenotype. *Journal of Immunology,* **139**, 313–18.

Gregory, C. D., Rowe, M. & Rickinson, A. B. (1990). Different Epstein–Barr virus–B cell interactions in phenotypically distinct clones of a Burkitt's lymphoma cell line. *Journal of General Virology,* **71**, 1481–95.

Gunven, P., Klein, G., Klein, E., Norin, T. & Singh, S. (1980). Surface immuno-globulins on Burkitt's lymphoma biopsy cells from 91 patients. *International Journal of Cancer,* **25**, 711–19.

Gutierrez, M. I., Bhatia, K., Barriga, F. *et al.* (1992). Molecular epidemiology of Burkitt's lymphoma from South America: differences in breakpoint location and Epstein–Barr virus association from tumours in other world regions. *Blood,* **79**, 3261–6.

Haluska, F. G., Finver, S., Tsujimoto, T. & Croce, C. M. (1986). The f(8;14) chromosomal translocation occurring in B-cell malignancies results from mistakes in V-D-J joining. *Nature,* **324**, 158–61.

Hammerschmidt, W. & Sugden, B. (1989). Genetic analysis of immortalizing functions of Epstein–Barr virus in human B lymphocytes. *Nature,* **340**, 393–7.

Harris, A. W., Pinkert, C. A., Crawford, M., Langdon, W. Y., Brinster, R. L. & Adams, J. M. (1988). The Eμ-myc Transgenic Mouse: A model for high-incidence spontaneous lymphoma and leukemia of early B cells. *Journal of Experimental Medicine,* **167**, 353–71.

Haupt, Y., Alexander, W. S., Barri, G., Klinken, S. P., Adams, J. M. (1991). Novel zinc finger gene implicated as myc collaborator by retrovirally accelerated lympho-magenesis in E*μ-myc* transgenic mice. *Cell,* **65**, 753–63.

Henderson, S., Rowe, M., Gregory, C. *et al.* (1991). Induction of bcl-2 expression by Epstein–Barr virus latent membrane protein 1 protects infected B cells from programmed cell death. *Cell,* **65**, 1107–15.

Hollstein, M., Sidransky, D., Vogelstein, B. & Harris, C. C. (1991). p53 mutations in Human Cancers. *Science,* **253**, 49–53.

Honigberg, S. M., McCarroll, R. M. & Esposito, R. E. (1993). Regulatory mechanisms in meiosis. *Current Opinion in Cell Biology,* **5**, 219–25.

Huen, D. S., Grand, R. J. & Young, L. S. (1988). A region of the Epstein–Barr virus nuclear antigen leader protein and adenovirus E1A are identical. *Oncogene,* **3**, 729–30.

Hunter, T. & Pines, J. (1991). Cyclins and cancer. *Cell,* **66**, 1071–4.

Hurley, E. A., Agger, S., McNeil, J. A. *et al.* (1991). When Epstein–Barr virus persistently infects B-cell lines, it frequently integrates. *Journal of Virology,* **65**, 1245–54.

Jacquemin, M. G., Sinclair, A. J., Shanahan, F. & Farrell, P. J. (1993). Analysis of p53, Rb, pim-1 and glucocorticoid receptor in Burkitt's lymphoma cell lines. *Proceedings of the Vth International Symposium on Epstein–Barr Virus and Associated Diseases*, in press.

Jansson, A., Masucci, M. & Rymo, L. (1992). Methylation of discrete sites within the enhancer region regulates the activity of the Epstein–Barr virus BamHI W promoter in Burkitt's lymphoma lines. *Journal of Virology,* **66**, 62–9.

Jiang, W-Q., Szekely, L., Wendel-Hausen, V., Ringertz, N., Klein, G. & Rosen, A. (1991). Colocalisation of the retinoblastoma protein and the Epstein–Barr virus encoded nuclear antigen EBNA-5. *Experimental Cellular Research*, **197**, 314–18.

Joos, S., Haluska, F. G., Falk, M. H. *et al.* (1992). Mapping chromosomal breakpoints of Burkitt's t(8:14) translocations far upstream of c-myc. *Cancer Research*, **52**, 6547–52.

Jucker, M., Schaadt, M., Diehl, V., Poppena, S., Jones, D. & Tesch, H. (1990). Heterogeneous expression of proto-oncogenes in Hodgkin's disease derived cell lines. *Hematological Oncolology*, **8**, 191–204.

Jucker, M., Roebroek, A. J. M., Mautner, J. *et al.* (1992). Expression of truncated transcripts of the proto-oncogene *c-fps/fes* in human lymphoma and lymphoid leukemia cell lines. *Oncogene*, **7**, 943–54.

Karran, L., Gao, Y., Smith, P. R. & Griffin, B. E. (1992). Expression of a family of complementary-strand transcripts in Epstein–Barr virus-infected cells. *Proceedings of the National Academy of Sciences, USA*, **89**, 8058–62.

Kieff, E. & Liebowitz, D. (1990). In *Virology*, 2nd edn, ed. B. N. Fields & D. M. Knipe, pp. 1889–1920.

Klein, G. (1991). Comparative action of myc and bcl-2 in B cell malignancy. *Cancer Cells*, **3**, 141–3.

Klein, G. (1983). Specific chromosomal translocations and the genesis of B cell derived tumours in mice and men. *Cell*, **32**, 311–15.

Klein, G. (1987). In defence of the 'old' Burkitt lymphoma scenario. In *Advances in Viral Oncology*, 7, ed. G. Klein, pp. 207–211. New York: Raven Press.

Klein, G. (1989). Virus latency and transformation: the strategy of Epstein–Barr virus. *Cell*, **58**, 5–8.

Klein, E., Klein, G., Nadkarni, J. S., Nadkarni, J. J., Wigzell, H. & Clifford, P. (1967). Surface IgM specificity on cells derived from a Burkitt's lymphoma. *Lancet*, **ii**, 1068–70.

Knutson, J. C. (1990). The level of c-fgr RNA is increased by EBNA-2, an Epstein–Barr virus gene required for B-cell immortalisation. *Journal of Virology*, **64**, 2530–6.

Krumm, A., Meulia, T., Brunvard, M. & Groudine, M. (1992). The block to transcriptional elongation within the human c-myc gene is determined in the promoter-proximal region. *Genes & Development*, **6(11)**, 2201–13.

Kupfer, S. R. & Summers, W. C. (1990). Identification of a glucocorticoid-responsive element in Epstein–Barr virus. *Journal of Virology*, **64**, 1984–90.

Lacy, J., Summers, W. P. & Summers, W. C. (1989). Post-transcriptional mechanisms of deregulation of MYC following conversion of a human B cell line by Epstein–Barr virus. *EMBO Journal*, **8**, 1973–80.

Lam, K. M. C., Syed, N., Whittle, H. & Crawford, D. H. (1991). Circulating Epstein–Barr virus carrying B-cells in acute malaria. *Journal American Medical Association*, **265**, 876–9.

Lenoir, G. M. & Bornkamm, G. W. (1987). Burkitt's Lymphoma, a human cancer model for the multistep development of cancer: proposal for a new scenario. In *Advances in Viral Oncology*, ed. G. Klein, pp. 173–206. New York: Raven Press.

Ling, N. R., Hardie, D., Lowe, J., Johnson, G. D., Khan, M. & MacLennan, I. C. M. (1989). A phenotypic study of cells from Burkitt lymphoma and EBV-B-lymphoblastoid lines and their relationship to cells in normal lymphoid tissues. *International Journal of Cancer*, **43**, 112–18.

Lombardi, L., Grinani, F., Sternas, L., Cechova, K., Inghiraci, G. & Dalla-Favera, R. (1990). Mechanism of negative feed-back regulation of c-myc gene expression in B-cells and its inactivation in tumour cells. *Current Topics in Microbiology and Immunology,* **166**, 293–301.

McDonnell, T. J. & Korsmeyer, S. J. (1991). Progression from lymphoid hyperplasia to high grade malignant lymphoma in mice transgenic for the t(14:18). *Nature,* **349**, 254–6.

MacLennan, I. C. M. & Gray, D. (1986). Antigen-driven selection of virgin and memory cells. *Immunology Review,* **91**, 61–85.

Magrath, I. (1990). The pathogenesis of Burkitt's lymphoma. *Advances in Cancer Research,* **55**, 132–270.

Mannick, J. B., Cohen, J. I., Birkenbach, M., Marchini, A. & Kieff, E. (1991). The Epstein–Barr virus nuclear protein encoded by the leader of the EBNA RNAs is important in B-lymphocyte transformation. *Journal of Virology,* **65**, 6826–37.

Masucci, M. G., Contreras-Salazar, B., Ragnar, E. *et al.* (1989). 5-Azacytidine up regulates the expression of Epstein–Barr virus nuclear antigen 2 (EBNA-2) through EBNA-6 and latent membrane protein in the Burkitt's lymphoma line Rael. *Journal of Virology,* **63**, 3135–41.

Masucci, M. G., Torsteinsdottir, S., Colombani, B. J., Brautbar, C., Klein, E. & Klein, G. (1987). Down regulation of class I HLA antigens and of the Epstein–Barr virus-encoded latent membrane protein in Burkitt lymphoma lines. *Proceedings of the National Academy of Sciences, USA,* **84**, 4567–71.

Meeker, T. C., Loeb, J., Ayres, M. & Sellers, W. (1990). The human *Pim*-1 gene is selectively transcribed in different hemato-lymphoid cell lines in spite of a G+C-rich housekeeping promoter. *Molecular and Cellular Biology,* **10**, 1680–8.

Motokura, T. & Arnold, A. (1993). Cyclin D and oncogenesis. *Current Opinion in Genetics and Development,* **3**, 5–10.

Murray, R. J., Kurilla, M. G., Brooks, J. M. *et al.* (1992). Identification of target antigens for the human cytotoxic T cell response to Epstein–Barr virus (EBV): Implications for the immune control of EBV-positive malignancies. *Journal of Experimental Medicine,* **176**, 157–68.

Neri, A., Barriga, F., Knowles, D. M., Magrath, I. T. & Dalla-Favera, R. (1988). Different regions of the immunoglobulin heavy chain locus are involved in chromosomal translocations in distinct pathogenetic forms of Burkitt's lymphoma. *Proceedings of the National Academy of Sciences, USA,* **85**, 2748–52.

Neri, A., Barriga, F., Inghirami, G. *et al.* (1991). Epstein–Barr virus infection precedes clonal expansion in Burkitt's and acquired immunodeficiency syndrome-associated lymphoma. *Blood,* **77**, 1092–5.

Nudelman, E., Kannagi, R., Hakomori, S. *et al.* (1983). A glycolipid antigen associated with Burkitt's lymphoma defined by a mono-clonal antibody. *Science,* **220**, 509–11.

Palmero, I., Holder, A., Sinclair, A. J., Dickson, C. & Peters, G. (1993). Cyclins D1 and D2 are differentially expressed in human B-lymphoid cell lines. *Oncogene,* **8**, 1049–54.

Pelicci, P. G., Knowles, D. M., Magrath. I. & Dalla-Favera, R. (1986). Chromosomal breakpoints and structural alterations of the c-myc locus differ in endemic and sporadic forms of Burkitt lymphoma. *Proceedings of the National Academy of Sciences, USA,* **83**, 2984–8.

Pettersson, S., Cook, G. P., Bruggemann, M., Williams, G. T. & Neuberger, M. S. (1990). A second B cell specific enhancer 3′ of the immunoglobulin heavy-chain locus. *Nature,* **344**, 165–8.

Polack, A., Strobl, L., Feederle, R. *et al.* (1991). The intron enhancer of the immunoglobulin kappa gene activates c-myc but does not induce the Burkitt-specific promoter shift. *Oncogene,* **6**, 2033–40.

Pope, J. H., Horne, M. K. & Scott, W. (1968). Transformation of foetal human leukocytes *in vitro* by filtrates of a human leukaemia cell line containing herpes-like virus. *International Journal of Cancer*, **3**, 857–66.

Raab-Traub, N. & Flynn, K. (1986). The Structure of the termini of the Epstein–Barr virus as a marker of clonal cellular proliferation. *Cell*, **47**, 883–9.

Rabbitts, T. H., Hamlyn, P. H. & Baer, R. (1983). Altered nucleotide sequences of a translocated c-myc gene in Burkitt lymphoma. *Nature*, **306**, 760–5.

Ramqvist, T., Magnusson, K. P., Wang, Y., Szekely, L., Klein, G. & Wiman, K. G. (1993). Wild-type p53 induces apoptosis in a Burkitt lymphoma (BL) line that carried mutant p53. *Oncogene*, **8**, 1495–500.

Rickinson, A. B., Young, L. S. & Rowe, M. (1987). Influence of the Epstein–Barr virus nuclear antigen EBNA2 on the growth phenotype of virus-transformed B cells. *Journal of Virology*, **61**, 1310–17.

Rooney, C. M., Gregory, C. D., Rowe, M. *et al.* (1986). Endemic Burkitt's lymphoma: phenotypic analysis of tumour biopsy cells and of derived tumour cell lines. *Journal National Cancer Institute*, **77**, 681–7.

Rowe, M. & Gregory, C. (1989). Epstein–Barr virus and Burkitt's lymphoma. *Advances in Viral Oncology*, **8**, 237.

Rowe, M., Young, L. S., Cadwallader, K., Petti, L. & Kieff, E. (1989). Distinction between Epstein–Barr virus type A (EBNA 2A) and type B (EBNA 2B) isolates extends to the EBNA 3 family of nuclear proteins. *Journal of Virology*, **63**, 1031–9.

Rowe, M., Rooney, C. M., Edwards, C. F., Lenoir, G. M. & Rickinson, A. B. (1986). Epstein–Barr virus status and tumour cell phenotype in sporadic Burkitt's lymphoma. *International Journal of Cancer*, **37**, 367–73.

Rowe, M., Rooney, C. M., Rickinson, A. B. *et al.* (1985). Distinctions between endemic and sporadic forms of Epstein–Barr virus-positive Burkitt's lymphoma. *International Journal of Cancer*, **35**, 435–41.

Rowe, M., Rowe, D. T., Gregory, C. D. *et al.* (1987). Differences in B cell growth phenotype reflect novel patterns of Epstein–Barr virus latent gene expression in Burkitt's lymphoma cells. *EMBO Journal*, **6**, 2743–51.

Rustgi, A. K., Dyson, N. & Bernards, R. (1991). Amino-terminal domains of *c-myc* and N-*myc* proteins mediate binding to the retinoblastoma gene product. *Nature*, **352**, 541–4.

Sample, J., Brooks, L., Sample, C. *et al.* (1991). Restricted Epstein–Barr virus protein expression in Burkitt lymphoma is due to a different Epstein–Barr nuclear antigen 1 transcriptional initiation site. *Proceedings of the National Academy of Sciences, USA*, **88**, 6343–7.

Sample, J., Young, L., Martin, B., Chatman, T., Rickinson, A. & Keiff, E. (1990). Epstein–Barr virus types 1 and 2 differ in their EBNA-3A, EBNA-3B and EBNA-3C genes. *Journal of Virology*, **64**, 4084–92.

Sandlund, J. T., Neckers, L. M., Schneller, H. E., Woodruff, L. S. & Magrath, I. T. (1993). Theophylline induced differentiation provides direct evidence for the deregulation of c-myc in Burkitt's lymphoma and suggests participation of immunoglobulin enhancer sequences. *Cancer Research*, **53**, 127–32.

Schaefer, B. C., Woisetschlaeger, M., Strominger, J. L. & Speck, S. H. (1991). Exclusive expression of Epstein–Barr virus nuclear antigen 1 in Burkitt lymphoma arises from a third promoter, distinct from the promoters used in latently infected lymphocytes. *Proceedings of the National Academy of Sciences, USA*, **88**, 6550–4.

Schmidt, E. V., Pattengale, P. K., Weir, L. & Leder, P. (1988). Transgenic mice bearing the human c-myc gene activated by an immunoglobulin enhancer: A pre-B-cell lymphoma model. *Proceedings of the National Academy of Sciences, USA*, **85**, 6047–51.

Schuster, C., Chasserot-Golaz, S., Urier, G., Beck, G., & Sergeant, A. (1991). Evidence for a functional glucocorticoid responsive element in the Epstein–Barr virus genome. *Molecular Endocrinology*, **5**, 267–72.

Sculley, T. B., Apolloni, A., Stumm, R. S. *et al.* (1989). Expression of Epstein–Barr virus nuclear antigens 3, 4 and 6 are altered in cell lines containing B-type virus. *Virology*, **171**, 401–8.

Shiramizu, B., Barriga, F., Neequaye, J. *et al.* (1991). Patterns of chromosomal breakpoint locations in Burkitt's lymphoma: relevance to geography and Epstein–Barr virus association. *Blood*, **77**, 1516–26.

Shirodkar, S., Ewen, M., DeCaprio, Morgan, J., Livingston, D. M. & Chittenden, T. (1992). The transcription factor E2F interacts with the retinoblastoma product and a p107-cyclin A complex in a cell cycle-regulated manner. *Cell*, **68**, 157–66.

Showe, L. C., Ballantine, M., Nishikura, K., Erikson, J., Kaji, H., Croce, C. M. (1985). Cloning and sequencing of a c-myc oncogene in a Burkitt's lymphoma cell line that is translocated to a germ line alpha switch region. *Molecular and Cellular Biology*, **5**, 501–9.

Sinclair, A. J., Brimmell, M. & Farrell, P. J. (1992). Reciprocal antagonism of steroid hormones and BZLF1 in switch between Epstein–Barr virus latent and productive cycle gene expression. *Journal of Virology*, **66**, 70–7.

Sinclair, A. J. & Farrell, P. J. (1992). Epstein–Barr virus transcription factors. *Cell Growth & Differentiation*, **3**, 557–63.

Sinclair, A. J., Jacquemin, M. G., Brooks, L. *et al.* (1994). Reduced signal transduction through corticoid receptor in Burkitt's lymphoma cell lines. *Virology*, in press.

Smith, P. R. & Griffin, B. E. (1992). Transcription of the Epstein-Barr virus gene EBNA-1 from different promoters in nasopharyngeal carcinoma and B-lymphoblastoid cells. *Journal of Virology*, **66**, 706–14.

Smith, P. R., Gao, Y., Karran, L., Jones, M. D., Snudden, D. & Griffin, B. E. (1993). Complex nature of the major viral polyadenylated transcripts in Epstein–Barr virus-associated tumours. *Journal of Virology*, **67**, 3217–25.

Spencer, C. A., LeStrange, R. C., Novak, U., Hayward, W. S. & Groudine, M. (1990). The block to transcription elongation is promoter dependent in normal and Burkitt's lymphoma *c-myc* alleles. *Genes & Development*, **4**, 75–88.

Spencer, C. A. & Groudine, M. (1991). Control of c-myc regulation in normal and neoplastic cells. *Advances in Cancer Research*, **56**, 1–48.

Strobl, L. J. & Eick, D. (1992). Hold back of RNA polymerase II at the transcription start site mediates down-regulation of c-myc *in vivo*. *EMBO Journal*, **11**, 3307–14.

Strobl, L. J., Kohlhuber, F., Mautner, J., Polak, A. & Eick, D. (1993). Absence of a paused transcription complex from the c-myc P2 promoter of the translocation chromosome in Burkitt's lymphoma cells: implication for the c-myc P1/P2 promoter shift. *Oncogene*, **8**, 1437–47.

Sugden, W. & Warren, N. (1989). A promoter of Epstein–Barr virus that can function during latent infection can be transactivated by EBNA-1, a viral protein required for viral DNA replication during latent infection. *Journal of Virology*, **63**, 2644–9.

Swaminathan, S., Tomkinson, B. & Kieff, E. (1991). Recombinant Epstein–Barr virus with small RNA (EBER) genes deleted transforms lymphocytes and replicates *in vitro*. *Proceedings of the National Academy of Sciences, USA*, **88**, 1546–50.

Swendeman, S. & Thorley-Lawson, D. A. (1987). The activation antigen BLAST-2, when shed is autocrine BCGF for normal and transformed B-cells. *EMBO Journal*, **6**, 1637–42.

Szekely, L., Selivanova, G., Hagnusson, K. P., Klein, G. & Wiman, K. G. (1993).

EBNA-5, an EBV encoded nuclear antigen, binds to the RB and p53 proteins. *Proceedings of the National Academy of Sciences, USA,* **90**, 5455–9.

Thorley-Lawson, D. A., Swendeman, S. L. & Edson, C. M. (1986). Biochemical analysis suggests distinct functional roles for the Blast-1 and Blast-2 antigens. *Journal of Immunology,* **136**, 1745–51.

Thorley-Lawson, D. A. & Mann, K. (1985). Early events in Epstein–Barr virus infection provide a model for B-cell activation. *Journal of Experimental Medicine,* **162**, 45–59.

Tomkinson, B., Robertson, E. & Kieff, E. (1993). Epstein–Barr virus nuclear proteins EBNA-3A and EBNA-3C are essential for B-lymphocyte growth transformation. *Journal of Virology,* **67**, 2014–54.

van Lohuizen, M., Verbeek, S., Scheijen, B., Wientjens, E., van der Gulden, H. & Berns, A. (1991). Identification of cooperating oncogenes in Eμ-*myc* transgenic mice by provirus tagging. *Cell,* **65**, 737–52.

Vousden, K. H., Crook, T. R. & Farrell, P. J. (1993). Biological activities of p53 mutants in Burkitt's lymphoma cells. *Journal of General Virology,* **74**, 803–10.

Vousden, K. H. (1991). Human papillomavirus transforming genes. *Seminars in Virology,* **2**, 307–17.

Wang, F., Gregory, C., Sample, C. *et al.* (1990). Epstein–Barr virus latent membrane protein (LMP1) and nuclear proteins 2 and 3C are effectors of phenotypic changes in B lymphocytes: EBNA-2 and LMP1 cooperatively induce CD23. *Journal of Virology,* **64**, 2309–18.

Wang, F., Kikutani, H., Tsang, S-T, Kishimoto, T. & Kieff, E. (1991). Epstein–Barr virus nuclear protein 2 transactivates a cis-acting CD23 DNA element. *Journal of Virology,* **65**, 4101–6.

Whittle, H. C., Brown, J., Marsh, K. *et al.* (1984). T-cell control of Epstein–Barr virus-infected B cells is lost during *P. falciparum* malaria. *Nature,* **312**, 449–50.

Wiman, W. G., Magnusson, K. P., Ramqvist, T. & Klein, G. (1991). Mutant *p53* detected in a majority of Burkitt lymphoma cell lines by monoclonal antibody PAb240. *Oncogene,* **6**, 1633–8.

Woisetschlaeger, M., Strominger, J. L. & Speck, S. H. (1989). Mutually exclusive use of viral promoters in Epstein–Barr virus latently infected lymphocytes. *Proceedings of the National Academy of Sciences, USA,* **86**, 6498–502.

Woisetschlaeger, M., Yandava, C. N., Furmanski, L. A., Strominger, J. L. & Speck, S. H. (1990). Promoter switching in Epstein–Barr virus during the initial stages of infection of B lymphocytes. *Proceedings of the National Academy of Sciences, USA,* **87**, 1725–9.

Yao, Q. Y., Rickinson, A. B. & Epstein, M. A. (1985). A re-examination of the Epstein–Barr virus carrier state in healthy seropositive individuals. *International Journal of Cancer,* **35**, 35–42.

Young, L. S., Yao, Q. Y., Rooney, C. M. *et al.* (1987). New type B isolates of Epstein–Barr virus from Burkitt's lymphoma and from normal individuals in endemic areas. *Journal of General Virology,* **68**, 2853–62.

Zimber, U., Addlinger, H. K., Lenoir, G. M. *et al.* (1986). Geographical prevalance of two types of Epstein–Barr virus. *Virology,* **154**, 56–66.

Zimber-Strobl, U., Suentzenich, K-O., Laux, G. *et al.* (1991). Epstein–Barr virus nuclear antigen 2 activates transcription of the terminal protein gene. *Journal of Virology,* **65**, 415–23.

SPECIFICALLY MUTATED EPSTEIN–BARR VIRUS RECOMBINANTS: DEFINING THE MINIMAL GENOME FOR PRIMARY B LYMPHOCYTE TRANSFORMATION

E. KIEFF, K. IZUMI, K. KAYE,
R. LONGNECKER, J. MANNICK, C. MILLER,
E. ROBERTSON, S. SWAMINATHAN,
B. TOMKINSON, X. TUNG AND
R. YALAMANCHILI

Program in Virology and Department of Microbiology and Molecular Genetics and Medicine, Harvard University, 75 Francis St., Boston, MA 02115, USA

INTRODUCTION

In the past 30 years since the discovery of Epstein–Barr Virus (EBV) in Burkitt lymphoma (BL) cells growing in culture, the epidemiology and role of this Herpes virus in benign and malignant disease, and the relevant molecular biological properties of the virus in establishing latent infection in B lymphocytes, in causing proliferation of the latently infected cells *in vitro*, and in replicating in B lymphocytes, have been extensively investigated. However, relatively little has been done to correlate the molecular biological observations with EBV recombinant molecular genetics which could provide a critical and independent approach to defining the role of viral genes in specific aspects of virus infection and its effects on cell growth. For the most part, the lag in recombinant EBV molecular genetics arose because of the poor permissivity of B lymphocytes for EBV replication and the lack of an alternative, more permissive host cell. Recently, however, substantial progress has been made in the development of strategies for specifically mutating EBV genes and for creating EBV recombinants. These strategies have resulted in significant advances in understanding the role of specific EBV genes in lymphocyte infection and in cell growth transformation. This review will focus on latent EBV infection and B lymphocyte growth transformation. The relevant biology and biochemistry of EBV infection, the general strategies for constructing EBV recombinants, and the significant results of the application of these strategies to the study of the role of specific EBV genes in latent infection and growth transformation will be described.

BACKGROUND

First, the relevant medical and molecular biology (for more extensive reviews and primary references see Liebowitz & Kieff, 1993; Miller, 1989; Epstein & Achong, 1979, 1986; Farrell, 1992; Henle & Henle, 1979; Klein, 1987; Middleton *et al.*, 1991; Moss *et al.*, 1992; Rogers, Strominger & Speck, 1992; Thomas, Allday & Crawford, 1991; Magrath, 1990; Amen *et al.*, 1986, and also contributions by Rickinson, Farrell and Klein in this volume). EBV is transmitted in saliva. Infection begins in the oropharyngeal epithelium which is permissive for virus replication. In the course of primary infection, as many as 10% of the circulating B lymphocytes become infected. These cells are largely non-permissive for virus replication and are probably the important site of long-term latent EBV infection and of periodic reactivation for reseeding of virus on to the oropharyngeal epithelium. Initially, natural killer cells, and later, $CD8^+$ cytotoxic T lymphocytes eliminate almost all virus infected B lymphocytes. In adolescence, this phase of infection and immune response results in the manifestation of acute infectious mononucleosis. After primary EBV infection, only about 1 in 10^5–10^6 peripheral blood B lymphocytes is latently infected with EBV. When these latently infected cells are cultured *in vitro*, or when normal B lymphocytes are infected with EBV *in vitro*, the latently infected cells will continuously proliferate. These latently infected cells are tumorigenic in nude mouse brain or in SCID mice. The effect on B lymphocyte proliferation is also seen in cotton top tamarins where inoculation of a large dose of EBV results in an acute polyclonal lymphoproliferative disease. A related disease occurs in humans whose cytotoxic T lymphocytes specific for EBV transformed lymphocytes have been destroyed or rendered functionally incompetent or in humans with X linked or sporadic genetic susceptibility to EBV infection. Thus, although latent infection of B lymphocytes is associated with the expression of virus genes which can cause cell proliferation, in the normal host the proliferating latently infected cells are well contained and provide a long term reservoir of virus from which infection can reactivate, can spread to the oropharyngeal epithelium, and can infect non-immune hosts.

When primary B lymphocytes are infected with EBV *in vitro*, the virus expresses six nuclear proteins or EBNAs, two integral membrane proteins or LMPs and two small RNAs or EBERs. The EBNAs, LMPs and EBERs are expressed in some latently infected cells *in vivo* since normal infected humans have continuously circulating T lymphocytes which specifically recognize peptides derived from each of the EBNAs and LMPs (with the possible exception of EBNA 1) in the context of class I histocompatibility molecules. EBNA or LMP mRNAs have also been detected in the peripheral blood of normal seropositive people. Further, these genes are expressed in latently infected cells in EBV associated lymphoproliferative disease. However, the full complement of EBNAs and LMPs may not be expressed

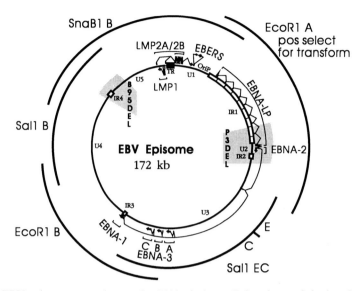

Fig. 1. EBV episome, transcripts, and mRNAs in latent B lymphocyte infection. Largely unique (U1-U5) and highly repetitive internal (IR1–4) or terminal (TR) repeat DNA domains, the episome DNA replication origin and termination site (ori P) and the location of exons encoding EBV nuclear proteins (EBNA 1, 2, 3A, 3B, 3C, LP) or membrane proteins (LMP 1, 2A or 2B) genes are indicated by dark boxes. The EBNA mRNAs are spliced as indicated from a single transcript. Transcription is shown as it initiates from the IR1 or Bam W promoter. In many infected lymphocytes, after expression of EBNA LP and EBNA 2, the Bam C promoter in U1 dominates EBNA transcription. Alternative splicing then leads to expression of all of the EBNAs. EBNA-LP is encoded by repeating exons (W1 or W2) from IR1 and two short unique exons from U2. Translation of EBNA LP is dependent on a splice between the exons downstream of the W or C promoters and the first W1 exon. RNAs with a 5b shorter first W1 exon have an ATG at the beginning of the EBNA LP open reading frame and express EBNA LP. RNAs with a longer W1 exon are not translated to EBNA LP. These RNAs have a long untranslated leader and EBNA 2, 3A, 3B, 3C or EBNA 1 from alternative open reading frames at the end of their respective mRNAs. The EBERs are two small, non-polyadenylated RNAs which are transcribed in latent or productive infection. LMP 1 and LMP 2B are transcribed in opposite directions under control of promoters separated by only 200 bp. The promoters share the same EBNA 2 response element and are both up regulated by EBNA 2 expression. Another clockwise promoter downstream of the LMP 1 polyadenylation site transcribes LMP 2A. LMP 2A also has an upstream EBNA 2 response element. The cosmids used in one cosmid marker rescue of transformation experiments (EcoR1 A) or in the first second site marker rescue experiments (Sal1 C) or in the five cosmid genome reconstruction experiments are shown.

in all latently infected cells. Another state of latent infection has been noted in BL cells where, frequently, EBNA 1 is the only EBNA which is expressed. If a similar state of latency can occur in a normal B lymphocyte, the infected cell would not be recognized as foreign by most EBV immune CD8[+] cytotoxic B lymphocytes.

The key steps in initiation of latent infection in primary B lymphocytes *in vitro* (Fig. 1) are:

1. Within 8 hours after infection, the viral genome circularizes, and RNA

is transcribed under control of the Bam W or IR1 latency promoter. The RNA is processed into two multiply spliced mRNAs which differ only in the second exon acceptor site and which encode either EBNA LP from the multiply spliced first exon or EBNA 2 from a terminal exon. Immediate early or other viral genes expressed in the course of lytic infection are not expressed during latent infection.

2. Within 24 hours, EBNA 2 turns on EBV promoters for LMP1 and LMP2 and cell promoters for CD21, CD23 and c-fgr. The EBNA latency promoter in some cells shifts upstream to a Bam C or U1 latency promoter (probably due to EBNA 2 responsive elements around the Bam C promoter). EBNA transcription now extends through at least two-thirds of the R strand of the viral genome and the EBNA transcript is processed into RNAs encoding EBNA LP, EBNA 2, EBNA 3A, EBNA 3B, EBNA 3C and EBNA 1.

3. Within 48 hours, all the EBNAs and LMPs are expressed. The expression of many B lymphocyte genes is induced by EBNA 2 or LMP 1 or by other EBNAs. EBNA 3A, 3B and 3C are three distantly related genes which evolved from a single progenitor since the three are distantly homologous, are tandemly located in the EBV genome, are similarly processed into a short and long exon at the end of EBNA mRNAs, and encode proteins of approximately 1000 amino acids. EBNA 3C can induce CD21 expression in B lymphocytes. LMP1 and LMP2 localize to a patch in the plasma membrane. LMP1 has transforming effects on immortalized rodent fibroblasts including loss of contact inhibition, serum independence, anchorage independence and nude mouse tumourigenicity. LMP1 also induces expression of many B lymphocyte activation and adhesion molecules and of Bcl-2 and A20, two proteins which have anti apoptotic effects in B lymphocytes. Much of this activation is probably mediated by NF-kb interaction. LMP2 interacts with src family tyrosine kinases (Burkhardt *et al.*, 1992).

4. By 36–48 hours, cell DNA replication begins. Cell division follows. The EBERs are then expressed. The EBV episome is maintained and amplified through cell division in response to EBNA 1 binding to the EBV ori-p element and the recruitment of cellular factors. Viral DNA synthesis initiates at the dyad symmetry component of the origin and terminates at the family of repeats component of the origin. (Both components being composed of repeats of the same EBNA 1 cognate sequence).

5. The EBNAs, LMPs and EBERs continue to be expressed and the cell continues to proliferate.

OBJECTIVES

Recombinant EBV-based molecular genetic experiments were initiated in order to evaluate the relative importance of the EBNAs, LMPs and EBERs in latent infection and cell growth transformation, and to define critical domains and specific sites in these molecules which can be used to further delineate essential biochemical interactions.

Background of the experimental strategy

The approach is based on previous studies of herpes simplex and pseudorabies virus molecular genetics, on the unique ability of EBV to establish latent infection and to transform B lymphocytes into cell lines (LCLs), on strategies for transfection of B lymphocytes, and on strategies for the induction of lytic EBV infection in latently infected cells. The important principles of those previous experiments are:

1. The transfection of herpes simplex virus (HSV) DNA or of overlapping clones of pseudorabies virus (PRV) DNA into cells can result in virus replication.
2. Transfection of HSV DNA into cells along with a mutagenized fragment of HSV DNA results in a high frequency of incorporation of the transfected DNA fragment into the correct site in the progeny viral genomes. Transfection of cells with a DNA fragment and infection with HSV results in a lower frequency of incorporation of the mutagenized fragment into the correct site in the viral genome.
3. Lymphoblasts can be transfected by electroporation.
4. Lytic EBV infection can be induced with a low efficiency in latently infected cells with phorbal esters, or butyrate, or by transfection of cells with an expression vector for the z immediate early transactivator of lytic EBV infection.

Thus, the strategies which evolved were based on lymphoblast transfection, on induction of partial permissivity for EBV replication, on the expectation that a transfected EBV DNA fragment would recombine with the replicating EBV genome in lytically infected cells and on the ability to identify EBV infected primary B lymphocytes by their continuous proliferation into LCLs. This last point raises the most important theoretical barrier at the start of this investigation. How can mutation be studied in a gene which is essential to the processes of latent infection and cell growth transformation, when the only way of obtaining recombinants is through infection of primary B lymphocytes and the clonal derivation of LCLs?

'Gain of function', transforming, EBV recombinants

The first recombinant EBV molecular genetic experiments avoided the conceptual barrier of inability to isolate non-transforming mutants by creating 'gain of function' transforming mutants, using the P3HR-1 EBV strain which has a deletion involving two nuclear proteins. There was considerable precedent for this approach. P3HR-1 is a BL derived cell line which is one of the most spontaneously permissive cell lines for lytic EBV infection. The P3HR-1 EBV is replication competent, but the resulting virus is unable to transform primary B lymphocytes. The P3HR-1 EBV genome is deleted for a DNA segment which includes the last two exons of EBNA LP and the entire EBNA 2 exon. A favourite working hypothesis was that the deletion involving EBNA LP and EBNA 2 was the basis for the inability of P3HR-1 to growth transform primary B lymphocytes. P3HR-1 EBV can induce lytic EBV infection in another Burkitt lymphoma-derived cell line, Raji, which carries a replication defective endogenous EBV genome containing two deletions. The resulting virus stocks could transform primary B lymphocytes. The resultant LCLs contain EBV genomes which have markers characteristic of the P3HR-1 genome except for restoration of the deleted DNA segment that encodes EBNA LP and EBNA 2.

The first recombinant EBV molecular genetic experiments were done by transfecting into P3HR-1 cells a cloned wild type (WT) EBV DNA fragment which spanned the P3HR-1 deletion, inducing replication and infecting primary B lymphocytes with the resultant virus (Hammerschmidt & Sugden, 1989; Cohen et al., 1989). The infected primary B lymphocytes were then either plated in soft agarose over fibroblast feeder layers or seeded into multiple micro well plates. In both sets of experiments, the conditions for the growth of newly transformed primary B lymphocytes were near optimal, and 'gain of function', transforming, EBV recombinants were clonally derived in clones of resultant LCLs. The recombinants had the restriction endonuclease digestion products expected for EBV P3HR-1 DNA which had incorporated the WT DNA at the former deletion site. The procedures were standardized so that the number of transformants obtained following transfection with a WT EBV DNA fragment spanning the deletion was similar among repeated experiments. To evaluate specifically the roles of EBNA LP and EBNA 2 in the rescue of transformation competence, mutations were made in the EBNA LP and EBNA 2 open reading frames within the cloned transfected DNA fragments. Transfection of P3HR-1 cells with cloned EBV DNA fragments which span the site of the P3HR-1 deletion and which had been specifically mutated by insertion of a stop codon or by deletion of substantial parts of the EBNA 2 open reading frame gave no recombinants which could induce B lymphocytes to grow into LCLs. An interesting correlative series of experiments was based on the observation that there are two EBV strains which have diverged in their EBNA 2, EBNA 3A, 3B and 3C genes. Type 1 (T1 or A type) strains have a much greater ability to transform primary B lymphocytes. EBV recombinants constructed by transfection of P3HR-1 cells with WT DNA having the T1 EBNA 2 exhibited the T1 EBV transforming phenotype; while replacement of the EBNA 2 open reading frame with T2 EBNA 2 results in a T2 transforming phenotype. The consistent failure of at least two DNA clones with specifically mutated EBNA 2 open reading frames to yield LCLs in multiple experiments, the consistent positive results with WT constructs transfected in parallel and the type specific effects of EBNA 2 prove that EBNA 2 is critical to the ability of EBV to transform primary B lymphocytes.

An extended series of these marker rescue experiments demonstrated that the procedures were sufficiently reproducible to determine the role of specific sites and domains of EBNA 2 in enabling EBV to transform primary B lymphocytes and in transactivating LMP1 expression (Cohen, Wang & Kieff, 1991). Two independent constructs of 11 linker insertions or 15 deletions were tested and found to fall into four groups. Ten deletions and one linker insertion resulted in inactivation of transformation marker rescue following transfection of P3HR-1 with the specifically mutated DNAs. Since several of these deletions overlap with each other, or with the inactivating linker insertion, only four domains are defined by these experiments as being essential. Two deletions and one linker insertion resulted in a marked reduction in transformation marker rescue. One deletion and one linker insertion resulted in a small reduction in transformation marker rescue. Two deletions and eight linker insertions had no effect. All mutations which affected marker rescue of transformation also had a similar decreased activity in transactivation of LMP1 expression in transient transfection assays. In these transient assays, the level of EBNA 2 protein expression was similar to WT EBNA 2. These results are consistent with the hypothesis that the essential role of EBNA 2 in enabling EBV to transform B lymphocytes is mediated by gene transactivation.

Subsequent biochemical studies have focused on the essential EBNA 2 domains (numbered 1–4 from amino to carboxy termini). Region 4 includes an acidic domain which, when fused to the gal 4 DNA binding domain, can transactivate a promoter with upstream gal 4 binding sites in transient assays in B lymphocytes (Cohen & Kieff, 1991). Mutation of tryptophan 454 in the acidic domain to a serine results in loss of the transactivating and transforming activity (Cohen, 1992). The acidic transactivating domain of HSV VP16 can substitute for the EBNA 2 domain with similar transformation and transactivation efficiencies. Region 3 includes a gly–arg repeat domain which interacts with many proteins and with nucleic acids. Among nucleic acids, the arg–gly element interacts preferentially with poly-rG. Among proteins there is preferential interaction with histone, H1. Curiously, deletion of the arg–gly repeat element markedly reduces transformation efficiency but substantially increases the efficiency of transient transactivation by EBNA 2 of the LMP1 promoter. These data are consistent with the arg–gly repeat having regulatory effects on transactivating activity which are important to the role of EBNA 2 in cell growth transformation.

Similar studies of EBNA LP indicate that the last two exons are critical to efficient transformation of primary B lymphocytes (Mannick *et al.*, 1991). Although an initial experiment with a deletion encompassing the last two exons of EBNA LP resulted in only a modest effect on the efficiency of transformation marker rescue from P3HR-1, subsequent experiments with similar deletions or with a stop codon inserted at the beginning of the penultimate or ultimate exons resulted in a reduction of more than 95% in the efficiency of transformation marker rescue. The transformants which were obtained grew poorly, tended to differentiate toward Ig secretion and tended to be more permissive for EBV replication. When virus was recovered from the LCLs, the recombinant virus was deficient in ability to transform primary B lymphocytes, and feeder layers facilitated LCL out growth. These results are most compatible with the hypothesis that WT EBNA LP is a transactivator of a lymphokine which has autocrine B cell growth factor activity. However, a mechanism of this kind has yet to be confirmed, and other data suggest that the amino terminal repeat domains of EBNA LP can interact inefficiently with Rb or p53 *in vitro*, raising the possibility that EBNA LP may have a low affinity interaction with a tumour suppressor gene (Szekely *et al.*, 1993).

Marker rescue of transformation as a positive selection strategy for EBV recombinants

Marker rescue of transformation by transfection of WT EBV DNA into P3HR-1 cells enables EBV recombinants to be selected specifically by their newly acquired ability to cause infected primary B lymphocytes to proliferate into LCLs. This is a sensitive and powerful selection strategy. Even with primary B lymphocytes from seropositive humans, the background outgrowth of LCLs is far less than 1%. When infected cells are plated at less than 0.5 transforming unit per micro well, most LCLs result from a single transforming event and are infected with a single recombinant. Since the transforming recombinants arise from homologous recombination between the transfected DNA and P3HR-1 EBV DNA, specific mutations in the DNA outside of the P3HR-1 deletion might be carried into the recombinant EBV genomes, depending on the precise point of recombination between the transfected DNA and the P3HR-1 EBV DNA. Several studies have used this strategy of making a mutation outside of the P3HR-1 deletion in the transfected, marker rescuing, DNA fragment, in the expectation that some of the recombinants will have incorporated the physically linked mutated DNA segment.

The first EBV recombinant molecular genetic experiment constructing a mutation outside of the P3HR-1 deletion was designed to evaluate the role of the EBERs (Fig. 1, Swaminathan, Tomkinson & Kieff, 1991). For these experiments, the EBER genes were deleted from a 50kb WT EBV DNA fragment which extends from 6 kb 5' to the EBERs through the P3HR-1 deletion (39–46 kb 3' to the EBERs) and ends 4 kb 3' to the deletion. The mutant fragment was used to transfect P3HR-1 cells and the resultant virus was used to infect primary B lymphocytes in a clonal transformation assay. Approximately 20% of the resultant LCLs were infected with a recombinant which was deleted for the EBERs. These LCLs did not differ from their WT-infected counterparts in their transformed cell growth characteristics, EBNA or LMP expression, or spontaneous lytic replication. EBER deleted recombinants also did not differ from WT recombinants in their sensitivity to interferons or in the permissivity of their interferon-treated LCLs for vesicular stomatitis virus replication (Swaminathan et al., 1992). Thus, the role of the EBERs in EBV infection remains an enigma.

Other pertinent recombinants that have been made using mutations linked to the P3HR-1 transformation marker include those with specific mutations in BHRF 1 (Marchini et al., 1991) or BCRF 1 (Swaminathan et al., 1993), two other genes which had been implicated in latent infection or cell growth transformation. BHRF 1 is distantly but colinearly homologous to bcl-2, which has an anti-apoptotic effect in B lymphocytes. Although BHRF 1 is expressed early in lytic infection, RNAs that include BHRF 1 have been identified in latently infected cells. Further, BHRF-1 is transiently expressed following serum refeeding of latently infected Raji cells. Thus, BHRF 1 could have an anti-apoptotic effect in latent or early lytically infected B lymphocytes. To investigate this possibility, EBV recombinants with specifically mutated BHRF 1 open reading frames were constructed. Since BHRF 1 maps less than 2 kb 3' to the P3HR-1 deletion, most transformation rescued recombinants incorporated the specific mutations in BHRF-1 that were introduced into the rescuing DNA fragment. EBV recombinants with a stop codon in the 24th codon of BHRF 1, or with a complete deletion of BHRF 1, were able to latently infect and transform primary B lymphocytes and the transformed cells were indistinguishable from WT recombinant-infected cells in their growth in vitro. A slightly different approach had yielded similar results (Lee & Yates, 1992). Since BHRF-1 is expressed at high levels early in lytic EBV infection when bcl-2 expression might be expected to be shut off, attempts were made to explore the hypothesis that BHRF 1 might protect cells from apoptotic death during lytic infection. Recombinants containing a stop codon in BHRF-1 did not, however, differ from wild type recombinants in their progression from latent through early lytic to late lytic infection, despite incubation of the infected cells in media with low serum, a condition which induces B lymphocyte apoptotic cell death. Thus, although the function of BHRF 1 remains an enigma, the recombinant EBV molecular genetic experiments exclude a role for BHRF 1 in latent, lytic or growth transforming EBV infection of primary B lymphocytes in vitro.

In contrast to the distant homology between BHRF 1 and bcl-2, BCRF 1 is

84% homologous to the human IL10 gene. IL10 can enhance the growth and differentiation of primary B lymphocytes *in vitro* and BCRF 1 also has this activity. Although BCRF1 is not expressed at high levels in latent infection, the BCRF1 gene is located between the EBERs and the EBNA promoters so that a low level of expression may be possible. To investigate the role of BCRF 1 in EBV infection, two specifically mutated EBV recombinants were constructed. The first contains a termination triplet in place of codon 116 of the BCRF 1 open reading frame, and the second contains a deletion of the entire open reading frame and surrounding sequence extending from bp 9535–12870. The frequency of incorporation of the BCRF 1 stop codon mutation into recombinant virus (as assayed in the resultant LCLs) was 28%. This approximates to the frequency expected for a mutation 30kb away from the marker rescuing DNA. (BCRF 1 is 3 kb closer to the positive selection marker than the EBERs, and EBER mutations were incorporated into 20% of the recombinants. BCRF 1 is 30 kb further away from the positive selection marker than BHRF 1, which is incorporated into >50% of the recombinants). The BCRF 1 stop codon or deletion mutant recombinants were not different from WT recombinants in latent or lytic infection or in cell growth transformation. Importantly, however, B lymphocytes lytically infected with BCRF 1 stop codon mutant recombinants induced very significant levels of gamma interferon in human peripheral blood mononuclear cell cultures, while lymphocytes infected with WT BCRF 1 recombinants induced little or no interferon. These results indicate that BCRF 1 has a more readily demonstrable role in inhibiting gamma interferon induction than in facilitating infected B lymphocyte outgrowth into LCLs. Other experiments with murine or human IL 10 or with BCRF 1 *in vitro* (Stewart & Rooney, 1992), or *in vivo* (Kurilla *et al.*, 1993) point to effects on interferon, NK and cytotoxic T cell responses which are likely to be biologically relevant to EBV infection *in vivo*.

Second site marker rescue enables the construction of mutations at any site in the EBV genome and determination of the role of other EBNAs and LMPs

Two observations made during the construction of EBV recombinants with mutations in DNA physically linked to the transformation marker led to new strategies for constructing and testing mutations in other EBV genes. First, the frequency with which transformation marker rescued recombinants had incorporated a linked mutation that was >30 kb away from the positive selection marker was 20% or more. The mutation was not only separated from the marker by a long DNA segment, but the intervening DNA consists almost entirely of multiple copies of a 3 kb repeat which would be expected to decrease genetic linkage to the selected marker. The hypothesis was therefore considered (and proven to be correct) that the high frequency of incorporation of the mutation was not completely dependent on (the distant) physical linkage to the selected marker. Secondly, the LCLs which arose following infection of B lymphocytes with recombinant viruses derived from P3HR-1 were frequently found to contain the parental P3HR-1 virus in addition to the transformation competent recombinant. This indicated that a sufficient excess of parental P3HR-1 EBV was released following transfection of P3HR-1 cells that many of the primary B lymphocytes which were infected with a recombinant EBV were also co-infected with P3HR-1 EBV (Fig. 2). Furthermore, this episome co-infection state frequently continued through many cell divisions, despite the lack of dependence on the co-infecting P3HR-1 EBV for infected cell proliferation. A potential consequence of this phenomenon was that mutations could be constructed in any gene, essential or not, outside of the P3HR-1 deletion because P3HR-1 co-infection of the transformed cells might provide the WT function. Since the P3HR-1 EBV titre approximated to the number of infected primary B lymphocytes used in a typical experiment, only about half of the cells infected with recombinant EBV were co-infected with P3HR-1. If the mutation were in a gene which is not critical for latency or growth transformation, co-infection would not be a uniform occurrence. If the mutation were in a gene essential for latency or growth transformation, then primary B lymphocytes infected with a mutant recombinant would grow into LCLs only if they were co-infected with P3HR-1 EBV (Fig. 2). The recovery of the mutant recombinant in the co-infected LCL would enable the recombinant to be characterized and then passaged to fresh primary B lymphocytes to further evaluate whether the specifically mutated recombinant genome required P3HR-1 co-infection for cell growth transformation (Fig. 2). A dependence on the WT gene from P3HR-1 could be verified by demonstrating that the B-lymphocytes infected with the second generation mutant EBV would grow into LCLs only when P3HR-1 EBV was provided exogenously. Additional proof of the role of the specific mutation in creating the dependence on P3HR-1 co-infection, as opposed to other random changes which might have occurred elsewhere in the genome, would derive from showing that independently derived specifically mutated recombinants consistently required P3HR-1 co-infection, while WT recombinants derived in parallel did not. A further line of evidence arises from the likelihood that induction of lytic EBV infection in the original co-infected LCL would result in recombination between the specifically mutated recombinant which would be WT for EBNA LP and EBNA 2 and the co-infecting P3HR-1 EBV which is WT for the specific gene and deleted for EBNA LP and EBNA 2 (Fig. 2). Secondary recombinants which are competent for latent infection would be able to transform primary B lymphocytes into LCLs and would be identified in the subsequent PCR analysis because they would not have the specifically mutated EBV DNA segment. If the defect were elsewhere in the mutated recombinant EBV genome, and not a

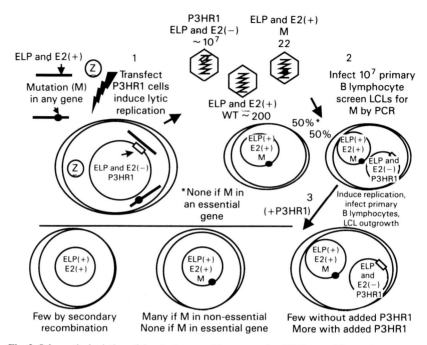

Fig. 2. Schematic depiction of the strategy used for generating EBV recombinants from P3HR-1 cells. In general, EBV recombinants are generated by transfecting latently infected B lymphocytes with cloned EBV DNA fragments and inducing lytic EBV infection in the transfected cells. The progeny virus is used to infect primary B lymphocytes or non EBV infected BL lymphoblasts. The experiments depicted in the diagram start with P3HR-1 cells which contain an EBV genome deleted for the last two exons of EBNA LP and the neighboring exon which encodes EBNA 2. The P3HR-1 EBV genome is defective and unable to initiate growth transformation of primary B lymphocytes unless the EBNA LP/2 deletion is restored by transfection with a wild type EBV DNA fragment which spans the deletion. When such a DNA fragment is transfected into P3HR-1 cells and lytic replication is induced, about 10^7 P3HR-1 virus and 10^2 restored recombinant virus is produced. In a typical experiment, the virus preparation is used to infect 1-2×10^7 primary B lymphocytes. The infected primary B lymphocytes are then plated into 1000 micro wells, The 100–200 wells which have a lymphocyte infected with a restored WT EBNA LP/2 recombinant are able to grow into lymphoblastoid cell lines. About half of the recombinant infected cells are initially co-infected with non recombinant P3HR-1 virus. The co-infected P3HR-1 genomes are lost as the resultant cell lines are maintained in culture over several months. A second, non-linked, cosmid EBV DNA fragment undergoes homologous recombination with the WT EBNA LP/2 recombinant genomes in lytically infected P3HR-1 cells with 10–12% efficiency. Thus, mutations engineered into a second non-linked cosmid EBV DNA fragment will wind up in 10–12% of the infected cell lines. If the mutation is in an essential transforming gene, the frequency of isolating the recombinant is only reduced by 50% since co-infecting P3HR-1 EBV will provide the WT gene in trans. To evaluate further whether the mutated recombinant EBV can infect and growth transform primary B lymphocytes, lytic infection can be induced in most co-infected LCLs and the virus plated onto primary B lymphocytes. Aggregates of virus can result in a small number of cells which are co-infected with both the specifically mutated recombinant and P3HR-1 EBVs. However, most primary B lymphocytes will be infected with the specifically mutated recombinant EBV or with P3HR-1 EBV or with a secondary recombinant. If the mutation affects the ability of the recombinant to establish latent infection or to growth transform primary B lymphocytes, no LCLs will grow out which are infected with the mutated recombinant alone. The mutated recombinant can, however, be demonstrated to be present in the virus preparation by adding sufficient exogenous P3HR-1 virus to infect most of the primary B lymphocytes so as to provide the WT gene function by complementation.

consequence of the specific mutation, phenotypically WT recombinants would be generated which would still have the specific mutation and would lack WT DNA at that site.

These principles were validated in a series of experiments in which a 30 kb WT T1 EBV DNA fragment which included the EBNA 3A, 3B and 3C genes was transfected into P3HR-1 cells along with the transformation marker rescuing DNA fragment and an expression plasmid for the z immediate early gene, to activate lytic EBV infection in the co-transfected cells (Tomkinson & Kieff, 1992*a*). Since P3HR-1 is a T2 EBV strain, and there are multiple type specific differences between the T1 and T2 EBNA 3A, 3B and 3C genes, incorporation of the T1 transfected DNA into P3HR-1 could be readily assayed by PCR. The resultant virus pool was plated on to primary B lymphocytes which were then distributed into a number of micro wells that exceeded by more than two-fold the number of expected transformation marker rescued recombinants. The surprising and important result was that 10–12% of the EBV recombinants had incorporated the T1 EBNA 3A and 3B genes. The EBNA 3C gene is near the end of the transfected DNA fragment and was incorporated into only 2% of the recombinants. Restriction endonuclease analysis revealed that most recombinants had the sites expected from a model in which the T1 EBNA 3 genes had been incorporated into the recombinant EBV genomes as a result of homologous recombination. As expected, about half of the recombinants were also co-infected with parental P3HR-1 EBV. When these co-infected LCLs were passaged for many months in culture, the co-infecting P3HR-1 EBV genomes were usually lost. When lytic infection was induced in the cells which were infected with only T1 EBNA 3A, 3B and 3C recombinant virus, the efficiency with which this virus replicated, established latent infection and growth transformed primary B lymphocytes was identical to T2 EBNA 3 control recombinants. Thus, the observed recombination results establish a baseline efficiency for incorporation of an unselected marker (T1 EBNA 3) into recombinant EBV genomes which had also incorporated the DNA fragments containing EBNA LP and EBNA 2.

Several parameters of these experiments have a bearing on the formulation of a model that might explain the finding that 10–12% of the transforming recombinants had incorporated the type 1 EBNA 3s (the unselected marker). First, a 5–10 fold molar excess of the z expression vector (relative to the marker rescuing EBNA LP-EBNA 2 and type 1 EBNA 3 DNA fragments) is used to induce lytic replication in the P3HR-1 cells. z expression is induced in about 5% of the transfected P3HR-1 cells. A cell which takes up the z expression plasmid is also likely to take up the other DNA fragments. However, since the marker rescuing and the type 1 EBNA 3 encoding DNA fragments are larger and less abundant than the z expression plasmid, <5% of the cells probably take up these plasmids. An origin for lytic DNA replication is located near EBNA 2 on the marker rescuing DNA fragment and that fragment is therefore likely to replicate (and recombine) preferentially in a transfected cell in which lytic replication is induced. In a typical experiment, 10^7 P3HR-1 cells are transfected and the virus which is produced includes approximately 10^7 parental P3HR-1 (which can be independently assayed by its ability to rescue transformation defective mutants in LMP1 or EBNA 3C by co-infection as will be described below). The number of recombinants which have incorporated the EBNA LP-EBNA 2 transformation marker fragment and, which therefore transform primary B lymphocytes into LCLs, is 1–2×10^2. The efficiency of incorporating this fragment is 1 in 10^5 (10^2 transforming recombinants among 10^7 P3HR-1 EBV). Of these, 10–12% have incorporated type 1 EBNA 3. Having inefficiently incorporated the first DNA fragment, 10–12% of these recombinants incorporated the second fragment, even though that fragment lacks an origin of DNA replication. Thus, an EBV genome which has recombined with one transfected DNA fragment is now particularly able to recombine with a second DNA fragment. This indicates that there is a limiting step in the generation of recombinants such as an enzyme, enzyme complex or site, which having been engaged by a viral genome tends to remain associated with that DNA through successive recombination events. This phenomenon is likely to extend to cells infected with any herpes virus; and may extend to other recombination events.

Exploiting the phenomenon of second site homologous recombination, specific mutated EBV recombinants were constructed with a stop codon inserted into codon 111 of EBNA 3B (Tomkinson & Kieff, 1992b) or into codon 20 of LMP2A (Longnecker et al., 1992). The truncation of EBNA 3B after 110 of the 938 EBNA 3B codons had no effect on the ability of the recombinant virus to latently infect, growth transform or lytically replicate in primary B lymphocytes in vitro. No parameter of latent infection, including other EBNA and LMP gene expression, episome copy number or spontaneous transition to lytic infection was affected.

Similar results were obtained with LMP2A. An LMP2A mutation was constructed in an 8 kbp EBV DNA fragment by inserting a stop codon after codon 19 of the LMP2A open reading frame. The efficiency of incorporation of this mutation into marker rescued recombinants was 6%, a significant reduction compared with that of WT T1 EBNA 3A or 3B genes which were incorporated into 10–12% of the recombinants from a 30kb EBV DNA fragment. All of the initial LCLs that were infected with the LMP2A mutant were also infected with non-recombinant P3HR-1. However, when lytic infection was induced in these co-infected LCLs, the resulting mutant progeny readily transformed primary B lymphocytes without added P3HR-1. The mutated recombinant virus could then be serially passaged without P3HR-1. Infection of primary B lymphocytes was unaffected by the mutation. LCLs transformed with LMP2A-mutated EBV recombinants grew as well as those infected in parallel with WT control recombinants. Lytic infection could be induced as readily in the LCLs infected with the LMP2A as in LCLs infected with the wild type recombinant following transfection with a z expression plasmid. Since LMP2A interacts with B lymphocyte src family tyrosine kinases and with LMP1, a lack of effect on cell growth transformation was unexpected and the growth properties of the cells infected with the mutant recombinant were extensively compared with those of cells infected in parallel with WT recombinants. No significant differences were observed in LCL growth, either *in vitro* or in SCID animals. Similarly, no difference was found with LMP2 mutated recombinants which were deleted and out of frame with regard to all of LMP2 after the first 120 codons or were truncated after the fifth transmembrane domain (Longnecker *et al.*, 1993*a,b*). Thus, LMP2 is completely non-critical for latent infection or growth of primary B lymphocytes *in vitro*.

However, a marked difference was observed when surface Ig was cross linked on cells infected with LMP2 mutants (Miller, Longnecker & Kieff, 1993*a*; Miller *et al.*, 1993*b*). Cells infected with recombinants that could not express LMP2A (the unique amino terminus of which is critical to the interaction with src family tyrosine kinases) were similar to primary B lymphocytes in their intracellular free calcium rise following surface Ig cross linking, while cells infected with WT recombinants had little or no response to sIg cross linking. The rise in intracellular free calcium in the cells infected with mutant recombinants was associated with induction of EBV replication, while EBV replication was only minimally induced in cells infected with WT recombinants. No difference was observed between mutant and WT infected cells when lytic infection was induced with calcium ionophore and TPA which bypass the signal transducing effects of sIg associated src family tyrosine kinases. Thus, LMP2A plays a key role in preventing activation of lytic EBV infection in response to sIg cross linking. This is quite likely to be critical to the survival of latently infected cells in the peripheral circulation where they may encounter cross linking of sIg or of other receptors in response to antigen or other activating ligands. Activation of lytic infection in that milieu would result in termination of latency, cell death as a consequence of lytic infection or of T cell cytotoxicity and virus death as a consequence of release into immune serum.

Second site homologous recombination has also been used to demonstrate that EBNA 3A, EBNA 3C and LMP1 are essential for latent infection and cell growth transformation. The experiments with EBNA 3A and 3C were done in parallel with those evaluating EBNA 3B (Tomkinson, Robertson & Kieff, 1993*a*), while the experiments with LMP1 used a DNA fragment similar in size and map coordinates to that used for the LMP2 mutations (Kaye, Izumi & Kieff, 1993). Over 600 EBV recombinants were obtained following eight independent transfections of P3HR-1 cells with either of two independently derived 30 kbp T1 EBV DNA fragments which had been specifically mutated by inserting a stop codon after EBNA 3A codon 302. The frequency of obtaining LCLs infected with EBNA 3A mutated recombinants was 1.4% among the 637 EBV recombinant infected LCLs compared with a 10% frequency of obtaining WT T1 EBNA 3A recombinants in control experiments done in parallel. The cells infected with T1 EBNA 3A mutants were all co-infected with P3HR-1. All lost the EBNA 3A mutation as they were expanded over the first several months in culture, with the exception of one in which the mutant EBNA 3A DNA had not replaced the WT EBNA 3A. Thus, against an expected number of 63 mutated EBNA 3A recombinants, and while obtaining the expected number of WT EBNA 3A recombinants in control experiments done in parallel, only one mutant EBNA 3A recombinant was stably maintained in LCLs and that mutant resulted from an unusual recombination event which would likely result in expression of the WT type 2 EBNA 3A gene. These experiments indicate that EBNA 3A is critical or essential to latent infection or growth transformation and that the amino terminal 302 amino acids of the 944 in WT T1 EBNA 3A had a dominant negative effect on the out growth of co-infected LCLs, resulting in continuous selection for a cell infected with a secondary recombinant which lacks the mutated T1 EBNA 3A.

Because EBNA 3C maps close to the end of the DNA fragment used to construct EBNA 3 recombinants, the expected frequency of obtaining recombinants which have incorporated WT EBNA 3C is only 2–3%. A stop codon insertion after codon 365 of the 992 EBNA 3C codons was incorporated into 2% of the resulting recombinants. Five independent LCLs were obtained containing T1 EBNA 3C mutants. Each LCL also carried the WT T2 EBNA 3C gene, presumably due to co-infection with P3HR-1 EBV. One of the five lost the EBNA 3C stop codon mutation within the first few months in culture, and two others lost the mutation over the next few months in culture. Of the two LCLs which stably maintained both mutant and WT EBNA 3C, one contained the mutated T1 EBNA 3C and WT T2 EBNA 3C genes in the same EBV genome; while the other was stably co-infected with a recombinant containing the mutated T1 EBNA 3C and WT EBNA LP-EBNA B2 genes and with P3HR-1 EBV which is WT for T2 EBNA 3C. When virus replication was induced in this latter LCL and the resultant 0.22u filtered virus used to infect primary B lymphocytes, all of the LCLs which grew out were infected with secondary recombinants which had the WT EBNA LP-EBNA 2 and the WT T2 EBNA 3C DNA. One of these progeny-infected LCLs had been infected with a secondary recombinant which had retained the WT T1 EBNA 3A gene and had replaced the mutated T1 EBNA 3C with the WT T2 EBNA 3C from the P3HR-1 genome. This specific replacement of the mutated EBNA 3C segment of the T1 EBNA 3 DNA fragment indicates that the fragment is not incidentally defective outside the specific EBNA 3C mutation. Furthermore, the recombinant containing the mutated EBNA 3C and WT EBNA LP-EBNA 2 genes was present in the virus preparation released from the co-infected LCL, but failed to transform primary B lymphocytes on its own since this recombinant was readily recovered in co-infected primary B lymphocytes when high titre exogenous P3HR-1 EBV was mixed with the virus preparation. Thus LCLs were never obtained in the absence of a WT EBNA 3C gene and these data indicate that WT EBNA 3C is critical to the establishment of latent growth-transforming infection in primary B lymphocytes.

Three types of specifically LMP1 mutated EBV recombinants have been constructed. Two have a stop codon inserted after codon 9 or 84 of the LMP1 open reading frame; while, the third has the same linker inserted into an LMP1 intron to serve as a WT control. Eight independent stop codon 9, three independent stop codon 84 and 5 WT stop codon control recombinants were made. Although four of the five LCLs infected with control recombinants were co-infected initially with P3HR-1, after eight months of continuous cell growth these cells had lost the co-infected P3HR-1 genomes, and had only the WT stop codon recombinant genome. In contrast, all 11 LCLs containing open reading frame stop codon recombinants were initially co-infected with P3HR-1; of the seven which could be grown continuously in culture for 18 months all retained P3HR-1 co-infection. When virus replication was induced in the LCLs that were co-infected with the stop codon control recombinants and with P3HR-1, and the progeny virus was used to transform primary B lymphocytes, many of the new LCLs contained the control recombinants alone. In contrast, when virus was passaged from the seven LCLs that carried the LMP1 mutants all LCLs (>300 were analysed) which arose from the infection of primary B lymphocytes were either co-infected with P3HR-1 or were infected with a secondary recombinant which had acquired the WT LMP1 gene from P3HR-1. Of 172 LCLs that were infected with secondary recombinants, none had retained the LPM1 open reading frame mutation. Four of the original open reading frame recombinant and P3HR-1 co-infected LCLs could be induced to make a sufficient amount of virus so that a complementation test could be done with exogenously added high titre P3HR-1. In each experiment, the added P3HR-1 virus complemented the open reading frame stop codon recombinant and primary B lymphocytes were transformed into LCLs which were co-infected with the open reading frame recombinant and with P3HR-1.

Interestingly, the recombinants containing a stop codon at position 9 expressed abundant cross reactive proteins that were products of reinitiation of translation at codons 44, 89 or 129. These proteins did not form patches in the plasma membrane or inhibit the growth of the co-infected LCLs, indicating that LMP1 initiated at codon 44 does not interact with a putative down stream effector of LMP1 which is in limiting abundance. Taken together, these data provide strong evidence for an essential role for LMP1 in primary B lymphocyte growth transformation. Further, the data point to the amino terminal cytoplasmic domain and the first two transmembrane domains as being critical to LMP1 interaction with the plasma membrane and to the function of LMP1 in transforming primary B lymphocytes to LCLs.

Subsequent experiments with mutated EBV recombinants which have alterations in the LMP1 amino terminal cytoplasmic amino acid sequence reveal that there is little, if any, sequence specificity in this part of LMP1, beyond a requirement for a hydrophilic sequence which can properly tether the first two transmembrane domains. These experiments strongly favour the model that the principal function of the amino terminus is to correctly position the first two transmembrane domains, and that correct positioning of the transmembrane domains is sufficient for transformation by the six transmembrane domains and the carboxy terminal cytoplasmic domain. Thus, a key component of LMP1's role in cell growth transformation probably derives from interaction with a membrane-associated growth regulating cell protein.

Non-EBV-infected B lymphoma cells as hosts for isolating and replicating EBV recombinants

All of the above experiments are dependent on the use of transformation marker rescue to identify and isolate EBV recombinants by their ability to cause primary B lymphocytes to grow into LCLs. While this is a powerful and easy selection, non-transforming mutations in EBNA LP or EBNA 2 cannot be derived; and non-transforming mutations in other genes can only be derived by providing the WT gene *in trans*. In almost all of the above experiments, the WT gene is provided by P3HR-1 co-infection which complicates the physical analysis of the recombinant genome and also complicates the genetic analysis. Attempts to transcomplement using simpler expression vectors for expression of the WT gene have so far not been successful (except in the case of EBNA LP), presumably because of the stringent regulation of EBNA and LMP gene expression in LCLs. Non-EBV-infected BL cells are infectable with EBV *in vitro* and were therefore a possible host for isolating EBV recombinants. These cells are not dependent on EBV for their growth, and transforming or non-transforming recombinants should therefore be obtainable. There were two obvious problems: First, how to identify or select for the cell infected with the EBV recombinant. Several positive selection markers which convey resistance to toxic drugs have been used for selection of drug resistant transfected BL cells. However, whether the usual promoters for expression of these genes would work in the context of the virus genome was uncertain. Secondly, there was considerable uncertainty about whether lytic EBV infection could be reactivated from an *in vitro* infected BL cell line. Several BL cell lines derived from EBV-positive BL can be induced to replicate EBV *in vitro*. However, EBV-negative BL cells, after infection with EBV *in vitro*, are notable for their integration of EBV DNA and their inability to be activated for lytic infection.

SV40 promoter and enhancer-driven hygromycin phosphotransferase (hyg) or guanine phosphoribosyltransferase (gpt) open reading frames proved to be expressed at effective levels from several different sites in the EBV genome; and to have minimal effects on EBV gene expression (Wang, Marchini & Kieff, 1991; Marchini *et al.*, 1992*a*; Marchini, Longnecker & Kieff, 1992*b*; Marchini, Kieff & Longnecker, 1993; Lee *et al.*, 1992; Lee & Yates, 1992; Longnecker *et al.*, 1993*b*). One effect noted is a local alteration in EBV gene expression so that a gene ordinarily expressed in lytic infection such as D1LMP1 is expressed in latent infection from EBV genomes which have the SV40 promoter and enhancer nearby. Most importantly, EBV genomes with expression cassettes for either hyg or gpt could be selected specifically in non-EBV infected BL cells or primary B lymphocytes using toxic drug resistance. A surprising finding was that the EBV genome persisted as an episome in the *in vitro* infected BL cells. Most of the infected cells expressed only EBNA 1. Some expressed all of the EBNAs and LMPs. These two types of latent EBV infection were noted among separate clones of the same cell line infected with cloned recombinant virus. BL cells varied in their efficiency for latent infection and conversion to toxic drug resistance. BJAB cells were most efficient for recovery of recombinant virus, BL41 and BL30 cells slightly less efficient and Louckes cells still less efficient. The titre of recombinant virus which was fully transforming and carried a positive selection marker was about 10 fold higher in primary B lymphocyte transformation assays than in BJAB conversion to toxic drug resistance. In some non-EBV infected BL cells, reactivation of lytic EBV infection was very rare. Even in response to z expression and phorbal ester treatment, which was the most effective inducer, lytic infection was hardly ever activatable in BJAB cells. BL41 and BL30 cells were more permissive and Louckes cells tended to be most permissive. Lytic infection could be induced in a substantial fraction of some clones of infected Louckes cells. EBV negative BL cells have been used to segregate and characterize non-transforming EBNA 2 null mutants of EBV which were created by transfection of P3HR-1 cells with DNA fragments carrying specific EBNA 2 mutations. Non-EBV infected BL30 cells have also been infected with the transformation defective P3HR-1 genome which had been altered by recombination with an EBV DNA fragment carrying a SV40-hyg cassette. The recombinant EBV-infected cells were then used as a host for creation of new EBV recombinants using a transformation marker rescue protocol similar to that used for marker rescue from P3HR-1 cells. In all these experiments, the principal limitation of the BL cells is the lower efficiency of obtaining cells infected with recombinant virus and the difficulty in inducing lytic EBV infection in the BJAB cells which are the most efficient for recovery of recombinant virus.

Reconstitution of EBV from sub-genomic DNA fragments

Based on the experience that transfection of susceptible cells with sub-genomic fragments of PRV results in lytic virus infection, overlapping cosmid libraries have been constructed from the B95-8 EBV strain and many attempts have been made to obtain lytic EBV replication in non-EBV infected BL cells by transfection with the cosmid DNAs and a z expression plasmid. These experiments have so far not resulted in the generation of transforming EBV.

In a parallel series of experiments, the five overlapping cosmids were transfected into P3HR-1 cells along with the z expression plasmid (Fig. 1, Tomkinson *et al.*, 1993*b*). Previous experiments had established that P3HR-1 cells can be transfected with cosmid size DNA fragments, that transfection with z expression plasmids can induce permissivity for replication of the endogenous P3HR-1 EBV genome, that intermolecular recombination can occur in these cells and that recombinant virus that contains the EBNA LP-EBNA 2 marker rescuing DNA (and parental P3HR-1 EBV) will be found in the LCLs which grow out following infection of primary B lymphocytes with the virus from the transfected P3HR-1 cells. When primary B lymphocytes were infected with the resulting virus and the resultant LCLs were screened for markers which distinguish between the transfected cosmid T1 EBV DNA and the T2 P3HR-1 EBV the important results were:

1. Approximately 10% of the recombinants had markers only of the transfected cosmid DNA; while the other 90% were derived in part from P3HR-1 DNA.
2. Overall, the T1 EBNA 3A gene was incorporated into 26% of the recombinants as opposed to only 10–12% when the T1 EBNA 3 carrying cosmid was transfected into cells with the transformation marker rescuing fragment alone. Of the 26% frequency of T1 EBNA 3 incorporation, 10/26 were due to the recombinants which consisted only of transfected DNA. The rest (16/26) were due to recombinants which had markers from only two transfected cosmid DNAs (4/26), markers from only three transfected cosmid DNAs (6/26) or markers from only four transfected cosmid DNAs (6/26). Of recombinants which had a marker from at least 1 EBV DNA fragment, 26% had a marker from a second DNA fragment, and more than 80% of those that had a marker from a second fragment had a marker from a third. Thus, these experiments provide further evidence that a molecule which has undergone one recombination event is likely to go on to a second, third, fourth or fifth event. Of important practical significance was the finding in these experiments that markers near the end of a cosmid fragment were incorporated into recombinants at a frequency of at least 12%, as was a large deletion in one of the fragments (the B95

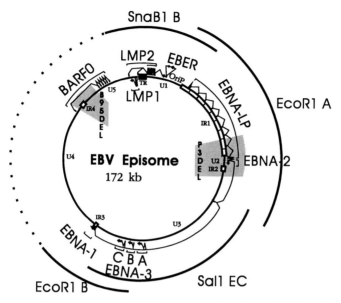

Fig. 3. Strategy for creating minimal EBV genomes sufficient for establishing latent growth transforming infection in primary B lymphocytes. Essential regions of the EcoR1 B and SnaB1 B fragments have been fused into a single cosmid and transfected into P3HR-1 cells along with the EcoR1 A and Sal1 E/C cosmids. P3HR-1 provides replication functions and some of the progeny virus is generated by recombination among the three cosmids. The BARFO transcript is shown on the schematic diagram of the EBV genome. The BARFO open reading frame is in the last exon.

-8 EBV genome from which the cosmid libraries were made is deleted for 14kb of DNA relative to other EBV genomes). The frequency of incorporation of the same large deletion in a two cosmid transfection experiment was nearly zero. Thus, this procedure offers the highest efficiency for producing recombinant EBV genomes with a specific mutation at any site.

Derivation of a minimal transforming EBV genome

The relatively high frequency of reconstruction of the EBV genome following five-cosmid transfections into P3HR-1 cells, including as one cosmid a cosmid with a large deletion, opened the possibility of making large deletions at any site in the EBV genome and thereby defining the minimal EBV genome for primary B lymphocyte growth transformation. Since all of the EBNAs and LMP1 are included in the part of the EBV genome between 168 kb and 110 kb in the EBV genome map, the rest of the genome between 110 kb and 168 kb should be dispensable for primary B lymphocyte growth transformation (Fig. 3). This smaller genome could potentially be reconstituted from three overlapping cosmids following transfection into

P3HR-1 cells which would provide the missing lytic replication functions. Such a small genome could fall below the minimal size for packaging and might be too large for packaging as a dimer. Furthermore, a transcript which includes a sizeable open reading frame (BARF0) is entirely encoded by DNA between 145 kb and 163 kb in latently infected lymphoid or nasopharyngeal carcinoma cells (Smith *et al.*, 1993). The appropriate constructs to test the possibility that this part of the genome could be deleted and be packaged as an extensively deleted EBV genome were made and the experiment was then attempted. More than 20% of the resulting LCLs were infected with an EBV recombinant consisting only of the three cosmids. As expected for LCLs established by infection with virus from a P3HR-1 cell transfection, about half the LCLs were co-infected with P3HR-1 EBV. Most importantly, half of the LCLs were infected only with a genome which resulted from recombination among the three transfected cosmids and lacked EBV DNA extending from 115–165 kb. These experiments demonstrate that the BARF0 transcript is not essential for primary B lymphocyte latent infection or cell growth transformation. Based on these experiments, further deletions have been made in the EBV genome using the new three cosmid transfection scheme for *in situ* construction of a transforming EBV. An additional 20 kb DNA segment has been deleted between the EBNA 2 and EBNA 3 encoding DNAs.

CONCLUSIONS

Over the past five years, initial forays in recombinant EBV molecular genetics have lead to the derivation and application of a number of strategies for constructing specifically mutated EBV recombinants. Aspects of these strategies, such as second site homologous recombination are likely to be useful in constructing mutations in other herpes virus genomes and may be useful in cell molecular genetics. This application of these strategies to the analysis of how EBV transforms primary B lymphocytes has revealed critical roles for EBNA LP, 2, 3A and 3C and LMP1, has led to the definition of key components of these genes and has demonstrated that most of the rest of the genome is not critical for primary B lymphocyte growth transformation *in vitro*. Mutants in EBNA 1, an important gene in maintaining the EBV episome in latent infection, have not as yet been evaluated. Genes which are expressed in latently infected B lymphocytes and which appear to be noncritical for latent infection or cell growth transformation *in vitro* are still likely to be important in latent infection *in vivo*. One of these, LMP2A, is likely to play a critical role in regulating reactivation from latency *in vivo*.

ACKNOWLEDGEMENTS

This research programme is supported by grant No. 47006 from the National Cancer Institute of the United States Public Health Service. K.I. and E.R.

are Fellows of the Cancer Research Institute. B.T. was a Fellow of the Cancer Research Institute. K.K., J.M. and S.S. are Physician Scientist Awardees of the National Cancer Institute or of the National Institute of Allergy and Infectious Disease of the United States Public Health Service. R.L. is a Fellow of the Leukaemia Society of America. Jeffrey Cohen, Andrew Marchini and Fred Wang, contributed substantially to these procedures. George Miller contributed several key reagents including B9 5–8 cells and a clone of P3HR-1 cells and demonstrated the use of the Z immediate early gene in inducing lytic EBV infection. The extensive contributions of the Arrand, Cheng, Crawford, deThe, Epstein, Farrell, Gerber, Griffin, Hardwick, zur Hausen, Hayward, Henle, Hutt-Fletcher, Joncas, Kenney, Klein, Levine, Macgrath, Menezes, Miller, Morgan, Moss, Osato, Pagano, Pearson, Perricaudet, Raab-Traub, Rawlins, Rickinson, Rooney, Rymo, Sairenji, Sample, Sargeant, Sixbey, Speck, Strominger, Studen, Takada, Thorley-Lawson, Tosato, Wolfe, Yates, Young and Zeng laboratories to the biology and molecular biology of EBV are more appropriately described in another review which is not focused on EBV recombinant molecular genetics (Leibowitz & Kieff, 1993).

REFERENCES

Amen, P., Lewin, N., Nordstrum, M. & Klein, G. (1986). EBV-activation of human B lymphocytes. *Current Topics in Microbiology and Immunology*, **132**, 266–71.

Burkhardt, A., Bolen, J., Kieff, E. & Longnecker, R. (1992). An Epstein–Barr Virus transformation associated membrane protein interacts with src family tyrosine kinases. *Journal of Virology*, **66**, 5161–67.

Cohen, J., Wang, F., Mannick, J. & Kieff, E. (1989). Epstein–Barr virus nuclear protein 2 is a key determinant of lymphocyte transformation. *Proceedings of the National Academy of Sciences, USA*, **86**, 9558–62.

Cohen, J., Wang, F. & Kieff, E. (1991). Epstein–Barr Virus nuclear protein-2 mutations define essential domains for transformation and transactivation. *Journal of Virology*, **65**, 2545–54.

Cohen, J. I. & Kieff, E. (1991). An Epstein–Barr Virus nuclear protein 2 domain essential for transformation is a direct transcriptional activator. *Journal of Virology*, **65**, 5880–5.

Cohen, J. I. (1992). A region of herpes simplex virus VP16 can substitute for a transforming domain of Epstein–Barr virus nuclear protein 2. *Proceedings of the National Academy of Sciences, USA*, **89**, 8030–4.

Epstein, M. & Achong, B. (1979). *The Epstein–Barr Virus*. Springer-Verlag, London.

Epstein, M. & Achong, B. (1986). *The Epstein–Barr Virus, Recent Advances*. Heinemann, London.

Farrell, P. J. (1992). Epstein–Barr virus. In *Genetic Maps*, 6th edn, ed. S. J. O'Brien, pp. 1.120–1.133. New York: Cold Spring Harbor Press.

Hammerschmidt, W. & Sugden, B. (1989). Genetic analysis of immortalizing functions of Epstein–Barr virus in human B lymphocytes. *Nature*, **340**, 393–7.

Henle, W. & Henle, G. (1979). Seroepidemiology of the virus. In *The Epstein–Barr Virus*, ed. M. Epstein & B. Achong, pp. 61–78. Berlin: Springer-Verlag.

Kaye, K., Izumi, K. & Kieff, E. (1993). Epstein–Barr virus latent membrane protein 1 is essential for B-lymphocyte growth transformation. *Proceedings of the National Academy of Sciences, USA,* **90**, In press.

Klein, G. (1987). *Advances in Viral Oncology.* New York: Raven Press.

Kurilla, M. G., Swaminathan, S., Welsh, R. M., Kieff, E. & Brutkiewicz, R. R. (1993). The effects of virally expressed interleukin-10 on vaccinia virus infection in mice. *Journal of Virology,* **67**, In press.

Lee, M. A., Kim, O. J. & Yates, J. L. (1992). Targeted gene disruption in Epstein–Barr virus. *Virology,* **189**, 253–65.

Lee, M. A. & Yates, J. L. (1992). BHRF1 of Epstein–Barr virus, which is homologous to human proto-oncogene bcl-2, is not essential for transformation of B cells or for virus replication *in vitro. Journal of Virology,* **66**, 1899–906.

Liebowitz, D. & Kieff, E. (1993). Epstein–Barr virus. In *The Human Herpesviruses*, ed. B. Roizmaan, R. J. Whitley & C. Lopez, pp. 107–172, New York: Raven Press.

Longnecker, R., Miller, C., Miao, X.-Q., Marchini, A. & Kieff, E. (1992). The only domain which distinguishes Epstein–Barr virus latent membrane protein 2A (LMP2A) from LMP2B is dispensable for lymphocyte infection and growth transformation *in vitro. Journal of Virology,* **66**, 6461–9.

Longnecker, R., Miller, C. L., Miao, X.-Q., Tomkinson, B. & Kieff, E. (1993*a*). The last seven transmembrane and carboxyterminal cytoplasmic domains of Epstein–Barr virus latent membrane protein 2 (LMP2) are dispensable for lymphocyte infection and growth transformation *in vitro. Journal of Virology,* **67**, 2006–13.

Longnecker, R., Miller, C. L., Tomkinson, B., Maio, X.-Q. & Kieff, E. (1993*b*). Deletion of DNA encoding the first five transmembrane domains of Epstein–Barr virus latent membrane proteins 2A and 2B. *Journal of Virology,* **67**, 5068–74.

Magrath, I. 1990. The pathogenesis of Burkitt's lymphoma. *Advances in Cancer Research,* **55**, 133–270.

Mannick, J. B., Cohen, J. I., Birkenbach, M., Marchini, A. & Kieff, E. (1991). The Epstein–Barr virus nuclear protein encoded by the leader of the EBNA RNAs (EBNA-LP) is important in B-lymphocyte transformation. *Journal of Virology,* **65**, 6826–37.

Marchini, A., Tomkinson, B., Cohen, J. & Kieff, E. (1991). BHRF1, the Epstein–Barr virus gene with homology to Bcl2, is dispensable for B-lymphocyte transformation and virus replication. *Journal of Virology,* **65**, 5991–6000.

Marchini, A., Cohen, J., Wang, F. & Kieff, E. (1992*a*). A selectable marker allows investigation of a non-transforming Epstein–Barr virus mutant. *Journal of Virology,* **66**, 3214–19.

Marchini, A., Longnecker, R. & Kieff, E. (1992*b*). Epstein–Barr virus (EBV) negative B-lymphoma cell lines for clonal isolation and replication of EBV recombinants. *Journal of Virology,* **66**, 4972–81.

Marchini, A., Kieff, E. & Longnecker, R. (1993). Marker rescue of a transformation negative Epstein–Barr virus (EBV) recombinant from an infected Burkitt lymphoma cell line: a method useful for analysis of genes essential for transformation. *Journal of Virology,* **67**, 606–9.

Middleton, T., Gahn, T. A., Martin, J. M. & Sugden, B. (1991). Immortalizing genes of Epstein–Barr virus. *Advances in Virus Research,* **40**, 19–55.

Miller, C. L., Longnecker, R. & Kieff, E. (1993*a*). The Epstein–Barr virus latent membrane protein 2A (LMP2A) blocks calcium mobilization in B lymphocytes. *Journal of Virology,* **67**, 3087–94.

Miller, C. L., Lee, J. H., Kieff, E. & Longnecker, R. (1993*b*). An integral membrane protein (LMP2) blocks reactivation of Epstein–Barr virus from latency. *Proceedings of the National Academy of Sciences, USA*, in press.

Miller, G. (1989). The Epstein–Barr virus. In *Fields' Virology*, ed. B. Fields & D. Knipe, pp. 1921–1958, New York: Raven Press.

Moss, D. J., Burrows, S. R., Khanna, R., Misko, I. S. & Sculley, T. B. (1992). Immune surveillance against Epstein–Barr virus. *Seminars in Immunology*, **4**, 97–104.

Rogers, R. P., Strominger, J. L. & Speck, S. H. (1992). Epstein–Barr virus in B lymphocytes: viral gene expression and function in latency. *Advances in Cancer Research*, **58**, 1–26.

Smith, P., Gao, Y. Karran, L., Jones, M., Snudden, D. & Griffin, B. (1993). Complex nature of the major viral polyadenylated transcripts in Epstein–Barr virus associated tumours. *Journal of Virology*, **67**, 3217–25.

Stewart, J. P. & Rooney, C. M. (1992). The interleukin-10 homolog encoded by Epstein–Barr virus enhances the reactivation of virus-specific cytotoxic T cell and HLA-unrestricted killer cell responses. *Virology*, **191**, 73–8.

Swaminathan, S., Tomkinson, B. & Kieff, E. (1991). Recombinant Epstein–Barr virus deleted for small RNA (EBER) genes transforms lymphocytes and replicates *in vitro*. *Proceedings of the National Academy of Sciences, USA*, **88**, 1546–50.

Swaminathan, S., Huneycutt, B., Reiss, C. & Kieff, E. (1992). Epstein–Barr virus encoded small RNAs (EBERs) do not modulate interferon effects in EBV infected lymphocytes. *Journal of Virology*, **66**, 5133–6.

Swaminathan, S., Hesselton, R., Sullivan, J. & Kieff, E. (1993). Epstein–Barr virus recombinants with specifically mutated BCRF1 genes. *Journal of Virology*, **67**, in press.

Szekely, L., Selivanova, G., Magnusson, K. P., Klein, G. & Wiman, K. G. (1993). EBNA-5, an Epstein–Barr virus-encoded nuclear antigen, binds to the retinoblastoma and p53 proteins. *Proceedings of the National Academy of Sciences, USA*, **90**, 5455–9.

Thomas, J. A., Allday, M. J. & Crawford, D. H. (1991). Epstein–Barr virus-associated lymphoproliferative disorders in immunocompromised individuals. *Advances in Cancer Research*, **57**, 329–80.

Tomkinson, B. & Kieff, E. (1992*a*). Second site homologous recombination in Epstein–Barr virus: insertion of type 1 EBNA-3 genes in place of type 2 has no effect on *in vitro* infection. *Journal of Virology*, **66**, 780–9.

Tomkinson, B. & Kieff, E. (1992*b*). Use of second site homologous recombination to demonstrate that Epstein–Barr Virus nuclear protein EBNA-3B is not important for B-lymphocyte infection or growth transformation *in vitro*. *Journal of Virology*, **66**, 2893–903.

Tomkinson, B., Robertson, E. & Kieff, E. (1993*a*). Epstein–Barr virus nuclear proteins (EBNA) 3A and 3C are essential for B lymphocyte growth transformation. *Journal of Virology*, **67**, 2014–25.

Tomkinson, B., Robertson, E., Yalamanchili, R., Longnecker, R. & Kieff, E. (1993*b*). Epstein–Barr virus recombinants from overlapping cosmid fragments. *Journal of Virology*, **67**, in press.

Wang, F., Marchini, A. & Kieff, E. (1991). Epstein–Barr Virus recombinants: use of positive selection markers to rescue mutants in EBV negative B lymphoma cells. *Journal of Virology*, **65**, 1701–9.

HEPATITIS VIRUSES AND LIVER CANCER

B. L. SLAGLE, S. A. BECKER AND J. S. BUTEL

Division of Molecular Virology, Baylor College of Medicine,
Houston, TX 77030, USA.

INTRODUCTION

Numerous seroepidemiologic studies have strongly implicated a viral aetiology for chronic liver disease and hepatocellular carcinoma (HCC) (Szmuness, 1978; Beasley *et al.*, 1981; Popper, 1988; Saito *et al.*, 1990; Miyamura *et al.*, 1991). At least five human hepatitis viruses have been identified (designated hepatitis A through E), but only the hepatitis B virus (HBV) and the hepatitis C virus (HCV; formerly known as non-A/non-B hepatitis virus) are associated with the development of HCC.

The mechanisms by which these viruses contribute to hepatocarcinogenesis have proven elusive. HBV and HCV are believed to have no direct oncogenic or cytopathic effect on the hepatocytes in which they replicate. Molecular studies of both human and animal HBVs during the past 20 years have implicated direct mechanisms of viral involvement in some HCCs and have suggested roles for both viral and cellular genes. After the discovery of HCV in 1989, a similar effort to understand its role in liver cancer began. It is known that diseases characterized by chronic liver injury and liver regeneration are associated with an increased risk of liver cancer (Melia *et al.*, 1984; Carlson & Eriksson, 1985; Niederau *et al.*, 1985; Eriksson, Carlson & Velez, 1986; Lieber *et al.*, 1986; Alter, 1988; Limmer *et al.*, 1988) and that the chronic liver disease associated with hepatotropic viruses is probably important in the development of HCC. Current models of hepatocarcinogenesis involving HBV and HCV predict that a combination of direct and indirect mechanisms for viral involvement contributes to the causation of HCC.

HEPATITIS B VIRUS

Discovered in 1965 as an antigen present in the serum of an Australian aborigine (Blumberg, Alter & Visnich, 1965), HBV is the prototype member of an unusual class of viruses, the Hepadnaviridae (Francki *et al.*, 1991). The hepadnaviruses are noted for their narrow host range and tissue tropism and, in addition to HBV, include woodchuck hepatitis virus (WHV) (Summers, Smolec & Snyder, 1978), ground squirrel hepatitis virus (GSHV) (Marion *et al.*, 1980), duck hepatitis B virus (DHBV) (Mason, Seal & Summers, 1980), and DHBV-related viruses from geese, grey heron, and

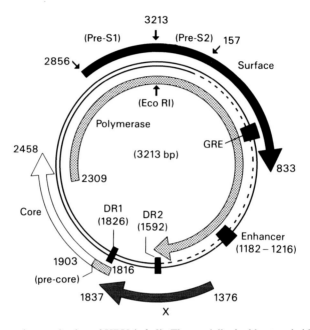

Fig. 1. Genomic organization of HBV (*adw*2). The partially double-stranded HBV genome, consisting of a long (−) and a short (+) strand of DNA, is represented by the inner circles. Four arrows depict the functional ORFs of the HBV genome which are designated surface, polymerase, core, and X. The boundaries of each ORF are designated by nucleotide number according to Valenzuela *et al*. (1980), with map position 1 at the unique *Eco*RI site present in the *adw* subtype. Small arrows at bp 2856 and bp 3213 within the surface ORF identify internal translational start sites utilized during virus replication. Two translational start sites within the core ORF (at bp 1816 and bp 1903) are also utilized to generate the core and pre-core viral proteins. The map positions for the major viral enhancer, the viral direct repeats (DR1 and DR2), and the glucocorticoid response element (GRE) are shown.

other species of ducks (Sprengel, Kaleta & Will, 1988). All hepadnaviruses share similar morphology, genome organization, and pathology (Marion, 1988; Schödel *et al*., 1989) and are associated with both acute and chronic liver disease (Hollinger, 1990). The mature HBV virion is a 42 nm, spherical, double-layered 'Dane' particle (Dane, Cameron & Briggs, 1970) containing an electron-dense 28 nm core structure. Treatment of Dane particles with detergents releases the nucleocapsid core of the virus which contains the HBV DNA and associated polymerase proteins.

The genome of hepadnaviruses is organized as a partially double-stranded, circular DNA approximately 3200 bp in length (Fig. 1); different isolates of HBV share approximately 90–98% nucleotide sequence homology (Valenzuela *et al*., 1980; Galibert, Chen & Mandart, 1982; Fujiyama *et al*., 1983; Ono *et al*., 1983). The full-length DNA minus strand is complementary to all HBV mRNAs, whereas the plus strand is 53–88% of unit length (Delius *et al.*, 1983). Molecular analyses of different HBV isolates have revealed four long open reading frames (ORFs; Fig. 1) that encode the proteins of the virus: surface antigen (HBsAg, the major envelope glycoprotein), core antigen (HBcAg, the virus nucleocapsid), polymerase [encompassing the virion terminal protein, reverse transcriptase, and RNase H activities (Seeger, Summers & Mason, 1991)], and X antigen [HBxAg, a *trans*-activator protein whose role in virus replication remains unclear (Schaller & Fischer, 1991; Rossner, 1992)].

Hepadnaviruses utilize several strategies to maximize their limited genetic information. Viral proteins are encoded from overlapping translational reading frames, allowing the virus to encode approximately 50% more protein than would be expected based on its genome size (Miller *et al.*, 1989). Additional protein diversity is achieved by the use of internal translational start sites both for the surface gene (resulting in the preS1, preS2 and S glycoproteins) and the core gene (generating the pre-core and core proteins). As a final illustration of compact organization of the hepadnaviruses, all of the regulatory signal sequences of the virus reside within protein-coding regions (Fig. 1). The different members of the hepadnavirus family share a similar genomic organization, with the exception that avian hepadnaviruses have a longer core gene and lack a separate X ORF (Ganem and Varmus, 1987).

Hepadnaviruses are associated with both acute and chronic liver disease. Age at the time of infection appears to be the major factor determining the frequency with which an infected person progresses from acute to chronic hepatitis (defined as circulating HBsAg for at least 6 months). Whereas chronic liver disease occurs in approximately 2–10% of adults infected with HBV (Beasley *et al.*, 1983), nearly 90% of those infected as infants or young children progress to chronic infection. The latter group, which may be chronically infected for decades, represents a population at a markedly increased risk (200-fold) of developing HCC (Beasley *et al.*, 1981). With an estimated 300 million chronically infected people worldwide, HBV-positive HCCs represent one of the most common cancers of humans. Studies in recent years have identified several possible roles for HBV in the development of liver cancer, and it appears that the mechanism for viral involvement will vary from tumour to tumour.

Potential oncogenic properties of HBV

HCC is associated with chronic, rather than acute, HBV infection. This suggests that molecular changes critical to tumour development must be favoured by long-term infection of the liver. During that time, both host and viral factors may interact to contribute to the risk of tumour development. Chronic HBV infection usually persists for several decades before the onset of HCC. Approximately 40% of Chinese males with chronic HBV infection will die owing to HBV-related HCC (Beasley *et al.*, 1981). Recent studies have identified several possible mechanisms by which HBV may contribute to tumour formation, and it appears that the role of the virus may vary among individual tumours.

Features of HBV integration

Although not part of the life-cycle of HBV, integration of viral DNA into the host chromosome occurs frequently during the course of chronic HBV infection. Almost 95% of HCCs from HBV-endemic areas (such as China) contain integrated HBV DNA, usually detected in DNA Southern hybridiz-ation experiments as one to four copies per cell (Bréchot *et al.*, 1981; Shafritz & Kew, 1981; Miller *et al.*, 1985; Zhou *et al.*, 1987). Although integration occurs at random locations in chromosomal DNA, a nonrandom selection is evident in HCCs for cells containing inserts from certain chromosomes (Slagle, Lee & Butel, 1992). The direct repeats of the virus (Fig. 1, DR1 and DR2) appear as preferred sites within the virus for recombination with the cell, as nearly 50% of all inserts contain one virus–cell junction that maps near this region in the viral genome (Shih *et al.*, 1987).

Gross chromosomal abnormalities are frequently apparent at the site of viral integration and include duplications, deletions, and chromosomal translocations (Dejean *et al.*, 1984; Mizusawa *et al.*, 1985; Hino, Shows & Rogler, 1986; Nagaya *et al.*, 1987; Rogler, Hino & Su, 1987; Tokino *et al.*, 1987; Yaginuma *et al.*, 1987; Zhou *et al.*, 1987; Pasquinelli *et al.*, 1988). Whether these chromosomal abnormalities are important in hepatocarcino-genesis depends on the identity and extent of disruption of nearby genes. The study of viral integrations cloned from HCCs led to the discovery of at least two previously unknown genes, the retinoic acid receptor gene (Dejean *et al.*, 1986) and the cyclin A gene (Wang *et al.*, 1990). Presumably, the altered expression of these growth regulatory genes in the two HCCs analysed provided a growth advantage to the affected hepatocytes. Certain integration-mediated chromosomal changes (such as inverted repeat struc-tures) may have an effect on genes located a distance from the actual integration site, and it is intriguing to consider the possibility of loss of tumour suppressor gene alleles mediated by HBV integration events (dis-cussed below).

Insertional activation of the myc family of genes

The role of integrated viral sequences in both human and woodchuck hepatocarcinogenesis has been the subject of intense investigation. There is strong evidence in the woodchuck system that viral DNA integration during chronic infection is directly related to hepatocarcinogenesis. Three independent woodchuck HCCs have been shown to contain WHV inserts within the c-*myc* gene (Hsu *et al.*, 1988; Möröy *et al.*, 1986). The increased levels of c-*myc* mRNA in those tumours is believed to result from integration of the WHV enhancer near the 5' end of c-*myc* or by stabilization of c-*myc* RNA by viral integration at the 3' end of c-*myc* (for review see Buendia, 1992). Even more striking is the finding of frequent WHV integration in N-*myc2*, an intron-less complementary DNA related to the woodchuck N-*myc1* gene that was originally discovered due to a WHV integration within the gene (Fourel *et al.*, 1990). Although not expressed in normal liver, N-*myc2* has retained N-*myc1*-like function and can cooperate with the protooncogene H-*ras* to achieve transformation of rat embryo fibroblasts (Fourel *et al.*, 1990). WHV integrations at N-*myc2*, either 5' of exon 1 or 3' of exon 3, are associated with increased expression of N-*myc2* (Wei *et al.*, 1992). The finding of WHV insertional activation of N-*myc2* expression in 13 of 19 WHV-infected woodchucks suggests that the increased levels of N-*myc2* provided some growth advantage that was selected for in the woodchuck tumours.

Insertional activation of the *myc* family of genes has been noted in about 50% of woodchuck HCCs examined (for review see Buendia, 1992). Involvement of the *myc* family of genes in liver cancer is not restricted to the woodchuck system. Amplification of c-*myc* has been detected in 6 of 14 liver tumours of GSHV-infected squirrels (Transy *et al.*, 1992). However, c-*myc* amplification occurs in the absence of specific GSHV integrations near the c-*myc* gene. Amplification of c-*myc* in human HCCs is rare (Trowbridge *et al.*, 1988), and there is no experimental evidence that insertional activation of the *myc* genes occurs in human HCCs.

Role for viral proteins in oncogenesis

The abnormal expression of certain cellular genes (due to mutation and/or altered or inappropriate expression) is known to sometimes contribute to cellular transformation (for review see Bishop, 1991). As both the HBV-encoded X protein and a 3'-truncated version of preS protein (preS/S^t) can activate the expression of different viral and cellular genes in *trans*, this suggests a possible role for HBV-encoded transactivating proteins in hepatocarcinogenesis (Takada & Koike, 1990; Lauer *et al.*, 1992; Rossner, 1992; Slagle *et al.*, 1993). Indeed, transgenic mice harbouring as a transgene bp 707–1856 of the HBV genome (including the entire X ORF; see Fig. 1) develop HCC (Kim *et al.*, 1991). In contrast, other transgenic mice carrying either the X ORF [bp 1376–1840; under the direction of the human alpha-1-antitrypsin promoter (Lee *et al.*, 1990)] or a 2.5 kb portion of HBV lacking the core ORF but containing both the S and the X ORFs (Dragani *et al.*, 1989) do not exhibit liver histopathologic changes. The latter mice do show an increased sensitivity to hepatocarcinogens, suggesting that HBx and/or

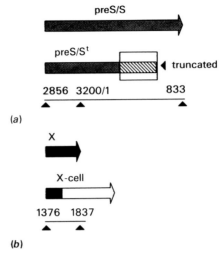

(a)

(b)

Fig. 2. HBV-encoded proteins with transactivator potential. At least two HBV-encoded proteins may transactivate the expression of viral and cellular promoters (described in text). (a) When the large (preS) surface protein is truncated within the region of bp 229 to bp ≈600 (shown in boxed region), the resulting preS/S^t protein gains transactivation ability (described by Lauer *et al.*, 1992). (b) During integration into the host genome, the transactivator X protein may become truncated and utilize adjacent cellular coding and polyadenylation signals (Takada & Koike, 1990; Hilger *et al.*, 1991). Such X-cellular hybrid proteins may also serve as transactivators of cellular gene expression. The viral DNA coding region for each transactivator protein is shown as a thin black line.

HBsAg expression may predispose cells to the effects of carcinogens that lead to the appearance of HCC (Dragani *et al.*, 1989).

Whether transactivation by HBx or preS/S^t is important in the development of human HCCs remains unknown. Most HBV inserts retain a portion of the X ORF, and the generation of preS/S^t during viral integration is a common phenomenon (Fig. 2). Thus, a significant proportion of HBV inserts may possess transactivation potential that could contribute to carcinogenesis. The transgenic mouse models harbouring all or defined regions of the HBV genome should continue to provide a powerful approach to understanding molecular effects of HBV infection and gene expression.

There is evidence that simple overexpression of viral proteins may contribute to the development of liver cancer. Studies in transgenic mice have clearly demonstrated that overexpression of the HBV preS1 protein correlates with cell death and triggers a set of events that culminate in the development of HCC (Chisari *et al.*, 1989). An accumulation of HBV core antigen may be similarly toxic to hepatocytes (Roingeard *et al.*, 1990). Presumably, during subsequent regeneration of damaged liver tissue, genetic and epigenetic abnormalities may occur that lead to the selection of cells at risk for accumulating additional genetic alterations (for review see Chisari, 1989).

Tumour suppressor genes and HCC

Both the activation of cellular protooncogenes and the inactivation of tumour suppressor genes are important in the development of human cancers (Bishop, 1991). Although activated oncogenes are not recovered with any regularity from chromosomal DNA of human HCCs, tumour-specific chromosome losses are frequently observed. Restriction fragment length polymorphism (RFLP) studies on DNA from matched normal and HCC patient tissues have identified tumour-specific allele losses (termed loss of heterozygosity, or LOH) on at least eight chromosome arms; these are chromosomes 4q, 5q, 8q, 11p, 13q, 16p, 16q and 17p (see Table 1). These losses have been interpreted to reflect a selection process which favoured cells that had lost growth-suppressing genes localized on those specific chromosome arms.

Current models of tumour suppressor gene inactivation involve a 2-hit process. The allele losses observed would represent one step of inactivation (deletion of the allele) for a target gene localized in those chromosome regions. The p53 tumour suppressor gene is believed to be a target gene for the chromosome 17p deletions, as the remaining p53 allele in those tumours frequently contains an inactivating point mutation (Bressac *et al.*, 1991; Hsu *et al.*, 1991; Scorsone *et al.*, 1992; Lai *et al.*, 1993). Although approximately 50% of HCCs from China and southern Africa contain p53 mutations, the frequency of p53 mutations is much less common in liver tumours from other geographical locations (Buetow *et al.*, 1992). The specificity of point mutations affecting codon 249 of the p53 protein in HCCs from regions of China and Africa has been interpreted as evidence for exposure to an environmental carcinogen(s) that may cooperate with HBV in the development of HCC (Ozturk *et al.*, 1991). The genes serving as targets for the remaining examples of chromosomal loss are unknown, and the role of HBV in mediating those allele losses remains unclear. Virus integration into chromosomal DNA in infected hepatocytes is random, as specific sequences in either cellular DNA or viral DNA are not involved. However, there is some selection for cells containing inserts into chromosomes 3, 11 and 17. These chromosomes have been over-represented when flanking cellular sequences next to viral inserts in HCCs have been analysed (Slagle *et al.*, 1993). Interestingly, allele losses from chromosomes 11 and 17 have been reported for tumours from several geographical locations (Table 1). The gross chromosomal abnormalities associated with HBV integration suggest that a single integration event might mediate biologically significant gene losses at distances far from the actual site of viral integration. Taken together, these data suggest that HBV integration may mediate tumour suppressor gene allele loss in some patients, with the biological consequences of viral integration events varying among subsets of tumours.

The direct association of viral proteins with cellular tumour suppressor

Table 1. *Chromosomal deletions associated with HCC*

Chromosome		Associated tumour suppressor gene[b]	Geographical origin of tumour samples	Reference
Arm	(% loss)[a]			
4q	(58)	—	mixed[c]	Buetow et al., 1989
	(50)	—	Japan	Zhang et al., 1990
5q	(44)	—	Japan	Fujimori et al., 1991
	(100)	—	NA[d]	Ding et al., 1991
8q	(44)	—	China[e]	Slagle et al., 1991
11p	(40)	—	NA	Wang & Rogler, 1988
	(67)	—	mixed[f]	Walker et al., 1991
13q	(50)	—	NA	Wang & Rogler, 1988
	(67)	—	mixed[f]	Walker et al., 1991
16p	(74)	—	China[e]	Slagle et al., 1993
16q	(57)	—	Japan	Zhang et al., 1990
	(52)	—	Japan	Tsuda et al., 1990
	(64)	—	Japan	Fujimori et al., 1991
	(40)	—	Japan	Nishida et al., 1993
	(85)	—	China[e]	Slagle et al., 1993
17p	(53)	p53[g]	China[e]	Slagle et al., 1991
	(54)	p53	Japan	Fujimori et al., 1991
	(71)	p53	China[e]	Scorsone et al., 1992
	(49)	p53	Japan	Nishida et al., 1993
	(50)	p53	NA	Ding et al., 1991
	(30)	p53	China[e]	Li et al., 1993

[a] Percentage loss (in parentheses) indicates number of patients with specific allele loss divided by total number of patients heterozygous at that locus ($\times$ 100). Only allele losses exceeding 40% are reported. Numbers of patients in each study population ranged from 12 to 70.

[b] Tumour suppressor gene listed is suspected to be involved in tumourigenesis because the remaining copy of the gene has also been mutated. Dashed lines (—) indicate that no tumour suppressor gene has been identified for that chromosomal region.

[c] Sample population includes four South African blacks, one American black, one Thai, three European caucasians, one Tonganese, one Japanese, and one Vietnamese.

[d] NA, not available. Geographical information on patient samples not reported.

[e] Sample population is from Quidong County.

[f] Sample population includes 13 Caucasians, one Indian, three Melanesians, one aborigine, and one Vietnamese.

[g] Remaining allele of p53 gene frequently contains a missense point mutation, as described by Scorsone et al. (1992).

gene products has been demonstrated for several DNA tumour viruses (for review see Levine & Momand, 1990). This protein–protein inter-action is believed to abrogate the normal function of the tumour suppres-sor gene, to the advantage of the virus, as the small DNA viruses must stimulate host cell DNA synthesis in order to complete their life cycles. One recent report suggests that the HBV-encoded X protein may directly interact with the p53 protein (Feitelson *et al.*, 1993).

Host immune response to virus-infected cells

Because not all HCCs from HBV-endemic areas contain integrated HBV sequences, an indirect role for the virus in the development of liver cancer must be considered. Replication of HBV is not considered directly cytotoxic to hepatocytes (Araki *et al.*, 1989). However, recent studies provide evidence consistent with a model in which the host immune response to virus-infected hepatocytes contributes to cell injury and/or cell death. The identification of HBcAg-specific cytotoxic T lymphocytes (CTL) in an HBV-infected patient supports an immune-mediated cytopathogenesis model (Guilhot *et al.*, 1992).

It is likely that all immunocompetent patients infected with HBV will mount a cell-mediated immune response to virus-infected hepatocytes. However, the severity of immune-mediated liver disease may be expected to vary among patients, and the contribution of these host factors to chronic liver disease (and eventually to hepatocarcinogenesis) remains unclear.

HEPATITIS C VIRUS

The introduction of a specific assay for HBV and the movement to a completely volunteer blood donor program uncovered cases of posttransfusion hepatitis which could not be attributed to either HBV or hepatitis A virus (HAV). This new agent, termed non-A, non-B hepatitis (NANBH) (for review see Houghton *et al.*, 1991), had a typical incubation period of 5 to 10 weeks and was not caused by any other known viruses (for review see Hollinger, 1990). Molecular studies indicated that at least one type of NANBH was caused by a virus less than 80 nm in diameter with a lipid-containing envelope (Bradley *et al.*, 1991; He *et al.*, 1987). In 1989, Choo and coworkers announced the identification of the NANBH agent, now known as hepatitis C virus (HCV) (Choo *et al.*, 1989).

Humans are thought to represent the natural reservoir for HCV, although the virus has been experimentally introduced into chimpanzees and marmosets. HCV infection is most often acquired in a manner similar to HBV infection (blood transfusions and intravenous drug abuse), although other routes of infection such as 'community acquired' appear to be important (Saito *et al.*, 1990; Chemello *et al.*, 1993; Serfaty *et al.*, 1993). In contrast to HBV, HCV does not appear to be efficiently transmitted by sexual contact (Serfaty *et al.*, 1993). Acute infection with HCV tends to be mild as compared to infection with HAV or HBV, and subclinical cases are common. However, approximately 50% of HCV infections progress to chronic viral infection (Plagemann, 1991), compared to 10% of HBV infections and none of HAV infections. The reason why HCV so readily establishes chronic infection is unknown; possible mechanisms include an inadequate immune response to the virus, an ability of the virus to adapt and successfully evade the host immune response, and the existence of a nonhepatic reservoir of HCV infection.

HCV has recently been cloned and sequenced and has been classified into a separate genus of the family *Flaviviridae* (Francki *et al.*, 1991) based on several biochemical criteria, including hydrophobicity analysis of putative proteins, genome organization, and characterization of the virion. The other two genera within the *Flaviviridae*, the flaviviruses [the prototype being yellow fever virus (Monath, 1990)] and the animal pestiviruses (bovine viral diarrhea virus and hog cholera virus), are only distantly related to HCV at the nucleic acid level (Miller & Purcell, 1990; Takamizawa *et al.*, 1991). All *Flaviviridae* contain similarly organized genomes ranging in size from 9.5–12.5 kb (Monath, 1990; Schlesinger & Schlesinger, 1990). The particles of HCV are between 30 and 60 nm in size (He *et al.*, 1987), and the buoyant density of plasma-derived HCV is similar to that of the pestiviruses ($1.088–1.107$ g/cm^3) (Horzinek, 1981; Bradley *et al.*, 1991).

Attempts to grow HCV in cell culture have met with limited success. Primary hepatocytes isolated from an acutely infected chimpanzee and placed in cell culture continued to produce both HCV antigens and appropriately sized virus particles (Jacob *et al.*, 1990). Recent data suggest that the liver may not be the sole site of virus replication, as HCV replication was detected in a human T-cell line (Shimizu *et al.*, 1992). Relative levels of infectious HCV in serum are measured using an endpoint dilution assay in chimpanzees or marmosets (Bradley *et al.*, 1991; Shindo *et al.*, 1992), and HCV is usually present during human infection in concentrations of 10^3 to 10^6 chimpanzee infectious doses (CID) per milliliter. Sensitive polymerase chain reaction (PCR) assays have been developed recently and are able to detect both positive-strand RNA (viral genomes) and negative-strand RNA (indicative of viral replication) in infected liver tissue (Takehara *et al.*, 1992). Serum HCV titres measured by PCR appear to correlate with CID titres.

Genomic organization of HCV

The single-strand, positive-sense RNA genome of HCV is organized similarly to that of other members of the *Flaviviridae* (Fig. 3). Non-translated regions (NTR) at both the 5′ and 3′ ends of the genome flank a single continuous ORF of 3010 amino acids. In contrast to portions of the long ORF which may vary considerably among HCV isolates, the 5′-NTR is highly conserved (>91% at the nucleic acid level; for review see Plagemann,

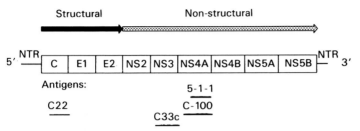

Fig. 3. Genomic organization of HCV. The positive-stranded RNA genome of HCV consists of approximately 9400 nucleotides, with a 341-base 5' non-translated region (NTR) and a 27- to 45-base 3' NTR (Takamizawa *et al.*, 1991). The remaining RNA is translated as a single polyprotein, and the relative order of the resulting processed proteins is shown in the boxes labelled C, E1, etc. The structural proteins of the virus (C, E1 and E2) cluster toward the 5' end of the genome and are followed by nonstructural proteins (NS2, NS3, NS4A, NS4B, NS5A and NS5B). Shown below the boxes are antigens of the virus (C-22, 5-1-1, C-100, and C-33c) that have been utilized in immunological assays for the detection of antibody to HCV (described in the text). Based on similarities to other known flaviviruses, HCV has been assigned to a separate genus within the family *Flaviviridae* (Francki *et al.*, 1991).

1991). By analogy to other viruses that contain 5'-NTRs, this 341 nucleotide region of HCV is expected to contain regulatory elements important in virus functions (for example, replication and viral gene expression). Indeed, a secondary stem–loop structure has been identified in the 5'-NTR (Han *et al.*, 1991; Brown *et al.*, 1992), and this region corresponds to an internal ribosome entry site (IRES) involved in the initiation of translation of the viral proteins (Tsukiyama-Kohara *et al.*, 1992). The 3'-NTR, which varies from 27 to 45 nucleotides in length among HCV isolates, may serve as a primer for negative-strand RNA synthesis during viral replication. The 3' end of the viral genome contains either a poly(A) or poly(U) sequence, depending on the isolate (reviewed by Houghton *et al.*, 1991).

The long ORF of HCV is translated as a single polyprotein of approximately 3011 amino acids, with the variation in length due to differences in the E2 region (Houghton *et al.*, 1991). The structural proteins of the virus are located at the amino terminus of the polyprotein (Fig. 3, ORFs designated C, E1 and E2) and are followed by the non-structural proteins (Fig. 3, designated NS2, NS3, NS4A, NS4B, NS5A and NS5B). As has been found with the flaviviruses and pestiviruses, the polyprotein of HCV is proteolytically processed to yield the mature viral proteins (Hijikata *et al.*, 1991; Grakoui *et al.*, 1993*b*). Expression of the polyprotein in cell culture has led to the identification of proteolytic cleavage products (Grakoui *et al.*, 1993*b*; Harada *et al.*, 1991; Inoue *et al.*, 1992; Kohara *et al.*, 1992; Kumar *et al.*, 1992; Matsuura *et al.*, 1992; Spaete *et al.*, 1992). Residing at the most amino-terminal portion is a highly basic core (C) protein believed to provide the capsid of the viral particle. The C protein has been expressed in monkey kidney cells (Harada *et al.*, 1991), and the resulting 22 kDa protein has been used in second-generation assays for the detection of anti-HCV antibodies (Schlipköter *et al.*, 1992; Todd *et al.*, 1992). The presumed envelope (E) region encodes the 31 kDa E1 and the 70 kDa E2 glycoproteins and contains variable and hypervariable regions (Weiner *et al.*, 1991) as well as several potential N-glycosylation sites. The hypervariable region of E2 has been suggested to be involved in the immune selection of HCV variants (Weiner *et al.*, 1992). Antisera to E2 and NS2 coprecipitate E1, suggesting a possible disulphide linkage of these proteins (Grakoui *et al.*, 1993*b*). The proteins NS2 through NS5B are thought to be non-structural proteins of the virus and, by analogy to flaviviruses, may provide functions that are important in virus replication. For example, the NS3 protein contains residues similar to putative NTP-binding helicases (Choo *et al.*, 1991) (Fig. 3), as well as a serine proteinase function that may be responsible for several cleavages of the viral polyprotein (Grakoui *et al.*, 1993*a*). Also, by analogy to the flaviviruses and pestiviruses, NS3 may participate with NS5B to form the viral replicase activity (Chambers *et al.*, 1990). The functions of NS2 and NS4 remain unclear.

There are extensive sequence differences among HCV isolates, and the virus is extremely heterogeneous both among infected individuals and within the same individual (Kato *et al.*, 1990; Cuypers *et al.*, 1991; Houghton *et al.*, 1991; Martell *et al.*, 1992; Tanaka *et al.*, 1992). A phylogenetic analysis of several HCV isolates suggests there are at least three HCV basic genotypes (Chan *et al.*, 1992) that are readily distinguishable by PCR amplification and DNA sequencing of portions of the 5'-NTR and NS3. All three genotypes are observed in Europe and Japan, whereas group I is the primary genotype detected in the United States (Houghton *et al.*, 1991). There is considerable variability among genotypes, from a low of 75% amino acid homology in several of the NS genes to a high of 98–100% homology in the C gene (Houghton *et al.*, 1991; Tanaka *et al.*, 1992). It is possible that the variable success observed in treatment of chronic HCV with interferon-α may be attributed to such variations in homology among HCV isolates. The quasi-species nature of HCV within an infected individual holds important implications for vaccine development and possibly for the ability of the virus to establish chronic infection.

Assays for the detection of HCV

Before specific tests for detection of HCV became available, blood screening was based on surrogate markers, namely elevated levels of alanine aminotransferase (ALT) and the presence of antibody to HBcAg (anti-HBc). A study comparing these surrogate markers found that screening for anti-HBc would prevent about 30% of post-transfusion hepatitis (PT-NANBH) cases, whereas screening for elevated ALT levels would prevent nearly 50% of such cases and required the discarding of fewer units of donor blood (Stevens et al., 1984).

There have been tremendous efforts dedicated to the development of assays for the detection of HCV. An antibody capture technique to detect antibodies to the nonstructural NS3 C-100 protein (Fig. 3) detected HCV antibodies in 80% and 71% of PT-NANBH patients from Japan and the United States, respectively (Kuo et al., 1989). An unacceptable level of false positive tests (Mortimer et al., 1989; McFarlane et al., 1990) led to the development of a 4-antigen recombinant immunoblot assay (4-RIBA), designed to confirm the C-100 protein assay (van der Poel et al., 1991). The antigens used were from both structural and non-structural proteins of the virus [Fig. 3, 5-1-1, C-100, C-33c, and C-22 (Lee et al., 1992)], and this permitted detection of anti-HCV antibodies earlier in the course of an infection. The assay has proven useful in the identification of C-100 assay false-positives (Serfaty et al., 1993). At present, a second-generation enzyme-linked immunosorbent assay utilizing C-22, C-33c, and C-100 is used to detect anti-HCV antibody (Schlipköter et al., 1992; Todd et al., 1992).

Immunological studies in humans and chimpanzees using the assays described above have shown that the time course of elevated ALT and detection of antibody to HCV varies considerably among individuals. Although nearly 80% of infected patients develop antibody to C-100, the average time before antibody detection is 22 weeks post-transfusion (for review see Plagemann, 1991). A similar slow appearance of the humoral immune response to C-100 has been noted in chimpanzees (Shimizu et al., 1990). More recent studies using assays to the HCV core antigen have detected antibody as soon as 8 weeks post-transfusion, whereas antibody to C-100 was detected 13 weeks post-transfusion in the same patient (Harada et al., 1991). Antibody to HCV may exist in the form of immune complexes during chronic liver disease (Hijikata et al., 1993).

Little is known about the cellular immune response to HCV infection. Patients chronically infected with HCV have CTL directed against all of the viral proteins (Botarelli et al., 1993). The non-structural NS4 protein appears to be most immunogenic, and a T-cell response to the HCV core antigen may be associated with less severe disease (Botarelli et al., 1993). The finding of HLA class I restricted CTLs within the hepatic parenchyma suggests that the cellular immune response may play a role in HCV chronic liver disease (Koziel et al., 1992).

PCR procedures permit the detection of HCV RNA much earlier after infection than antibodies to viral proteins. Reverse transcription of the HCV RNA using antisense-strand-specific and sense-strand-specific oligonucleotide primers allows the detection of minute amounts of both positive (genomic) strand and negative (replicating) HCV RNA (Takeuchi et al., 1990). Using the PCR assay, HCV RNA was detected in the serum of two

Table 2. *Possible mechanisms of HBV and HCV involvement in the development of HCC*

Characteristic	HBV	HCV
Establish chronic infection in adults[a]	Yes (5–10%)	Yes (50%)
Carry direct-acting oncogene	No	No
Encode viral transactivator protein(s)[b]	Yes	Unknown
Potential for insertional mutagenesis[c]	Yes	No
Replication of virus in tumour cells	No	No
Loss of tumour suppressor genes in virus-positive HCCs[d]	Yes	Unknown
Stimulate hepatocyte growth[e]	Yes	Yes

[a] Numbers shown represent the percentage of infected adults who develop chronic viral hepaitis, as described by Hollinger (1990).

[b] Proteins encoded by HBV that have the ability to transactivate viral and cellular genes include HBx (for review see Rossner, 1992) and preS/S^t (Lauer *et al.*, 1992).

[c] Integration of HBV sequences may alter the expression of nearby cellular growth control genes, as described for WHV and N-*myc2* (for review see Buendia, 1992). Because HCV does not replicate through a DNA intermediate, insertional mutagenesis is probably not involved in virus-mediated hepatocarcinogenesis.

[d] Tumour suppressor genes and chromosome allele losses (indicating potential sites for additional tumour suppressor genes) that have been reported for HBV-positive HCCs are summarized in Table 1.

[e] Both HBV and HCV cause liver damage, which induces liver regeneration. It is unknown if either virus directly stimulates DNA synthesis or cell division.

experimentally infected chimpanzees 3 days post-infection (Shimizu *et al.*, 1990). This technique has successfully detected HCV RNA in the sera and livers of chronically infected HCV patients as well (for review see Plagemann, 1991), and it is a valid alternative for the rapid and definitive diagnosis of HCV.

Potential oncogenic properties of HCV

Seroepidemiology studies have shown that chronic infection with HCV is an important risk factor for the development of HCC, particularly in areas of the world where HBV infection is less common (Bruix *et al.*, 1989; Colombo *et al.*, 1989; Di Bisceglie *et al.*, 1991) (Table 2). A recent increase of HCC in Japan has been attributed to a significant increase in HCV chronic liver disease in that country; nearly 80% of HBV-negative HCC cases were found to be HCV positive (Saito *et al.*, 1990; Miyamura *et al.*, 1991). In the United States, less than one third of HCC patients were positive for anti-HCV antibody (Yu *et al.*, 1990). The true incidence of HCV-related HCC is difficult to predict, as these first epidemiological studies utilized the C-100-based assay (which displayed a high rate of false-positive reactions). However, it appears that the relative role of HCV in HCC will vary between countries. It has been suggested that dual infection with HBV and HCV may lead to an increased risk of HCC (Yu *et al.*, 1990).

The molecular mechanism by which HCV might contribute to HCC is unclear. The virus does not carry a recognizable viral oncogene. Because the HCV replication cycle presumably does not include a DNA intermediate, there appears to be no opportunity for insertional mutagenesis via integration of viral DNA (as is observed with HBV). As very little is known of the functional properties of the viral proteins, transactivation of cellular growth regulatory genes and modifications of cell signalling pathways remain possible considerations. However, preliminary PCR studies failed to detect replicating viral RNA in HCC cells of anti-HCV-positive patients (Takeda *et al.*, 1992). By elimination of these direct mechanisms of viral involvement in hepatocarcinogenesis, it seems probable that the role for HCV is indirect. It is known that chronic liver inflammation and cirrhosis, regardless of aetiology, pose a risk for the development of HCC. One would predict that, if the role for HCV is indirect, dual infection with HCV and HBV might well elicit a more severe chronic liver disease.

CONCLUSIONS

Chronic infection with HBV or HCV is a major risk factor in the development of HCC. Both viral factors and the host immune response appear to contribute to the development of liver cancer. The threat posed by chronic HBV infection is multifaceted, and molecular events that occur during the course of chronic liver disease include insertional mutagenesis by viral DNA, transactivation of cellular regulatory genes by the viral transactivator proteins HBx and preS/S[t], tumour suppressor gene loss, and cell death and liver regeneration associated with the host immune response to HBV-infected cells. Importantly, not all liver cancers appear to have sustained the same virus-associated changes, suggesting that the role of HBV may differ from tumour to tumour.

Little is known about possible mechanisms by which HCV may contribute to hepatocarcinogenesis. As HBV and HCV may present complementary mechanisms of carcinogenesis, dual infections with these viruses would be predicted to incite more severe liver disease and potential for HCC development.

ACKNOWLEDGEMENTS

Work reported from the authors' laboratory was supported in part by grant CA54557 from the National Cancer Institute.

REFERENCES

Alter, H. J. (1988). Transfusion-associated non-A, non-B hepatitis: the first decade. In *Viral Hepatitis and Liver Disease*, ed. A. J. Zuckerman, pp. 537–542. New York: Alan A. Liss.

Araki, K., Miyazaki, J.-I., Hino, O., Tomita, N., Chisaka, O., Matsubara, K. & Yamamura, K.-I. (1989). Expression and replication of hepatitis B virus genome in transgenic mice. *Proceedings of the National Academy of Sciences, USA,* **86**, 207–11.

Beasley, R. P., Hwang, L.-Y., Lin, C.-C. & Chien, C.-S. (1981). Hepatocellular carcinoma and hepatitis B virus. A prospective study of 22,707 men in Taiwan. *Lancet,* **ii**, 1129–33.

Beasley, R. P., Hwang, L.-Y., Lin, C.-C., Ko, Y.-C. & Twu, S.-J. (1983). Incidence of hepatitis among students at a university in Taiwan. *American Journal of Epidemiology,* **117**, 213–22.

Bishop, J. M. (1991). Molecular themes in oncogenesis. *Cell,* **64**, 235–48.

Blumberg, B. S., Alter, H. J. & Visnich, S. (1965). A 'new' antigen in leukemia sera. *Journal of the American Medical Association,* **191**, 101–6.

Botarelli, P., Brunetto, M. R., Minutello, M. A. *et al.* (1993). T-lymphocyte response to hepatitis C virus in different clinical courses of infection. *Gastroenterology,* **104**, 580–7.

Bradley, D., McCaustland, K., Krawczynski, K., Spelbring, J., Humphrey, C. & Cook, E. H. (1991). Hepatitis C virus: buoyant density of the factor VIII-derived isolate in sucrose. *Journal of Medical Virology,* **34**, 206–8.

Bréchot, C., Hadchouel, M., Scotto, J. *et al.*, (1981). State of hepatitis B virus DNA in hepatocytes of patients with hepatitis B surface antigen-positive and -negative liver diseases. *Proceedings of the National Academy of Sciences, USA,* **78**, 3906–10.

Bressac, B., Kew, M., Wands, J. & Ozturk, M. (1991). Selective G to T mutations of p53 gene in hepatocellular carcinoma from southern Africa. *Nature,* **350**, 429–31.

Brown, E. A., Zhang, H., Ping, L.-H. & Lemon, S. M. (1992). Secondary structure of the 5' nontranslated regions of hepatitis C virus and pestivirus genomic RNAs. *Nucleic Acids Research,* **20**, 5041–5.

Bruix, J., Barrera, J. M., Calvet, X. *et al.* (1989). Prevalence of antibodies to hepatitis C virus in Spanish patients with hepatocellular carcinoma and hepatic cirrhosis. *Lancet,* **ii**, 1004–6.

Buendia, M. A. (1992). Hepatitis B viruses and hepatocellular carcinoma. *Advances in Cancer Research,* **59**, 167–226.

Buetow, K. H., Murray, J. C., Israel, J. L. *et al.* (1989). Loss of heterozygosity suggests tumour suppressor gene responsible for primary hepatocellular carcinoma. *Proceedings of the National Academy of Sciences, USA,* **86**, 8852–6.

Buetow, K. H., Sheffield, V. C., Zhu, M. *et al.* (1992). Low frequency of p53 mutations observed in a diverse collection of primary hepatocellular carcinomas. *Proceedings of the National Academy of Sciences, USA,* **89**, 9622–6.

Carlson, J. & Eriksson, S. (1985). Chronic 'cryptogenic' liver disease and malignant hepatoma in intermediate alpha-1-antitrypsin deficiency identified by a PiZ-specific monoclonal antibody. *Scandinavian Journal of Gastroenterology,* **20**, 835–42.

Chambers, T. J., Weir, R. C., Grakoui, A. *et al.* (1990). Evidence that the N-terminal domain of nonstructural protein NS3 from yellow fever virus is a serine protease responsible for site-specific cleavages in the viral polyprotein. *Proceedings of the National Academy of Sciences, USA,* **87**, 8898–902.

Chan, S. W., McOmish, F., Holmes, E. C. *et al.* (1992). Analysis of a new hepatitis C virus type and its phylogenetic relationship to existing variants. *Journal of General Virology,* **73**, 1131–41.

Chemello, L., Cavalletto, D., Pontisso, P. *et al.* (1993). Patterns of antibodies to hepatitis C virus in patients with chronic non-A, non-B hepatitis and their relationship to viral replication and liver disease. *Hepatology,* **17**, 179–82.

Chisari, F. V. (1989). Hepatitis B virus gene expression in transgenic mice. *Molecular Biology and Medicine,* **6**, 143–9.

Chisari, F. V., Klopchin, K., Moriyama, T. *et al.* (1989). Molecular pathogenesis of hepatocellular carcinoma in hepatitis B virus transgenic mice. *Cell,* **59**, 1145–56.

Choo, Q.-L., Kuo, G., Weiner, A. J., Overby, L. R., Bradley, D. W. & Houghton, M. (1989). Isolation of a cDNA clone derived from a blood-borne non-A, non-B hepatitis genome. *Science,* **244**, 359–61.

Choo, Q.-L., Richman, K. H., Han, J. H. *et al.* (1991). Genetic organization and diversity of the hepatitis C virus. *Proceedings of the National Academy of Sciences, USA,* **88**, 2451–5.

Colombo, M., Kuo, G., Choo, Q. L. *et al.* (1989). Prevalence of antibodies to hepatitis C virus in Italian patients with hepatocellular carcinoma. *Lancet,* **ii**, 1006–8.

Cuypers, H. T., Winkel, I. N., van der Poel, C. L. *et al.* (1991). Analysis of genomic variability of hepatitis C virus. *Journal of Hepatology,* **13**, suppl. 4, S15–19.

Dane, D. S., Cameron, C. H. & Briggs, M. (1970). Virus-like particles in serum of patients with Australia-antigen-associated hepatitis. *Lancet,* **i**, 695–8.

Dejean, A., Bougueleret, L., Grzeschik, K.-H., & Tiollais, P. (1986). Hepatitis B virus DNA integration in a sequence homologous to v-*erb*-A and steroid receptor genes in a hepatocellular carcinoma. *Nature,* **322**, 70–2.

Dejean, A., Sonigo, P., Wain-Hobson, S. & Tiollais, P. (1984). Specific hepatitis B virus integration in hepatocellular carcinoma DNA through a viral 11-base-pair direct repeat. *Proceedings of the National Academy of Sciences, USA,* **81**, 5350–4.

Delius, H., Gough, N. M., Cameron, C. H. & Murray, K. (1983). Structure of the hepatitis B virus genome. *Journal of Virology,* **47**, 337–43.

Di Bisceglie, A. M., Order, S. E., Klein, J. L. *et al.* (1991). The role of chronic viral hepatitis in hepatocellular carcinoma in the United States. *American Journal of Gastroenterology,* **86**, 335–8.

Ding, S.-F., Habib, N. A., Dooley, J., Wood, C., Bowles, L. & Delhanty, J. D. A. (1991). Loss of constitutional heterozygosity on chromosome 5q in hepatocellular carcinoma without cirrhosis. *British Journal of Cancer,* **64**, 1083–7.

Dragani, T. A., Manenti, G., Farza, H., Della Porta, G., Tiollais, P. & Pourcel, C. (1989). Transgenic mice containing hepatitis B virus sequences are more susceptible to carcinogen-induced hepatocarcinogenesis. *Carcinogenesis,* **11**, 953–6.

Eriksson, S., Carlson, J. & Velez, R. (1986). Risk of cirrhosis and primary liver cancer in alpha-1-antitrypsin deficiency. *New England Journal of Medicine,* **314**, 736–9.

Feitelson, M. A., Zhu, M., Duan, L.-X. & London, W. T. (1993). Hepatitis B x antigen and p53 are associated *in vitro* and in liver tissues from patients with primary hepatocellular carcinoma. *Oncogene,* **8**, 1109–17.

Fourel, G., Trepo, C., Bougueleret, L., Henglein, B., Ponzetto, A., Tiollais, P. & Buendia, M.-A. (1990). Frequent activation of N-*myc* genes by hepadnavirus insertion in woodchuck liver tumours. *Nature,* **347**, 294–8.

Francki, R. I. B., Fauquet, C. M., Knudson, D. L. & Brown, F. (1991). Classification and nomenclature of viruses: Fifth report of the International Committee on Taxonomy of Viruses. *Archives of Virology,* suppl. 2, 1–450.

Fujimori, M., Tokino, T., Hino, O. *et al.* (1991). Allelotype study of primary hepatocellular carcinoma. *Cancer Research,* **51**, 89–93.

Fujiyama, A., Miyanohara, A., Nozaki, C., Yoneyama, T., Ohtomo, N. &

Matsubara, K. (1983). Cloning and structural analyses of hepatitis B virus DNAs, subtype *adr*. *Nucleic Acids Research*, **11**, 4601–10.

Galibert, F., Chen, T. N. & Mandart, E. (1982). Nucleotide sequence of a cloned woodchuck hepatitis virus genome: comparison with the hepatitis B virus sequence. *Journal of Virology*, **41**, 51–65.

Ganem, D. & Varmus, H. E. (1987). The molecular biology of the hepatitis B viruses. *Annual Review of Biochemistry*, **56**, 651–93.

Grakoui, A., McCourt, D. W., Wychowski, C., Feinstone, S. M. & Rice, C. M. (1993*a*). Characterization of the hepatitis C virus-encoded serine proteinase: determination of proteinase-dependent polyprotein cleavage sites. *Journal of Virology*, **67**, 2832–43.

Grakoui, A., Wychowski, C., Lin, C., Feinstone, S. M. & Rice, C. M. (1993*b*). Expression and identification of hepatitis C virus polyprotein cleavage products. *Journal of Virology*, **67**, 1385–95.

Guilhot, S., Fowler, P., Portillo, G. *et al.* (1992). Hepatitis B virus (HBV)-specific cytotoxic T-cell response in humans: production of target cells by stable expression of HBV-encoded proteins in immortalized human B-cell lines. *Journal of Virology*, **66**, 2670–8.

Han, J. H., Shyamala, V., Richman, K. H. *et al.* (1991). Characterization of the terminal regions of hepatitis C viral RNA: identification of conserved sequences in the 5′ untranslated region and poly(A) tails at the 3′ end. *Proceedings of the National Academy of Sciences, USA*, **88**, 1711–15.

Harada, S., Watanabe, Y., Takeuchi, K. *et al.* (1991). Expression of processed core protein of hepatitis C virus in mammalian cells. *Journal of Virology*, **65**, 3015–21.

He, L.-F., Alling, D., Popkin, T., Shapiro, M., Alter, H. J. & Purcell, R. H. (1987). Determining the size of non-A, non-B hepatitis virus by filtration. *Journal of Infectious Diseases*, **156**, 636–40.

Hijikata, M., Kato, N., Ootsuyama, Y., Nakagawa, M. & Shimotohno, K. (1991). Gene mapping of the putative structural region of the hepatitis C virus genome by *in vitro* processing analysis. *Proceedings of the National Academy of Sciences, USA*, **88**, 5547–51.

Hijikata, M., Shimizu, Y. K., Kato, H. *et al.* (1993). Equilibrium centrifugation studies of hepatitis C virus: evidence for circulating immune complexes. *Journal of Virology*, **67**, 1953–8.

Hilger, C., Velhagen, I., Zentgraf, H. & Schröder, C. H. (1991). Diversity of hepatitis B virus X gene-related transcripts in hepatocellular carcinoma: a novel polyadenylation site on viral DNA. *Journal of Virology*, **65**, 4284–91.

Hino, O., Shows, T. B. & Rogler, C. E. (1986). Hepatitis B virus integration site in hepatocellular carcinoma at chromosome 17;18 translocation. *Proceedings of the National Academy of Sciences, USA*, **83**, 8338–42.

Hollinger, F. B. (1990). Hepatitis B virus. In *Fields' Virology*, ed. B. N. Fields, D. M. Knipe, R. M. Chanock, *et al.*, 2nd edn., pp. 2171–2236. New York: Raven Press.

Horzinek, M. C. (1981). *Non-Arthropod Borne Togaviruses*. Academic Press, London.

Houghton, M., Weiner, A., Han, J., Kuo, G. & Choo, Q.-L. (1991). Molecular biology of the hepatitis C viruses: implications for diagnosis, development and control of viral disease. *Hepatology*, **14**, 381–8.

Hsu, I. C., Metcalf, R. A., Sun, T., Welsh, J. A., Wang, N. J. & Harris, C. C. (1991). Mutational hotspot in the p53 gene in human hepatocellular carcinomas. *Nature*, **350**, 427–8.

Hsu, T.-Y., Möröy, T., Etiemble, J. *et al.* (1988). Activation of c-*myc* by woodchuck hepatitis virus insertion in hepatocellular carcinoma. *Cell*, **55**, 627–35.

Inoue, Y., Suzuki, R., Matsuura, Y. *et al.* (1992). Expression of the amino-terminal half of the NS1 region of the hepatitis C virus genome and detection of an antibody to the expressed protein in patients with liver diseases. *Journal of General Virology*, **73**, 2151–4.

Jacob, J. R., Burk, K. H., Eichberg, J. W., Dreesman, G. R. & Lanford, R. E. (1990). Expression of infectious viral particles by primary chimpanzee hepatocytes isolated during the acute phase of non-A, non-B hepatitis. *Journal of Infectious Diseases*, **161**, 1121–7.

Kato, N., Hijikata, M., Ootsuyama, Y. *et al.* (1990). Molecular cloning of the human hepatitis C virus genome from Japanese patients with non-A, non-B hepatitis. *Proceedings of the National Academy of Sciences, USA*, **87**, 9524–8.

Kim, C.-M., Koike, K., Saito, I., Miyamura, T. & Jay, G. (1991). *HBx* gene of hepatitis B virus induces liver cancer in transgenic mice. *Nature*, **351**, 317–20.

Kohara, M., Tsukiyama-Kohara, K., Maki, N. *et al.* (1992). Expression and characterization of glycoprotein gp35 of hepatitis C virus using recombinant vaccinia virus. *Journal of General Virology*, **73**, 2313–18.

Koziel, M. J., Dudley, D., Wong, J. T. *et al.* (1992). Intrahepatic cytotoxic T lymphocytes specific for hepatitis C virus in persons with chronic hepatitis. *Journal of Immunology*, **149**, 3339–44.

Kumar, U., Cheng, D., Thomas, H. & Monjardino, J. (1992). Cloning and sequencing of the structural region and expression of putative core gene of hepatitis C virus from a British case of chronic sporadic hepatitis. *Journal of General Virology*, **73**, 1521–5.

Kuo, G., Choo, Q.-L., Alter, H. J. *et al.* (1989). An assay for circulating antibodies to a major etiologic virus of human non-A, non-B hepatitis. *Science*, **244**, 362–4.

Lai, M. Y., Chang, H. C., Li, H. P. *et al.* (1993). Splicing mutations of the p53 gene in human hepatocellular carcinoma. *Cancer Research*, **53**, 1653–6.

Lauer, U., Weiss, L., Hofschneider, P. H. & Kekulé, A. S. (1992). The hepatitis B virus *pre-S/S'* transactivator is generated by 3' truncations within a defined region of the *S* gene. *Journal of Virology*, **66**, 5284–9.

Lee, S., McHutchinson, J., Francis, B. *et al.* (1992). Improved detection of antibodies to hepatitis C virus using a second generation ELISA. *Advances in Experimental Medicine and Biology*, **312**, 183–9.

Lee, T.-H., Finegold, M. J., Shen, R.-F., DeMayo, J. L., Woo, S. L. C. & Butel, J. S. (1990). Hepatitis B virus transactivator X protein is not tumorigenic in transgenic mice. *Journal of Virology*, **64**, 5939–47.

Levine, A. J. & Momand, J. (1990). Tumour suppressor genes: the p53 and retinoblastoma sensitivity genes and gene products. *Biochimica et Biophysica Acta*, **1032**, 119–36.

Li, D., Cao, Y., He, L., Wang, N. J. & Gu, J.-R. (1993). Aberrations of p53 gene in human hepatocellular carcinoma from China. *Carcinogenesis*, **14**, 169–73.

Lieber, C. S., Garro, A., Leo, M. A., Mak, K. M. & Worner, T. (1986). Alcohol and cancer. *Hepatology*, **6**, 1005–9.

Limmer, J., Fleig, W. E., Leupold, D., Bittner, R., Ditschuneit, H. & Beger, H. G. (1988). Hepatocellular carcinoma in type I glycogen storage disease. *Hepatology*, **8**, 531–7.

McFarlane, I. G., Smith, H. M., Johnson, P. J., Bray, G. P., Vergani, D. & Williams, R. (1990). Hepatitis C virus antibodies in chronic active hepatitis: pathogenetic factor or false-positive result? *Lancet*, **335**, 754–7.

Marion, P. L. (1988). Use of animal models to study hepatitis B virus. *Progress in Medical Virology*, **35**, 43–75.

Marion, P. L., Oshiro, L. S., Regnery, D. C., Scullard, G. H. & Robinson, W. S.

(1980). A virus in Beechey ground squirrels that is related to hepatitis B virus of humans. *Proceedings of the National Academy of Sciences, USA*, **77**, 2941–5.

Martell, M., Esteban, J. I., Quer, J. *et al.* (1992). Hepatitis C virus (HCV) circulates as a population of different but closely related genomes: quasispecies nature of HCV genome distribution. *Journal of Virology*, **66**, 3225–9.

Mason, W. S., Seal, G. & Summers, J. (1980). Virus of Pekin ducks with structural and biological relatedness to human hepatitis B virus. *Journal of Virology*, **36**, 829–36.

Matsuura, Y., Harada, S., Suzuki, R. *et al.* (1992). Expression of processed envelope protein of hepatitis C virus in mammalian and insect cells. *Journal of Virology*, **66**, 1425–31.

Melia, W. M., Johnson, P. J., Neuberger, J., Zaman, S., Portmann, B. C. & Williams, R. (1984). Hepatocellular carcinoma in primary biliary cirrhosis: detection by α-fetoprotein estimation. *Gastroenterology*, **87**, 660–3.

Miller, R. H., Kaneko, S., Chung, C. T., Girones, R. & Purcell, R. H. (1989). Compact organization of the hepatitis B virus genome. *Hepatology*, **9**, 322–7.

Miller, R. H., Lee, S.-C., Liaw, Y.-F. & Robinson, W. S. (1985). Hepatitis B viral DNA in infected human liver and in hepatocellular carcinoma. *Journal of Infectious Diseases*, **151**, 1081–92.

Miller, R. H. & Purcell, R. H. (1990). Hepatitis C virus shares amino acid sequence similarity with pestiviruses and flaviviruses as well as members of two plant virus supergroups. *Proceedings of the National Academy of Sciences, USA*, **87**, 2057–61.

Miyamura, T., Saito, I., Yoneyama, T. *et al.* (1991). Role of hepatitis C virus in hepatocellular carcinoma. In *Viral Hepatitis and Liver Disease*, ed. F. B. Hollinger, S. M. Lemon & H. Margolis, pp. 559–562. Baltimore: Williams & Wilkins.

Mizusawa, H., Taira, M., Yaginuma, K., Kobayashi, M., Yoshida, E. & Koike, K. (1985). Inversely repeating integrated hepatitis B virus DNA and cellular flanking sequences in the human hepatoma-derived cell line huSP. *Proceedings of the National Academy of Sciences, USA*, **82**, 208–12.

Monath, T. P. (1990). Flaviviruses. In *Fields Virology*, ed. B. N. Fields, D. M. Knipe, R. M. Chanock, *et al.*, 2nd edn., pp. 763–814. New York: Raven Press.

Möröy, T., Marchio, A., Etiemble, J., Trépo, C., Tiollais, P. & Buendia, M.-A. (1986). Rearrangement and enhanced expression of c-*myc* in hepatocellular carcinoma of hepatitis virus infected woodchucks. *Nature*, **324**, 276–9.

Mortimer, P. P., Cohen, B. J., Litton, P. A. *et al.* (1989). Hepatitis C virus antibody. *Lancet*, **ii**, 798–9.

Nagaya, T., Nakamura, T., Tokino, T. *et al.* (1987). The mode of hepatitis B virus DNA integration in chromosomes of human hepatocellular carcinoma. *Genes and Development*, **1**, 773–82.

Niederau, C., Fischer, R., Sonnenberg, A., Stremmel, W., Trampisch, H. J. & Strohmeyer, G. (1985). Survival and causes of death in cirrhotic and in noncirrhotic patients with primary hemochromatosis. *New England Journal of Medicine*, **313**, 1256–62.

Nishida, N., Fukuda, Y., Kokuryu, H. *et al.* (1993). Role and mutational heterogeneity of the p53 gene in hepatocellular carcinoma. *Cancer Research*, **53**, 368–72.

Ono, Y., Onda, H., Sasada, R., Igarashi, K., Sugino, Y. & Nishioka, K. (1983). The complete nucleotide sequences of the cloned hepatitis B virus DNA; subtype adr and adw. *Nucleic Acids Research*, **11**, 1747–57.

Ozturk, M. & collaborators. (1991). p53 mutation in hepatocellular carcinoma after aflatoxin exposure. *Lancet*, **338**, 1356–9.

Pasquinelli, C., Garreau, F., Bougueleret, L. *et al.* (1988). Rearrangement of a common cellular DNA domain on chromosome 4 in human primary liver tumours. *Journal of Virology*, **62**, 629–32.

Plagemann, P. G. W. (1991). Hepatitis C virus. *Archives of Virology*, **120**, 165–80.

Popper, H. (1988). Pathobiology of hepatocellular carcinoma. In *Viral Hepatitis and Liver Disease*, ed. A. J. Zuckerman, pp. 719–722. New York: Alan A. Liss.

Rogler, C. E., Hino, O. & Su, C.-Y. (1987). Molecular aspects of persistent woodchuck hepatitis virus and hepatitis B virus infection and hepatocellular carcinoma. *Hepatology*, **7**, 74S–8S.

Roingeard, P., Romet-Lemonne, J. L., Leturcq, D., Goudeau, A. & Essex, M. (1990). Hepatitis B virus core antigen (HBc Ag) accumulation in an HBV nonproducer clone of HepG2-transfected cells is associated with cytopathic effect. *Virology*, **179**, 113–20.

Rossner, M. T. (1992). Review: Hepatitis B virus X-gene product: a promiscuous transcriptional activator. *Journal of Medical Virology*, **36**, 101–17.

Saito, I., Miyamura, T., Ohbayashi, A. *et al.* (1990). Hepatitis C virus infection is associated with the development of hepatocellular carcinoma. *Proceedings of the National Academy of Sciences, USA*, **87**, 6547–9.

Schaller, H. & Fischer, M. (1991). Transcriptional control of hepadnavirus gene expression. *Current Topics in Microbiology and Immunology*, **168**, 21–39.

Schlesinger, S. & Schlesinger, M. J. (1990). Replication of togaviridae and flaviviridae. In *Fields Virology*, ed. B. N. Fields, D. M. Knipe, R. M. Chanock *et al.*, 2nd edn., pp. 697–711. New York: Raven Press.

Schlipköter, U., Gladziwa, U., Cholmakov, K. *et al.* (1992). Prevalence of hepatitis C virus infections in dialysis patients and their contacts using a second generation enzymed-linked immunosorbent assay. *Medical Microbiology and Immunology*, **181**, 173–80.

Schödel, F., Sprengel, R., Weimer, T., Fernholz, D., Schneider, R. & Will, H. (1989). Animal hepatitis B viruses. *Advances in Viral Oncology*, **8**, 73–102.

Scorsone, K. A., Zhou, Y.-Z., Butel, J. S. & Slagle, B. L. (1992). p53 mutations cluster at codon 249 in hepatitis B virus-positive hepatocellular carcinomas from China. *Cancer Research*, **52**, 1635–8.

Seeger, C., Summers, J. & Mason, W. S. (1991). Viral DNA synthesis. *Current Topics in Microbiology and Immunology*, **168**, 41–60.

Serfaty, L., Giral, P., Elghouzzi, M. H., Jullien, A. M. & Poupon, R. (1993). Risk factors for hepatitis C virus infection in hepatitis C virus antibody ELISA-positive blood donors according to RIBA-2 status: a case-control survey. *Hepatology*, **17**, 183–7.

Shafritz, D. A. & Kew, M. C. (1981). Identification of integrated hepatitis B virus DNA sequences in human hepatocellular carcinomas. *Hepatology*, **1**, 1–18.

Shih, C., Burke, K., Chou, M.-J. *et al.* (1987). Tight clustering of human hepatitis B virus integration sites in hepatomas near a triple-stranded region. *Journal of Virology*, **61**, 3491–8.

Shimizu, Y. K., Iwamoto, A., Hijikata, M., Purcell, R. H. & Yoshikura, H. (1992). Evidence for *in vitro* replication of hepatitis C virus genome in a human T-cell line. *Proceedings of the National Academy of Sciences, USA*, **89**, 5477–81.

Shimizu, Y. K., Weiner, A. J., Rosenblatt, J. *et al.* (1990). Early events in hepatitis C virus infection of chimpanzees. *Proceedings of the National Academy of Sciences, USA*, **87**, 6441–4.

Shindo, M., Di Bisceglie, A. M., Biswas, R., Mihalik, K. & Feinstone, S. M. (1992).

Hepatitis C virus replication during acute infection in the chimpanzee. *Journal of Infectious Diseases*, **166**, 424–7.

Slagle, B. L., Lee, T.-H. & Butel, J. S. (1992). Hepatitis B virus and hepatocellular carcinoma. *Progress in Medical Virology*, **39**, 167–203.

Slagle, B. L., Zhou, Y.-Z., Birchmeier, W. & Scorsone, K. A. (1993). Deletion of the E-cadherin gene in hepatitis B virus-positive Chinese hepatocellular carcinomas. *Hepatology*, **18**, 757–62.

Slagle, B. L., Zhou, Y.-Z. & Butel, J. S. (1991). Hepatitis B virus integration event in human chromosome 17p near the p53 gene identifies the region of the chromosome commonly deleted in virus-positive hepatocellular carcinomas. *Cancer Research*, **51**, 49–54.

Spaete, R. R., Alexander, D., Rugroden, M. E. *et al.* (1992). Characterization of the hepatitis C virus E2/NS1 gene product expressed in mammalian cells. *Virology*, **188**, 819–30.

Sprengel, R., Kaleta, E. F. & Will, H. (1988). Isolation and characterization of a hepatitis B virus endemic in herons. *Journal of Virology*, **62**, 3832–9.

Stevens, C. E., Aach, R. D., Hollinger, F. B. *et al.* (1984). Hepatitis B virus antibody in blood donors and the occurrence of non-A, non-B hepatitis in transfusion recipients. *Annals of Internal Medicine*, **101**, 733–8.

Summers, J., Smolec, J. M. & Snyder, R. (1978). A virus similar to human hepatitis B virus associated with hepatitis and hepatoma in woodchucks. *Proceedings of the National Academy of Sciences, USA*, **75**, 4533–7.

Szmuness, W. (1978). Hepatocellular carcinoma and the hepatitis B virus: evidence for a causal association. *Progress in Medical Virology*, **24**, 40–69.

Takada, S. & Koike, K. (1990). Trans-activation function of a 3' truncated X gene-cell fusion product from integrated hepatitis B virus DNA in chronic hepatitis tissues. *Proceedings of the National Academy of Sciences, USA*, **87**, 5628–32.

Takamizawa, A., Mori, C., Fuke, I. *et al.* (1991). Structure and organization of the hepatitis C virus genome isolated from human carriers. *Journal of Virology*, **65**, 1105–13.

Takeda, S., Shibata, M., Morishima, T. *et al.* (1992). Hepatitis C virus infection in hepatocellular carcinoma. *Cancer*, **70**, 2255–9.

Takehara, T., Hayashi, N., Mita, E. *et al.* (1992). Detection of the minus strand of hepatitis C virus RNA by reverse transcription and polymerase chain reaction: implications for hepatitis C virus replication in infected tissue. *Hepatology*, **15**, 387–90.

Takeuchi, K., Kubo, Y., Boonmar, S. *et al.* (1990). The putative nucleocapsid and envelope protein genes of hepatitis C virus determined by comparison of the nucleotide sequences of two isolates derived from an experimentally infected chimpanzee and healthy human carriers. *Journal of General Virology*, **71**, 3027–33.

Tanaka, T., Kato, N., Nakagawa, M. *et al.* (1992). Molecular cloning of hepatitis C virus genome from a single Japanese carrier: sequence variation within the same individual and among infected individuals. *Virus Research*, **23**, 39–53.

Todd, J., Kink, J., Leahy, D. *et al.* (1992). A novel semi-automated paramagnetic microparticle based enzyme immunoassay for hepatitis C virus: its application to serologic testing. *Journal of Immunoassay*, **13**, 393–410.

Tokino, T., Fukushige, S., Nakamura, T. *et al.* (1987). Chromosomal translocation and inverted duplication associated with integrated hepatitis B virus in hepatocellular carcinomas. *Journal of Virology*, **61**, 3848–54.

Transy, C., Fourel, G., Robinson, W. S., Tiollais, P., Marion, P. L. & Buendia, M.-A. (1992). Frequent amplification of c-*myc* in ground squirrel liver tumors associated with past or ongoing infection with a hepadnavirus. *Proceedings of the National Academy of Sciences, USA,* **89**, 3874–8.

Trowbridge, R., Fagan, E. A., Davison, F. *et al.* (1988). Amplification of the c-*myc* gene locus in a human hepatic tumour containing integrated hepatitis B virus DNA. In *Viral Hepatitis and Liver Disease,* ed. A. J. Zuckerman, pp. 764–768. New York: Alan R. Liss.

Tsuda, H., Zhang, W., Shimosato, Y. *et al.* (1990). Allele loss on chromosome 16 associated with progression of human hepatocellular carcinoma. *Proceedings of the National Academy of Sciences, USA,* **87**, 6791–4.

Tsukiyama-Kohara, K., Iizuka, N., Kohara, M. & Nomoto, A. (1992). Internal ribosome entry site within hepatitis C virus RNA. *Journal of Virology,* **66**, 1476–83.

Valenzuela, P., Quiroga, M., Zaldivar, J., Gray, P. & Rutter, W. J. (1980). The nucleotide sequence of the hepatitis B viral genome and the identification of the major viral genes. In *Animal Virus Genetics,* ed. B. N. Fields, and R. Jaenisch, pp. 57–70. New York: Academic Press.

van der Poel, C. L., Cuypers, H. T. M., Reesink, H. W. *et al.* (1991). Confirmation of hepatitis C virus infection by new four-antigen recombinant immunoblot assay. *Lancet,* **337**, 317–19.

Walker, G. J., Hayward, N. K., Falvey, S. & Cooksley, W. G. E. (1991). Loss of somatic heterozygosity in hepatocellular carcinoma. *Cancer Research,* **51**, 4367–70.

Wang, H. P. & Rogler, C. E. (1988). Deletions in human chromosome arms 11p and 13q in primary hepatocellular carcinomas. *Cytogenetics and Cell Genetics,* **48**, 72–8.

Wang, J., Chenivesse, X., Henglein, B. & Bréchot, C. (1990). Hepatitis B virus integration in a cyclin A gene in a hepatocellular carcinoma. *Nature,* **343**, 555–7.

Wei, Y., Fourel, G., Ponzetto, A., Silvestro, M., Tiollais, P. & Buendia, M.-A. (1992). Hepadnavirus integration: mechanisms of activation of the N-*myc*2 retrotransposon in woodchuck liver tumours. *Journal of Virology,* **66**, 5265–76.

Weiner, A. J., Brauer, M. J., Rosenblatt, J. *et al.* (1991). Variable and hypervariable domains are found in the regions of HCV corresponding to the flavivirus envelope and NS1 proteins and the pestivirus envelope glycoproteins. *Virology,* **180**, 842–8.

Weiner, A. J., Geysen, H. M., Christopherson, C. *et al.* (1992). Evidence for immune selection of hepatitis C virus (HCV) putative envelope glycoprotein variants: potential role in chronic HCV infections. *Proceedings of the National Academy of Sciences, USA,* **89**, 3468–72.

Yaginuma, K., Kobayashi, H., Kobayashi, M., Morishima, T., Matsuyama, K. & Koike, K. (1987). Multiple integration site of hepatitis B virus DNA in hepatocellular carcinoma and chronic active hepatitis tissues from children. *Journal of Virology,* **61**, 1808–13.

Yu, M. C., Tong, M. J., Coursaget, P., Ross, R. K., Govindarajan, S. & Henderson, B. E. (1990). Prevalence of hepatitis B and C viral markers in black and white patients with hepatocellular carcinoma in the United States. *Journal of the National Cancer Institute,* **82**, 1038–41.

Zhang, W., Hirohashi, S., Tsuda, H. *et al.* (1990). Frequent loss of heterozygosity on chromosomes 16 and 4 in human hepatocellular carcinoma. *Japanese Journal of Cancer Reseaech,* **81**, 108–11.

Zhou, Y.-Z., Butel, J. S., Li, P.-J., Finegold, M. J. & Melnick, J. L. (1987). Integrated state of subgenomic fragments of hepatitis B virus DNA in hepatocellular carcinoma from mainland China. *Journal of the National Cancer Institute,* **79**, 223–31.

HEPATITIS B VIRUSES AND LIVER CANCER: THE WOODCHUCK MODEL

M. A. BUENDIA

Unité de Recombinaison et Expression Génétique (INSERM U163), Institut Pasteur, 28 rue du Dr. Roux, 75724, Paris, Cedex 15, France.

INTRODUCTION

Hepatitis B virus (HBV) ranks among the few recognized human tumour viruses by virtue of a strong epidemiological association between long-lasting HBV infection and hepatocellular carcinoma (HCC) (Szmuness, 1978). It has been estimated that the life-time risk of developing HCC is increased by a factor of 30 to 100 for chronic carriers of the virus as compared to non-infected populations (Beasley *et al.*, 1981; Obata *et al.*, 1980). However, the mechanisms whereby HBV may induce tumour formation remain uncertain. The HBV genome carries no dominant oncogene, and HBV replication has no direct cytopathic effect in infected hepatocytes. The emergence of liver tumours after several decades of chronic liver disease might therefore result from indirect mechanisms, based on chronic liver injury and compensatory regeneration, which favour the occurrence of secondary genetic lesions (Chisari *et al.*, 1989). Alternatively, the virus might act as an insertional mutagen, as suggested by the finding of integrated viral forms in a majority of hepatocellular carcinomas (Bréchot *et al.*, 1981, 1982). HBV DNA integration might promote genomic instability (Hino, Tabata & Hotta, 1991), or deregulate the normal growth control 'in *trans*' through truncated viral *trans*-activators (Kekulé *et al.*, 1990; Wollersheim, Debelka & Hofschneider, 1988). Furthermore, potentially oncogenic targets for viral insertion have been identified in a few tumours (Dejean *et al.*, 1986; Wang *et al.*, 1990; Zhang *et al.*, 1992).

Further arguments linking HBV and HCC have been provided by studies of naturally occurring animal models. The human hepatitis B virus belongs to a small group of viruses called hepadnaviruses, which also include two rodent viruses, the woodchuck hepatitis virus (WHV) and the ground squirrel hepatitis virus (GSHV) and avian viruses, the duck hepatitis virus (DHBV) and the heron hepatitis virus (HHBV) (for reviews, see Buendia, 1992; Robinson, 1990; Schödel *et al.*, 1989). Wild woodchucks and ground squirrels infected by hepadnaviruses develop chronic liver disease and HCC at a high frequency (Marion *et al.*, 1986; Summers, Smolec & Snyder, 1978), and inoculation of newborn woodchucks with WHV results in the formation of HCC in all persistently infected animals (Popper *et al.*, 1987*a*).

Recent studies of woodchuck liver tumours have demonstrated that viral integration is a critical step in the oncogenic process triggered by WHV, and have shown that *myc* family oncogenes play a pivotal role in woodchuck hepatocarcinogenesis. In this chapter, the pathological properties of WHV and the known oncogenic events associated with WHV infection will be presented. The similarities and the differences between the woodchuck/ WHV model and other hepadnavirus systems will be discussed.

ONCOGENIC PROPERTIES OF WOODCHUCK HEPATITIS VIRUS

Woodchuck hepatitis virus was first identified as a hepatitis B-like agent associated with chronic hepatitis and hepatocellular carcinoma in captive woodchucks (*Marmotta monax*) kept at the Penrose Research Laboratory, Philadelphia (Snyder & Summers, 1980; Summers, Smolec & Snyder, 1978). In the wild, the prevalence of WHV infections varies considerably in different parts of the world: about 30% of woodchucks from the central eastern region of United States are chronic WHV carriers, whereas woodchucks from other countries, such as *Marmotta marmotta* in Europe, are apparently neither infected nor susceptible to infection with WHV (Chomel et al., 1984). The natural mode of spread of the virus has not been firmly established, but vertical or perinatal transmission from carrier dams to their offspring is supported by several lines of evidence. First, experimental inoculation of newborn animals with WHV virions usually leads to a high rate of persistent infections, whereas adult animals develop acute infections which are resolved (Popper et al., 1987a). This situation is highly reminiscent of human infection with HBV, in which the establishment of chronicity varies from 95% in neonates to less than 0.5% in adults. Secondly, it has been shown that WHV DNA is present in the testes and ovaries of adult woodchucks (Korba et al., 1988), and replicative forms of the virus have been detected in the livers and sera of several woodchuck feral litters (Kulonen & Millman, 1988), suggesting that WHV infections might start 'in utero'. Studies of the natural history of experimental WHV infection have revealed the presence of WHV DNA in the bone marrow first, followed by the liver, the spleen, peripheral blood lymphocytes, lymph nodes and thymus (Korba et al., 1989a). In addition, groups of WHV-positive cells have been detected in many tissues, including kidney, pancreas, testes and ovaries (Korba et al., 1990). Interestingly, high levels of WHV replication intermediates are found not only in the liver, but also in the spleen, at variance with human infections with HBV.

Persistent WHV infections are generally associated with chronic hepatitis of varying severity, characterized by mild to severe portal or periportal inflammation, with occasional necrosis, bile duct proliferation and ground-glass cells, mimicking the defined human 'healthy carrier' state (Popper et al., 1987b, 1981; Snyder & Summers, 1980; Toshkov et al., 1990). Foci of altered hepatocytes, resembling those described in the course of chemical carcinogenesis in small rodents, might represent precancerous lesions. The development of hepatocellular carcinoma in carrier animals occurs generally after a latency period of two to four years (Popper *et al.*, 1987*a*), and liver tumours also arise in about 20% of seroconverted woodchucks (Korba *et al.*, 1989*b*). Livers bearing malignant tumours are characterized by a marked necroinflammatory reaction, but liver cirrhosis has never been observed, even in aged animals. WHV-infected woodchucks develop essentially well-differentiated hepatocellular carcinoma which lacks the ability to form distant metastases. It is worth noticing that this species is not prone to spontaneous malignancies: more than 600 autopsies of wild-caught and colony-born woodchucks have revealed non-hepatic cancers in only 14 aged animals (Gerin *et al.*, 1991).

The incidence of liver tumours in WHV-infected woodchucks is the highest reported incidence of HCC following carcinogenic treatment (Popper *et al.*, 1987*a*), designating WHV as the most efficient oncogenic agent among known hepatocarcinogens. In particular, WHV appears to be more oncogenic than other mammalian hepadnaviruses. The frequency of tumour incidence in humans and rodents is generally correlated with the fractional life-span in a similar manner. The average life-span of captive healthy woodchucks, which never develop HCCs, is about 10 years. Hepatocarcinogenesis appears therefore much more rapid in woodchucks than in humans. Comparative analyses of HCC formation in WHV- versus GSHV-infected woodchucks have shown that tumours arise more rapidly and more frequently in WHV infections (Seeger *et al.*, 1991*a*). Thus, viral determinants, which remain presently unknown, rather than host specificities, appear to be responsible for the different oncogenicities of mammalian hepadnaviruses. Moreover, the strong oncogenic efficiency of WHV has been demonstrated in the absence of dietary or environmental carcinogenic co-factors (Popper *et al.*, 1987*a*). It has been shown that aflatoxin B1 (AFB1) is hepatocarcinogenic for woodchucks, but that AFB1 treatment neither alters the frequency of persistent WHV infections nor accelerates HCC development in WHV-infected woodchucks (Tennant *et al.*, 1991).

STATE OF WHV DNA IN CHRONICALLY INFECTED LIVERS AND HCCs

Molecular cloning and nucleotide sequencing of several WHV isolates have revealed that the genome of WHV is highly similar to that of HBV (Cohen *et al.*, 1988; Etiemble *et al.*, 1986; Galibert, Chen & Mandart, 1982; Girones *et al.*, 1989; Kodama *et al.*, 1985) (Fig. 1). These viruses share a similar small size of their DNA genomes, a common genetic organization, and the amino acid sequences of the viral genes are highly conserved, showing an overall 54% identity.

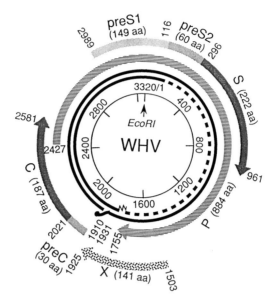

Fig. 1. The WHV genome. In the virion, WHV DNA (represented by a thick line) is partially double stranded (dotted line). The long viral DNA strand of fixed length (around 3.3 kb) encodes seven proteins from four open reading frames, shown as large arrows. A protein is covalently linked to the 5' end of the long strand and a short oligoribonucleotide at the 5' end of the short strand. The viral surface (S), core (C), polymerase (P), and X genes are represented.

Studies of the state of WHV DNA in woodchuck livers during acute and chronic infections have shown abundant replicative forms in infected hepatocytes (over 1000 genome units per cell) (Fig. 2). The life-cycle of hepadnaviruses has been mainly characterized using the DHBV model, but WHV has also been extensively studied (Seeger, Summers & Mason, 1991*b*). Although the replication pathway of hepadnaviruses is entirely extrachromosomal, it presents some analogies with the replication cycle of retroviruses, mainly by a step of reverse transcription of an RNA intermediate called pregenome (Summers & Mason, 1982). Similar patterns of transcription, associated with active viral replication, have been described for the WHV and the HBV genomes. Two abundant transcripts of 2.1 and 3.7 kb, specifying the major structural proteins and the viral polymerase, are produced in infected cells (Möröy *et al.*, 1985). The 3.7 kb transcript is also the pregenome RNA. Minor spliced transcripts of unknown function have been detected (Hantz *et al.*, 1992; Ogston & Razman, 1992). Recent data indicate that the large surface protein and the X gene product are made from discrete mRNAs present at very low levels (Y. Wei & M. A. Buendia, unpublished results).

Besides normal replication intermediates, integrated forms of WHV DNA have been found in chronically infected livers (Rogler & Summers,

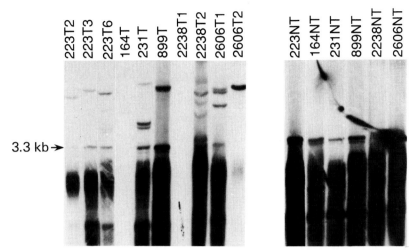

Fig. 2. State of viral DNA in chronically infected woodchuck livers and in HCCs. Genomic DNA from woodchuck livers (lanes NT) and tumours (lanes T), was digested with PvuII, an enzyme which does not cut in the WHV genome, and analysed by Southern blotting and hybridization with WHV DNA. Abundant replicative forms are shown in the livers by an intense smear beneath 3.3 kb. The presence of integrated WHV sequences in tumours is indicated by discrete bands of high molecular weight.

1984). In addition, large and highly rearranged WHV sequences, called 'novel forms' are also present in nonintegrated state during persistent infections (Rogler & Summers, 1982). Woodchuck HCCs generally develop in livers supporting active viral replication (Fig. 2), and most tumour samples contain free replicative forms of WHV at reduced levels. Whether these forms are really present in transformed cells, or originate from surrounding liver tissues contaminating the tumour samples, remain to be determined.

Southern blot analysis of genomic DNA from a large number of wood-chuck HCCs have shown the presence of integrated WHV sequences in a large majority (about 90%) of cases, both in chronic carriers and in seroconverted animals (Hansen *et al.*, 1993; Hsu *et al.*, 1990; Korba *et al.*, 1989*a*,*b*; Ogston *et al.*, 1982; Wei *et al.*, 1992*a*). The presence of discrete bands of high molecular weight that annealed with a WHV DNA probe indicate the clonal origin of the tumours. One or several distinct integration sites may be detected in the same tumour (Fig. 2). In animals developing simultaneously several tumour masses, a distinct viral integration pattern was observed for each tumour (Wei *et al.*, 1992*a*), indicating a different clonal origin. Therefore, like in human HBV-related HCCs, integration of viral DNA into the host genome precedes the clonal outgrowth of trans-formed liver cells in woodchuck HCC.

Analysis of the structure of cloned viral inserts have also pointed out a remarkable similarity between integrated WHV and HBV sequences. WHV inserts are made of continuous subgenomic fragments, or of complex, rearranged sequences; no complete genome has been observed (Fourel *et al.*, 1990; Hsu *et al.*, 1988; Ogston *et al.*, 1982; Wei *et al.*, 1992*a*). Both major and minor alterations may occur in flanking cellular DNA at the viral insertion site, including microdeletions, short duplications, or large re-arrangements which might reflect chromosomal deletions or translocations (Fourel *et al.*, 1990; Hsu *et al.*, 1988). In a tumour, integration of a large, recombined WHV sequence was not associated with gross alterations in cellular DNA, indicating that the viral DNA was rearranged prior to integration (Hsu *et al.*, 1988). The viral junctions are scattered all over the WHV genome, with a higher than average rate at or near the direct repeat DR1, corresponding to a preferred integration site in the HBV genome (Nagaya *et al.*, 1987). Sequences covering the viral S and X genes and the WHV enhancers are present in most WHV inserts (Wei *et al.*, 1992*a*), whereas C gene sequences are less represented. Little is known about the mechanisms leading to the integration of hepadnavirus DNA in cellular chromosomes, in the absence of an integrase gene in the hepadnavirus genome. The presence of patch homology between HBV or WHV DNA and cellular flanking DNA has suggested that hepadnaviruses might integrate through illegitimate recombination, like other DNA viruses (Matsubara and Tokino, 1990). It has been proposed that HBV might integrate its DNA through invasion of cellular DNA by free virus DNA ends from replication intermediates (Shih *et al.*, 1987). A cellular enzyme, the topoisomerase I (topoI) has been implicated in the process of WHV DNA integration (Wang & Rogler, 1991).

INSERTIONAL ACTIVATION OF *MYC* GENES BY WHV DNA

Search for transcriptional activation of known proto-oncogenes and for viral integration sites in woodchuck HCCs has revealed that WHV acts mainly as an insertional mutagen of *myc* family genes.

In an initial study of nine woodchuck tumours, rearrangement and overexpression of c-*myc* were observed in three HCCs (Möröy *et al.*, 1986). In the first tumour analysed in detail, the woodchuck c-*myc* gene was recombined with a cellular sequence termed 'hcr' with no apparent linkage to WHV integration (Etiemble *et al.*, 1989). An identical rearrangement between c-*myc* and hcr has recently been described in an independent woodchuck tumour (Hino *et al.*, 1992). In the two remaining cases, WHV DNA integrations in c-*myc* were demonstrated by molecular cloning of the mutated *myc* gene (Hsu *et al.*, 1988). Viral sequences were integrated either 600 bp upstream of the first c-*myc* exon, in the opposite transcriptional orientation relative to c-*myc*, or in the 3′ untranslated region of the gene, 70 bp downstream of the translation termination codon, in the same transcriptional orientation. In these tumours, c-*myc* RNA accumulated at enhanced steady-state levels, consisting of either normal c-*myc* transcripts initiated at the P1 and P2 promoters of the gene, or chimeric c-*myc*-WHV transcripts in which the coding region of c-*myc* was unaltered and WHV sequences replaced the c-*myc* 3' untranslated region. Since c-*myc* expression was controlled by its normal promoters, and viral sequences spanning the region homologous to the HBV enhancer were present in the inserts, it seems likely that a mechanism of viral enhancer insertion was associated with activation of the oncogene in these HCCs. This situation is similar to numerous examples of retroviral integration events (Payne, Bishop & Varmus, 1982; Selten *et al.*, 1984). The patterns of WHV DNA insertion in c-*myc* in these tumours share many common aspects with those of Moloney murine leukemia virus (MoMuLV) and its MCF derivatives in murine T-cell lymphomas. Deregulated expression of c-*myc* has been attributed to the insertion of potent transcriptional enhancers of viral origin together with the deletion of cellular sequences exerting negative regulatory effects on c-*myc* expression (Corcoran *et al.*, 1984). Indeed, it has recently been shown that the WHV genome contains an efficient enhancer, capable of activating transcription at heterologous promoters from different positions and orientations (Murakami *et al.*, 1990; Wei *et al.*, 1992*a*). A survey of 40 additional woodchuck HCCs for c-*myc* rearrangements has shown only one supplementary case in which WHV DNA was inserted into the c-*myc* locus, and provoked elevated expression of aberrant c-*myc* transcripts (Wei *et al.*, 1992*b*). Presently, it may be estimated that insertional activation of c-*myc* is involved in less than 10% of woodchuck tumours.

Further studies have outlined a higher frequency of viral integrations in the woodchuck N-*myc* genes (Fig. 3). Molecular cloning of a viral insertion site has led to the discovery of a second functional N-*myc* gene in the woodchuck genome (Fourel *et al.*, 1990). This gene, termed N-*myc*2, presents typical features of a processed pseudogene and probably evolved from the parental N-*myc*1 gene by retrotransposition of a mature N-*myc* transcript (Fig. 4). N-*myc*2 is devoid of introns, flanked by short direct repeats and carries the remnants of a polyA tail at its 3' extremity. Unlike most described pseudogenes, N-*myc*2 has retained extensive coding and transforming homology with N-*myc*. In particular, functionally important domains such as the leucine zipper and basic helix–loop–helix motifs are conserved, and N-*myc*2 can complement an activated *ras* gene in the rat embryo fibroblast (REF) transformation assay. Promoter sequences have recently been mapped in the retrotransposed unit, at the 5' end of the second N-*myc* exon, and N-*myc*2 is faintly expressed in the brain of normal adult woodchucks (Fourel *et al.*, 1992). N-*myc*2 represents by far the most frequent target for WHV DNA integration. Studies from two different groups have shown that N-*myc*2 was activated by nearby insertion of WHV DNA in 40% of woodchuck tumours, whereas the parental N-*myc* gene was

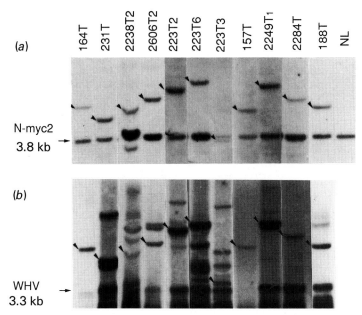

Fig. 3. Viral integration near N-*myc2* in woodchuck tumours. Southern blots of genomic DNA digested with PvuII were sequentially hybridized with a N-*myc* probe (*a*) and total WHV DNA (*b*). Rearranged N-*myc* bands co-migrating with WHV-specific bands are indicated by arrowheads.

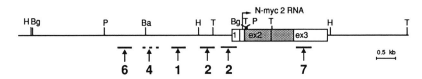

Fig. 4. Schematic illustration of WHV integration sites in the woodchuck N-*myc2* gene in a study of 52 HCCs.

mutated in only one case (Fourel *et al.*, 1990; Hansen *et al.*, 1993). Furthermore, it has been estimated that about 25% of the total integration events detected in woodchuck tumours analysed occurred in the N-*myc2* locus. Comparative analysis of woodchuck HCCs from wild-caught, naturally infected and colony-born, experimentally infected animals revealed comparable patterns of viral integration. The reasons for the strong clustering of WHV integration sites near N-*myc2* are not yet clear. In retroviral models, common integration sites in cellular proto-oncogenes have been documented in many lymphomas and leukaemias (Cuypers *et al.*, 1984; Moreau-Gachelin, Tavitian & Tambourin, 1988; Payne, Bishop & Varmus, 1982; Selten *et al.*, 1984), reflecting selection by clonal outgrowth of transformed cells. A particular chromatin structure at the N-*myc2* locus might favour the access of viral sequences; alternatively, the juxtaposition of WHV DNA and N-*myc2* gene might be particularly efficient in triggering hepatocyte proliferation. First detected in the 3′ non-coding region of N-*myc2* in 5/30 HCCs examined (Fourel *et al.*, 1990), the viral insertion sites were later found to be clustered in the vicinity of N-*myc2*, within a 3 kb region upstream of the gene or immediately downstream of the N-*myc2* coding domain, in 22/52 tumours, as illustrated in Fig. 4. In these tumours, increased levels of N-*myc2* mRNA or N-*myc2*/WHV co-transcripts resulted from enhanced transcriptional activity of the normal promoter of the gene. This suggests again a role for the WHV enhancers, which are present in all viral inserts lying near N-*myc2*. Enhanced production of N-*myc2* RNA of normal size was also observed in a majority of WHV-related woodchuck HCCs in the absence of any detectable rearrangement of the gene. In these cases, integration of WHV DNA in a hotspot located at distance of N-*myc2* in the same chromosomal region might contribute in N-*myc2* activation (G. Fourel and M. A. Buendia, unpublished results).

The importance of insertional activation of *myc* genes in woodchuck HCCs has been further demonstrated by the development of primary liver cancer in transgenic mice carrying the mutated c-*myc* gene and nearby integrated WHV sequences (Etiemble *et al.*, in press). In two different mouse strains, enforced, liver-specific expression of c-*myc* was correlated with the appearance of HCCs within 8 to 12 months of birth in a large majority of transgenic animals. These animals develop tumours in a stochastic fashion consistent with the notion that Myc is necessary but not sufficient for malignancy to develop, and provide interesting tools with which to study the oncogenic steps of the malignant process associated with persistent hepadnavirus infection.

CONCLUSIONS

The high oncogenic efficiency of WHV compared with other hepadnaviruses may be related to a specific ability of this virus to integrate its DNA near *myc* family genes and to activate transcription from the *myc* promoters by a direct, *cis*-acting effect of integrated viral sequences. Integration of HBV DNA near *myc* genes has never been described in human HCCs. The N-*myc2* retroposon is limited to a small number of mammalian species, including woodchucks, but absent in humans, and might therefore represent a species specificity accounting for the differences observed between HBV and WHV integration patterns. However, a gene homologous to the woodchuck N-*myc2* retroposon is also present in the ground squirrel genome (Transy *et al.*, 1992). The squirrel and woodchuck N-*myc2* sequences are highly conserved both in coding and regulatory regions, showing 98% identity at the nucleotide level (C. Transy and M. A. Buendia, unpublished results). Although WHV and GSHV are closely related viruses that differ no more than two HBV subtypes between each other, integration of GSHV DNA within *myc* family genes has not been observed in a study of 24 ground squirrel HCCs (Transy *et al.*, 1992). Furthermore, GSHV also failed to provoke insertional activation of *myc* genes in chronically infected woodchucks (Hansen *et al.*, 1993). In the two rodent species, GSHV-induced carcinogenesis is associated with a high rate of genetic amplifications at the c-*myc* locus, which may not, to our present knowledge, be linked to a direct effect of the virus.

The reasons for this marked discrepancy between the oncogenic strategies of closely related hepadnaviruses are not clear. A reduced capacity of GSHV DNA to integrate into the host cellular genome has been suggested by the finding of integrated GSHV sequences in only 20% of ground squirrel tumours (Transy *et al.*, 1992). However, GSHV integration events are frequent in woodchuck HCCs induced by persistent GSHV infection (Hansen *et al.*, 1993), and HBV DNA insertions, that occur in more than 80% of human HCCs, do not apparently disrupt the c-*myc* or N-*myc* genes. Although the human genome contains no N-*myc2* gene, the human c-*myc* and N-*myc* genes represent potentially accessible target sites, as indicated by the finding of human papilloma virus (HPV) insertions near these genes in cervical carcinomas (Couturier *et al.*, 1991). In a second hypothesis, the regulatory sequences of WHV, in particular the enhancer elements, might be more efficient in activating the *myc* promoters than their GSHV and HBV counterparts. Thus, viral integration near *myc* might provide a selective growth advantage to the target cell only in the case of WHV insertion. The regulatory regions are highly conserved among mammalian hepadnaviruses, and further studies are required to test this possibility. Finally, WHV might differ by other oncogenic properties: continuous WHV replication might specifically trigger early, preneoplastic events in infected hepatocytes; these potential alterations, which remain presently unknown, might in turn favour the occurrence of insertional activation of *myc* as a secondary oncogenic step.

Evidence has been provided that the *myc* family oncogenes lie at the meeting point of the carcinogenic process induced by rodent hepadnaviruses. The Myc oncoproteins are implicated in the control of normal cell proliferation in response to external stimuli, and their inappropriate expression may either contribute to neoplastic transformation of a variety of cell types, or activate a programmed cell death pathway (DePinho, Schreiber-Agus & Alt, 1991; Evan *et al.*, 1992). Overexpression of c-*myc* is commonly observed in rodent hepatocarcinogenesis (Chandar, Lombardi & Locker, 1989; Makino *et al.*, 1984). In humans, increased expression of c-*myc* has been described in a majority of HCCs, and it also occurs in cirrhosis (Gu *et al.*, 1986; Himeno *et al.*, 1988). In rare cases, it was associated with significant amplification of the c-*myc* locus (Gu *et al.*, 1986; Trowbridge *et al.*, 1988). The c-*myc* promoter has been shown to be activated 'in *trans*' by the HBV X gene product in transient transfections of human hepatoma cell lines (Balsano *et al.*, 1991), suggesting that, *in vivo*, the HBV transactivator might activate intracellular pathways leading to oncogenic deregulation of *myc* expression. Lessons from the animals models, and particularly the woodchuck model, might contribute to a better understanding of the mechanisms leading to liver cell transformation upon chronic HBV infections in humans.

ACKNOWLEDGEMENTS

I would like to thank P. Tiollais and my collaborators who have contributed to a large part of the work reported here, and provided stimulating discussions and comments while these data were generated, with special thanks to G. Fourel, C. A. Renard, C. Transy and Y. Wei.

REFERENCES

Balsano, C., Avantaggiati, M. L., Natoli, G. *et al.* (1991). Transactivation of c-*fos* and c-*myc* protooncogenes by both full-length and truncated versions of the HBV-X protein. In *Viral Hepatitis and Liver Disease*, ed. F. B. Hollinger, S. M. Lemon & H. S. Margolis, pp. 572–76. Baltimore: The Williams and Wilkins Co.

Beasley, R. P., Lin, C. C., Hwang, L. Y. & Chien, C. S. (1981). Hepatocellular carcinoma and hepatitis B virus: a prospective study of 22,707 men in Taiwan. *Lancet*, **ii**, 1129–33.

Bréchot, C., Hadchouel, M., Scotto, J. *et al.* (1981). Detection of hepatitis B virus DNA in liver and serum: a direct appraisal of the chronic carrier state. *Lancet*, **ii**, 765–8.

Bréchot, C., Pourcel, C., Hadchouel, M. *et al.* (1982). State of hepatitis B virus DNA in liver diseases. *Hepatology*, **2**, 27S–34S.

Buendia, M. A. (1992). Hepatitis B viruses and hepatocellular carcinoma. *Advances in Cancer Research*, **59**, 167–226.

Chandar, N., Lombardi, B. & Locker, J. (1989). C-*myc* gene amplification during hepatocarcinogenesis by a choline-devoid diet. *Proceedings of the National Academy of Sciences, USA*, **86**, 2703–7.

Chisari, F. V., Klopchin, K., Moriyama, T. *et al.* (1989). Molecular pathogenesis of hepatocellular carcinoma in hepatitis B virus transgenic mice. *Cell*, **59**, 1145–56.

Chomel, B., Trépo, C. H., Pichoud, C. H., Jacquet, C., Boulay, P. & Joubert, L. (1984). Infection spontanée et expérimentale de la marmotte alpine (*Marmota marmota*) par le virus de l'hépatite de la marmotte nord-américaine (*Marmota monax*). *Comparative Immunology and Microbiology of Infectious Diseases*, **7**, 179–94.

Cohen, J. I., Miller, R. H., Rosenblum, B., Denniston, K., Gerin, J. L. & Purcell, R. H. (1988). Sequence comparison of woodchuck hepatitis virus replicative forms shows conservation of the genome. *Virology*, **162**, 12–20.

Corcoran, L. M., Adams, J. M., Dunn, A. R. & Cory, S. (1984). Murine T lymphomas in which the cellular myc oncogene has been activated by retrovirus insertion. *Cell*, **37**, 113–22.

Couturier, J., Sastre-Garau, X., Schneider-Maunoury, S., Labib, A. & Orth, G. (1991). Integration of papillomavirus DNA near myc genes in genital carcinomas and its consequences for proto-oncogene expression. *Journal of Virology*, **65**, 4534–8.

Cuypers, H. T., Selten, G., Quint, W. *et al.* (1984). Murine leukemia virus-induced T-cell lymphomagenesis: Integration of proviruses in a distinct chromosomal region. *Cell*, **37**, 141–50.

Dejean, A., Bougueleret, L., Grzeschik, K. H. & Tiollais, P. (1986). Hepatitis B virus DNA integration in a sequence homologous to v-*erb*A and steroid receptor genes in a hepatocellular carcinoma. *Nature*, **322**, 70–2.

DePinho, R. A., Schreiber-Agus, N. & Alt, F. W. (1991). Myc family oncogenes in the development of normal and neoplastic cells. *Advances in Cancer Research*, **57**, 1–45.

Etiemble, J., Degott, C., Renard, C. A. *et al.* (1994). Liver specific expression and high oncogenic efficiency of a *c-myc* transgene activated by woodchuck hepatitis virus insertion. *Oncogene*, in press.

Etiemble, J., Möröy, T., Jacquemin, E., Tiollais, P. & Buendia, M. A. (1989). Fused transcripts of c-*myc* and a new cellular locus, *hcr*, in a primary liver tumour. *Oncogene*, **4**, 51–7.

Etiemble, J., Möröy, T., Trépo, C., Tiollais, P. & Buendia, M. A. (1986). Nucleotide sequence of the woodchuck hepatitis virus surface antigen mRNAs and the variability of three overlapping viral genes. *Gene*, **50**, 207–14.

Evan, G. I., Wyllie, A. H., Gilbert, C. S. *et al.* (1992). Induction of apoptosis in fibroblasts by c-myc protein. *Cell*, **69**, 119–28.

Fourel, G., Transy, C., Tennant, B. C. & Buendia, M. A. (1992). Expression of the woodchuck N-myc2 retroposon in brain and in liver tumours is driven by a cryptic N-myc promoter. *Molecular and Cellular Biology*, **12**, 5336–44.

Fourel, G., Trépo, C., Bougueleret, L. *et al.* (1990). Frequent activation of N-*myc* genes by hepadnavirus insertion in woodchuck liver tumours. *Nature*, **347**, 294–8.

Galibert, F., Chen, T. N. & Mandart, E. (1982). Nucleotide sequence of a cloned woodchuck hepatitis virus genome: comparison with the hepatitis B virus sequence. *Journal of Virology*, **41**, 51–65.

Gerin, J. L., Cote, P. J., Korba, B. E., Miller, R. H. Purcell, R. H. & Tennant, B. C. (1991). Hepatitis B virus and liver cancer: the woodchuck as an experimental model of hepadnavirus-induced liver cancer. In *Viral Hepatitis and Liver Disease*, ed. F. B. Hollinger, S. M. Lemon & H. Margolis, pp. 556–59, Baltimore: The Williams and Wilkins Co.

Girones, R., Cote, P. J., Hornbuckle, W. E. *et al.* (1989). Complete nucleotide sequence of a molecular clone of woodchuck hepatitis virus that is infectious in the natural host. *Proceedings of the National Academy of Sciences, USA*, **86**, 1846–9.

Gu, J. R., Hu, L. F., Cheng, Y. C. & Wan, D. F. (1986). Oncogenes in human primary hepatic cancer. *Journal of Cell Physiology*, **4** (**suppl.**), 13–20.

Hansen, L. J., Tennant, B. C., Seeger, C. & Ganem, D. (1993). Differential activation of myc gene family members in hepatic carcinogenesis by closely related hepatitis B viruses. *Molecular and Cellular Biology*, **13**, 659–67.

Hantz, O., Baginski, I., Fourel, I., Chemin, I. & Trépo, C. (1992). Viral spliced RNA are produced, encapsidated and reverse transcribed during in vivo woodchuck hepatitis virus infection. *Virology*, **190**, 193–200.

Himeno, Y., Fukuda, Y., Hatanaka, M. & Imura, H. (1988). Expression of oncogenes in human liver disease. *Liver*, **8**, 208–12.

Hino, O., Kitagawa, T., Nomura, K. *et al.* (1992). Comparative molecular pathogenesis of hepatocellular carcinomas. In *Progress in Clinical and Biological Re-*

search: *Comparative Molecular Carcinogenesis*, ed. A. J. P. Klein-Szanto, M. W. Anderson, J. C. Barrett & T. J. Slaga, pp. 173–85, New York: Wiley-Liss, Inc.

Hino, O., Tabata, S. & Hotta, Y. (1991). Evidence for increased *in vitro* recombination with insertion of human hepatitis B virus DNA. *Proceedings of the National Academy of Sciences, USA*, **88**, 9248–52.

Hsu, T. Y., Fourel, G., Etiemble, J., Tiollais, P. & Buendia, M. A. (1990). Integration of hepatitis virus DNA near c-myc in woodchuck hepatocellular carcinoma. *Gastroenterology Japan*, **25**, 43–48.

Hsu, T. Y., Möröy, T., Etiemble, J., Louise, A., Trépo, C., Tiollais, P. & Buendia, M. A. (1988). Activation of c-*myc* by woodchuck hepatitis virus insertion in hepatocellular carcinoma. *Cell*, **55**, 627–35.

Kekulé, A. S., Lauer, U., Meyer, M., Caselmann, W. H., Hofschneider, P. H. & Koshy, R. (1990). The pre-S2/S region of integrated hepatitis B virus DNA encodes a transcriptional transactivator. *Nature*, **343**, 457–61.

Kodama, K., Ogasawara, N., Yoshikawa, H. & Murakami, S. (1985). Nucleotide sequence of a cloned woodchuck hepatitis virus: Evolutional relationship between hepadnaviruses. *Journal of Virology*, **56**, 978–86.

Korba, B. E., Brown, T. L., Wells, F. V. *et al.* (1990). Natural history of experimental woodchuck hepatitis virus infection: molecular virologic features of the pancreas, kidney, ovary and testis. *Journal of Virology*, **64**, 4499–506.

Korba, B. E., Cote, P. J., Wells, F. V. *et al.* (1989*a*). Natural history of woodchuck hepatitis virus infections during the course of experimental viral infection: molecular virologic features of the liver and lymphoid tissues. *Journal of Virology*, **63**, 1360–70.

Korba, B. E., Wells, F. V., Baldwin, B. *et al.* (1989*b*). Hepatocellular carcinoma in woodchuck hepatitis virus-infected woodchucks: presence of viral DNA in tumour tissue from chronic carriers and animals serologically recovered from acute infections. *Hepatology*, **9**, 461–70.

Korba, B. E., Gowans, E. J., Wells, F. V., Tennant, B. C., Clarke, R. & Gerin, J. L. (1988). Systemic distribution of woodchuck hepatitis virus in the tissue of experimentally infected woodchucks. *Virology*, **165**, 172–81.

Kulonen, K. & Millman, I. (1988). Vertical transmission of woodchuck hepatitis virus. *Journal of Medical Virology*, **26**, 233–42.

Makino, R., Hayashi, K., Sato, S. & Sugimura, T. (1984). Expressions of the c-Ha-*ras* and c-*myc* genes in rat liver tumour. *Biochemical and Biophysical Research Communications*, **119**, 1096–102.

Marion, P. L., Van Davelaar, M. J., Knight, S. S. *et al.* (1986). Hepatocellular carcinoma in ground squirrels persistently infected with ground squirrel hepatitis virus. *Proceedings of the National Academy of Sciences, USA*, **83**, 4543–6.

Matsubara, K. & Tokino, T. (1990). Integration of hepatitis B virus DNA and its implications for hepatocarcinogenesis. *Molecular Biological Medicine*, **7**, 243–60.

Moreau-Gachelin, F., Tavitian, A. & Tambourin, P. (1988). Spi-1 is a putative oncogene in virally induced murine erythroleukemias. *Nature*, **331**, 277–80.

Möröy, T., Etiemble, J., Trépo, C., Tiollais, P. & Buendia, M. A. (1985). Transcription of woodchuck hepatitis virus in the chronically infected liver. *EMBO Journal*, **4**, 1507–14.

Möröy, T., Marchio, A., Etiemble, J., Trépo, C., Tiollais, P. & Buendia, M. A. (1986). Rearrangement and enhanced expression of c-*myc* in hepatocellular carcinoma of hepatitis virus infected woodchucks. *Nature*, **324**, 276–9.

Murakami, S., Uchijima, M., Shimoda, A., Kaneko, S., Kobayashi, K. & Hattori, N. (1990). Hepadnavirus enhancer and its binding proteins. *Gastroenterology Japan*, **25**, 11–19.

Nagaya, T., Nakamura, T., Tokino, T. *et al.* (1987). The mode of hepatitis B virus DNA integration in chromosomes of human hepatocellular carcinoma. *Genes and Development*, **1**, 773–82.

Obata, H., Hayashi, N., Motoike, Y. *et al.* (1980). A prospective study on the development of hepatocellular carcinoma from liver cirrhosis with persistent hepatitis B virus infection. *International Journal of Cancer*, **25**, 741–7.

Ogston, C. W., Jonak, G. J., Rogler, C. E., Astrin, S. M. & Summers, J. (1982). Cloning and structural analysis of integrated woodchuck hepatitis virus sequences from hepatocellular carcinomas of woodchucks. *Cell*, **29**, 385–94.

Ogston, C. W. & Razman, D. G. (1992). Spliced RNA of woodchuck hepatitis Virus. *Virology*, **189**, 245–52.

Payne, G. S., Bishop, J. M. & Varmus, H. E. (1982). Multiple arrangements of viral DNA and an activated host oncogene in bursal lymphomas. *Nature*, **295**, 209–14.

Popper, H., Roth, L., Purcell, R. H., Tennant, B. C. & Gerin, J. L. (1987*a*). Hepatocarcinogenicity of the woodchuck hepatitis virus. *Proceedings of the National Academy of Sciences, USA*, **84**, 866–70.

Popper, H., Shafritz, D. A. & Hoofnagle, J. H. (1987*b*). Relation of the hepatitis B virus carrier state to hepatocellular carcinoma. *Hepatology*, **7**, 764–72.

Robinson, W. S. (1990). Hepadnaviridae and their replication. In *Fields Virology*, ed. B. N. Fields, D. M. Knipe, R. M. Chanock *et al.*, pp. 2137–69, New York: Raven Press.

Rogler, C. E. & Summers, J. (1982). Novel forms of woodchuck hepatitis virus DNA isolated from chronically infected woodchuck liver nuclei. *Journal of Virology*, **44**, 852–63.

Rogler, C. E. & Summers, J. (1984). Cloning and structural analysis of integrated woodchuck hepatitis virus sequences from a chronically infected liver. *Journal of Virology*, **50**, 832–7.

Schödel, F., Sprengel, R., Weimer, T., Fernholz, D., Schneider, R. & Will, H. (1989). Animal hepatitis B viruses. In *Advances in Viral Oncology*, ed. G. Klein, pp. 73–102, New York: Raven Press, Ltd.

Seeger, C., Baldwin, B., Hornbuckle, W. E. *et al.* (1991*a*). Woodchuck hepatitis virus is a more efficient oncogenic agent than ground squirrel hepatitis virus in a common host. *Journal of Virology*, **65**, 1673–9.

Seeger, C., Summers, J. & Mason, W. S. (1991*b*). Viral DNA synthesis. In *Current Topics in Microbiology and Immunology*, ed. W. S. Mason & C. Seeger, pp. 40–60. Berlin–Heidelberg: Springer-Verlag.

Selten, G., Cuypers, H. T., Zijlstra, M., Melief, C. & Berns, A. (1984). Involvement of c-*myc* in MuLV-induced T cell lymphomas in mice: frequency and mechanisms of activation. *EMBO Journal*, **3**, 3215–22.

Shih, C., Burke, K., Chou, M. J. *et al.* (1987). Tight clustering of human hepatitis B virus integration sites in hepatomas near a triple-stranded region. *Journal of Virology*, **61**, 3491–8.

Snyder, R. L. & Summers, J. (1980). Woodchuck hepatitis virus and hepatocellular carcinoma. In *Cold Spring Harbor Conferences on Cell Proliferation VII. Viruses in Naturally Occurring tumours*, ed. M. Essex, G. Todaro & H. zur Hausen, pp. 447–57, New York: Cold Spring Harbor Laboratory Press.

Summers, J. & Mason, W. S. (1982). Replication of the genome of a hepatitis B-like virus by reverse transcription of an RNA intermediate. *Cell*, **29**, 403–15.

Summers, J., Smolec, J. M. & Snyder, R. (1978). A virus similar to human hepatitis B virus associated with hepatitis and hepatoma in woodchucks. *Proceedings of the National Academy of Sciences, USA*, **75**, 4533–7.

Szmuness, W. (1978). Hepatocellular carcinoma and the hepatitis B virus: evidence for a causal association. *Progress in Medical Virology,* **24**, 40–69.

Tennant, B. C. H., Tennant, W. E., Yeager, A. E. *et al.* (1991). Effects of aflatoxin B1 on experimental hepatitis virus infection and hepatocellular carcinoma. In *Viral Hepatitis and Liver Disease*, ed. F. B. Hollinger, S. M. Lemon & H. S. Margolis, pp. 599–600, Baltimore: The Williams and Wilkins Co.

Toshkov, I., Hacker, H. J., Roggendorf, M. & Bannasch, P. (1990). Phenotypic patterns of preneoplastic and neoplastic hepatic lesions in woodchucks infected with woodchuck hepatitis virus. *Journal of Cancer Research and Clinical Oncology,* **116**, 581–90.

Transy, C., Fourel, G., Robinson, W. S., Tiollais, P., Marion, P. L. & Buendia, M. A. (1992). Frequent amplification of c-*myc* in ground squirrel liver tumours associated with past or ongoing infection with a hepadnavirus. *Proceedings of the National Academy of Sciences, USA,* **89**, 3874–8.

Trowbridge, R., Fagan, E. A., Davison, F. *et al.* (1988). Amplification of the c-*myc* gene locus in a human hepatic tumour containing integrated hepatitis B virus DNA. In *Viral Hepatitis and Liver Disease*, ed. A. J. Zuckerman, pp. 764–68, New York: Alan R. Liss, Inc.

Wang, H. P. & Rogler, C. E. (1991). Topoisomerase I-mediated integration of hepadnavirus DNA *in vitro. Journal of Virology,* **65**, 2381–92.

Wang, J., Chenivesse, X., Henglein, B. & Bréchot, C. (1990). Hepatitis B virus integration in a cyclin A gene in a human hepatocellular carcinoma. *Nature,* **343**, 555–7.

Wei, Y., Fourel, G., Ponzetto, A., Silvestro, M., Tiollais, P. & Buendia, M. A. (1992*a*). Hepadnavirus integration: mechanisms of activation of the N-*myc*2 retrotransposon in woodchuck liver tumours. *Journal of Virology,* **66**, 5265–76.

Wei, Y., Ponzetto, A., Tiollais, P. & Buendia, M. A. (1992*b*). Multiple rearrangements and activated expression of c-*myc* induced by woodchuck hepatitis virus integration in a primary liver tumour. *Research in Virology,* **143**, 89–96.

Wollersheim, M., Debelka, U. & Hofschneider, P. H. (1988). A transactivating function encoded in the hepatitis B virus X gene is conserved in the integrated state. *Oncogene,* **3**, 545–52.

Zhang, X. K., Egan, J. O., Huang, D. P., Sun, Z. L., Chien, V. K. Y. & Chiu, J. F. (1992). Hepatitis B virus DNA integration and expression of an erb-B-like gene in human hepatocellular carcinoma. *Biochemica et Biophysica Research Communications,* **188**, 344–51.

MECHANISMS OF HTLV LEUKAEMOGENESIS

S. A. STEWART, B. POON AND I. S. Y. CHEN,

Division of Hematology–Oncology, Departments of Medicine, and Microbiology & Immunology, UCLA School of Medicine and Jonsson Comprehensive Cancer Center, Los Angeles, CA 90024-1678, USA.

INTRODUCTION

The study of retroviruses and their potential to cause disease in man has been an area of great interest since the discovery of the first oncogenic retrovirus in chickens, and subsequently, in most other animals. The initial identification of human T-cell leukaemia virus types I (HTLV-I) and II (HTLV-II) in patients diagnosed with cutaneous T-cell lymphoma (Poiesz *et al.*, 1980) and hairy-cell leukaemia (Saxon, Stevens & Golde, 1978), respectively, was the start of the ongoing research into the mechanisms of pathogenesis of these human retroviruses. HTLV-I has been subsequently established by seroepidemiology and molecular analysis as the aetiological agent of adult T-cell leukaemia (ATL), which is characterized by a malignant proliferation of CD4+ T-lymphocytes. Clinical features of the disease include hypercalcaemia, lymphoadenopathy, skin lesions due to leukaemic cell infiltration, involvement of the spleen or liver and immunodeficiency (Kondo *et al.*, 1987; Murphy *et al.*, 1989). Prognosis for patients with acute ATL is poor, with an average survival, after diagnosis, of months. HTLV-I infection is endemic to regions of southern Japan, the Caribbean basin, central Africa, northeastern South America, as well as the southeastern United States (Biggar *et al.*, 1985; Blattner *et al.*, 1982; Blayney *et al.*, 1983; Bunn *et al.*, 1983; Catovsky *et al.*, 1982; Merino *et al.*, 1984; Saxinger *et al.*, 1984; Su *et al.*, 1985). Recent reports have indicated a high rate of HTLV seropositivity among intravenous drug abusers (IVDA) in the United States and Europe (Jason *et al.*, 1985; Lee *et al.*, 1989; Robert-Guroff *et al.*, 1986; Sandler, 1986; Tedder *et al.*, 1984).

Only 1–4% of individuals infected with HTLV-I progress to development of ATL (Kondo *et al.*, 1987; Murphy *et al.*, 1989), and among this small minority, disease occurs after a latency period of 20–30 years post-infection. The majority of asymptomatic carriers characteristically have few infected lymphocytes, as HTLV proviral DNA is only detectable by the highly sensitive use of polymerase chain reaction (PCR). These infected cells exhibit a polyclonal pattern of integration, which is in marked contrast to pre-leukaemic and acute ATL patients, where the tumour cells are mono-clonal or oligoclonal with respect to the pattern of HTLV integration (Seiki *et al.*, 1984; Yoshida, Miyoshi & Hinuma, 1982), suggesting an outgrowth of individual T-cell clones. One characteristic feature of leukaemic cells is the lack of detectable expression of viral genes within these cells, suggesting that viral products may not be needed for maintenance of the tumourigenic phenotype. The block to viral expression is not due to an intrinsic defect in the viral genome, as culturing of fresh ATL cells leads to expression and detection of viral mRNA and proteins (Seiki, Hattori & Yoshida, 1982; Yoshida *et al.*, 1982). Twenty to 30% of ATL tumours possess defective viral genomes, so the lack of expression is due to both genetic and epigenetic mechanisms. The reasons for the lack of viral expression are currently unexplained, although viral latency may serve as an escape from immune surveillance.

HTLV-I is also associated with a neurological disorder termed HTLV-I-associated myelopathy (HAM) or tropical spastic paraparesis (TSP) (Bha-gavati *et al.*, 1988; Gessain *et al.*, 1985; Jacobson *et al.*, 1988*b*; Osame *et al.*, 1987*a*), a chronic demyelinating disease characterized by weakness and spasticity of the extremities, hyperflexia and mild peripheral loss (Osame *et al.*, 1987*b*; Vernant *et al.*, 1987). HAM/TSP has been described in all areas of the world known to be endemic for HTLV-I. The presence of reactive HTLV-I antibodies as well as molecular identification of HTLV-I sequences in HAM/TSP patients have further correlated HTLV-I infection with this disorder (Kwok *et al.*, 1988; Nishimura *et al.*, 1988; Osame *et al.*, 1987*b*). There is some debate as to whether the HTLV-I virus responsible for HAM/TSP is identical to the virus that causes ATL. Some biological differences were noted between an ATL viral isolate and virus produced by T-cell lines derived from peripheral blood and cerebrospinal fluid (CSF) of TSP patients (Jacobson *et al.*, 1988*a*). One group found, using restriction enzyme mapping, a HTLV-I clone from the CSF of a HAM/TSP patient to be related but distinct from HTLV-I isolated from ATL patients (Sarin *et al.*, 1989). Another group, also using restriction enzyme mapping analysis, reported that T-cell lines established from the CSF of HAM/TSP patients produced viruses that were indistinguishable from ATL viruses (Nishimura *et al.*, 1988).

The pathogenesis of HAM/TSP appears to be distinct from ATL. Disease can develop within a few years of infection, and afflicted individuals are usually infected in adulthood. Infected cells are polyclonal, as determined by the integration pattern within these cells (Greenberg *et al.*, 1989*b*; Yoshida *et al.*, 1987). There have been reports linking development of HAM with certain HLA haplotypes (Elovaara *et al.*, 1993; Ijichi *et al.*, 1989; Kannagi *et al.*, 1991), as well as detection of higher levels of both HTLV proviral DNA and percentage of infected cells in HAM patients as compared with asymptomatic carriers (Gessain *et al.*, 1990; Kira *et al.*, 1991*b*; Yoshida *et al.*, 1989).

Conflicting evidence has been reported regarding a possible link between HTLV-I and multiple sclerosis (MS), a neurological disorder that has some similarities to HAM/TSP. Some groups have detected HTLV sequences in MS patients using PCR (Greenberg *et al.*, 1989*a*, Reddy *et al.*, 1989), while other groups, using the same technique, have failed to amplify any HTLV-related sequences (Bangham *et al.*, 1983; Chen *et al.*, 1990; Merelli *et al.*, 1993; Richardson *et al.*, 1989).

HTLV-II is highly related to HTLV-I, but its association with disease is more tenuous. Only two isolates have been associated with an atypical form of hairy-cell leukaemia involving T-cells (Rosenblatt *et al.*, 1986; Saxon *et al.*, 1978). In one of these patients, HTLV-II was shown to be integrated in a monoclonal population of cells, and these leukemic cells did not express viral RNA (Rosenblatt *et al.*, 1988), a phenotype consistent with the properties of HTLV-I-infected ATL patients. If HTLV-II has characteristics of leukaemogenesis similar to HTLV-I, such as low percentage of infected individuals progressing to disease and a long latency period, it may take years before sufficient information from cases accumulates to establish HTLV-II with a particular disease. Since there is a high degree of antibody cross-reactivity between HTLV-I and -II, PCR was used to identify a significant ratio of HTLV-II infection in populations of HTLV-infected IVDA in New Orleans (Lee *et al.*, 1989), San Francisco (Feigal *et al.*, 1991), Italy (Zella *et al.*, 1990) and Argentina (Bouzas *et al.*, 1991). A population of native Americans in New Mexico is endemic for HTLV-II, and among this population, two patients have been reported to exhibit a chronic neuro-degenerative syndrome similar to HAM (Hjelle *et al.*, 1990, 1992). Some groups have reported incidences of HAM/TSP in HTLV-II-infected individuals (Hjelle *et al.*, 1992; Murphy *et al.*, 1993), as well as detection of HTLV-II sequences and viruses from HAM/TSP patients (Jacobson *et al.*, 1993; Kira *et al.*, 1991*a*), while other groups have not been able to correlate the presence of HTLV-II with onset of any neurological disorder (Kiyokawa *et al.*, 1991; Miyamoto *et al.*, 1992). There is evidence for a high degree of co-infection with HIV and HTLV-II among IVDA (Ehrlich *et al.*, 1989; Khabbaz *et al.*, 1991), and it is of interest to determine whether the presence of HTLV affects HIV disease progression.

Due to the lack of viral gene expression in HTLV-infected patients and the long latency period before onset of disease, a great deal of knowledge about HTLV transformation has necessarily been obtained from *in vitro* studies. *In vitro* infection of cells by HTLV is generally accomplished by co-cultivating the cells to be infected with irradiated virus-infected cells, as infection by cell-free virus is highly inefficient. Both HTLV-I and HTLV-II have the ability to transform normal human T-lymphocytes *in vitro* (Chen, Quan & Golde, 1983; Miyoshi *et al.*, 1983*a*; Popovic *et al.*, 1983). Infection

Table 1. *Comparison of in vivo vs. in vitro HTLV transformed T-cells*

Adult T-cell leukaemia	*In vitro* transformed T-cells
Generally CD4+ and IL-2R positive	Generally CD4+ and IL-2R positive
No common integration site	No common integration site
Clonal	Not clonal
Little or no expression of provirus	Provirus is always expressed
Difficult to propagate in culture	Cultures in absence of IL-2

of peripheral blood or cord blood T-cells leads to the emergence of CD4+-transformed cells which are initially polyclonal in phenotype, and eventually, upon continuous passage, develop into an oligoclonal population with respect to integration sites. T-lymphocytes from other animals such as monkeys (Miyoshi *et al.*, 1982), rabbits (Miyoshi *et al.*, 1983*b*), cats (Hoshino *et al.*, 1984) and rats (Tateno *et al.*, 1984) can also be immortalized by HTLV-I *in vitro*. It should be noted that cells transformed *in vitro* differ in several important aspects from cells isolated from ATL patients (Table 1). Culturing of ATL cells *in vitro* results in the outgrowth of a clonal population which is usually distinct from the clone that predominates *in vivo*. This is likely the result of *in vitro* culturing selection pressures, which almost certainly are different from the selection conditions occurring *in vivo* to generate the leukaemic clone. Cells transformed *in vitro* are capable of production of replication-competent virus and expression of viral messages and proteins. This contrasts with the lack of HTLV gene expression in ATL tumour cells (Seiki *et al.*, 1984; Yoshida *et al.*, 1982).

The ability of HTLV-I and -II to immortalize T-cells *in vitro* has led to an in-depth investigation of their mechanism of transformation to better understand their *in vivo* leukaemogenicity. Other animal retroviruses have been shown to induce tumours via insertional mutagenesis, where integration occurs near vital cellular genes and disrupts their normal function (Weiss *et al.*, 1984). Analysis of HTLV-I-induced lymphomas do not indicate a common or specific site of integration (Seiki *et al.*, 1984), nor does the HTLV genome contain genes that resemble a typical cellular oncogene (Seiki *et al.*, 1982; Shimotohno *et al.*, 1985). The role of tumour suppressor genes in HTLV disease progression is still under investigation, as recent reports have detected disruption of the p53 gene in ATL tumours (Nagai *et al.*, 1991; Sakashita *et al.*, 1992; Sugito *et al.*, 1991; Yamato *et al.*, 1993). However, the association between p53 and HTLV-I-induced leukaemogenesis is rather inconsistent, as p53 abnormalities are evidenced in only 1–5% of ATL tumours. In addition to the three essential viral genes present in all retroviruses, *gag, pol* and *env*, the HTLV genome encodes additional genes

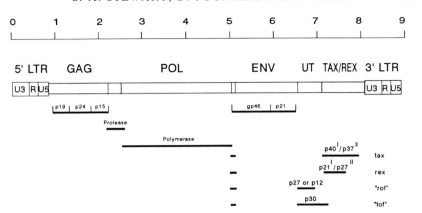

Fig. 1. Genomic organization of HTLV. The structure and organization of HTLV provirus, including the long terminal repeats (LTR) and genes encoding the known proteins, are shown at the top of the figure, drawn to scale. Enumeration above the genome depicts length in kilobases. Sizes and positions of the individual proteins encoded by the provirus are shown below the genome. Sizes of HTLV-I and HTLV-II Tax and Rex protein products are indicated by superscript 'I' and 'II', respectively. Reported sizes of the potentially encoded proteins, Rof and Tof (Ciminale et al., 1992; Koralnik et al., 1993), are also indicated.

that have the ability to modulate the viral life-cycle and possibly affect normal cellular function (Fig. 1). The expression of these viral products may contribute to the leukaemogenic and transforming potential of HTLV.

TRANSFORMATION

The tax gene encodes a 40 kilodalton (kDa) and 37 kDa protein in HTLV-I and -II, respectively (Miwa et al., 1984). Tax is a nuclear phosphoprotein which specifically trans-activates the viral long terminal repeat (LTR), as well as a variety of cellular genes in the absence of direct DNA binding, as determined by DNA binding assays (Cann et al., 1985; Felber et al., 1985; Fujisawa et al., 1985; Lee et al., 1984; Slamon et al., 1985; Sodroski et al., 1985). The rex gene encodes a protein of 21 kDa and 27 kDa in HTLV-I and HTLV-II, respectively. Like Tax, Rex is a phosphoprotein which is localized to the nucleus. Rex functions post-transcriptionally by increasing the ratio of unspliced or singly spliced to doubly spliced viral mRNAs. Rex, therefore, specifically promotes the appearance of viral structural proteins while inhibiting its own message (Hidaka et al., 1988; Inoue et al., 1986a,b, 1987). In addition, Rex has been implicated in stabilizing certain cellular mRNAs (White et al., 1991). Rex will be discussed in more detail later in this review.

Tax specifically trans-activates the HTLV-I LTR, which contains a variety of transcriptional regulatory sites, including three imperfectly conserved 21-base pair (bp) repeats located in the U3 region. Deletional studies of the LTR indicate that at least two of these repeats are required for tax *trans*-activation; however, direct Tax binding to the LTR has not been observed (Brady *et al.*, 1987; Kitada *et al.*, 1987; Shimotohno *et al.*, 1986). It is therefore thought that Tax activates the viral LTR through protein–protein interactions, modification of pre-existing cellular proteins, or induction of new cellular proteins. To determine which cellular transcription factors might be involved in binding the LTR and driving viral transcription, many investigators began precise dissection of the LTR. The 21 bp repeats within the LTR have been subdivided into three motifs, referred to as A, B and C, based on their binding properties. Mutagenesis of the central B motif results in elimination of *tax* activation, while mutagenesis of either the A or C motif alone results in a marked decrease in *tax* activation (Fujisawa *et al.*, 1989; Montagne *et al.*, 1990; Suzuki *et al.*, 1993). Fujisawa *et al.* (1989) identified a 12 bp element within the B motif, which shares a consensus binding sequence with the activating transcription factor/CRE-binding protein (ATF/CREB) family of transcription factors. This cAMP response element (CRE) is specifically bound by various members of the ATF/CREB family. Binding to this 12-bp element alone is not, however, sufficient for LTR *trans*-activation.

Muchardt *et al.* (1992*a*,*b*) recently showed that AP-2 is able to bind the A motif, excluding binding of the LTR by other members of the ATF/CREB family. AP-2 binding specifically activates expression of the viral LTR, as assayed by expression of a reporter gene in transfection assays. Interestingly, AP-2 and Tax appear to function antagonistically with regards to the viral LTR. Gel retardation assays demonstrated that Tax specifically inhibited the ability of AP-2 to bind the 21-bp repeats (Muchardt *et al.*, 1992*a*,*b*). These observations raise the possibility of a Tax-independent as well as a Tax-dependent pathway of viral *trans*-activation.

In addition to ATF/CREB binding, HTLV enhancer binding protein types 1 (HEB 1) and 2 (HEB 2) have also been shown to specifically bind the LTR through the A and B or C and B motifs, respectively. Montagne *et al.* (1990) reported that HEB 2 strongly bound the LTR in the presence of Tax, while Tax seemed to exhibit little influence on the weak binding of the ATF/CREB members. These data suggest that some cellular factors are active regardless of Tax presence, while others require Tax. Tax may therefore function by altering the DNA binding specificity of particular transcription factors. Collectively, these studies indicate that the LTR is capable of undergoing both positive and negative transcriptional regulation. This regulation may be dependent on both signals delivered to the cell, resulting in the activation of different transcription factors, as well as the intracellular presence of Tax. The exact role of tax in the transcriptional regulation of the viral LTR remains to be determined.

In addition to the binding of two 21 bp repeats, Bosselut *et al.* (1990) identified a site in the viral LTR located between nucleotides (nt) -155 and

-117, which is required for viral transcription (Giltin *et al.*, 1991). Members of the Ets family of transcription factors, which are lymphocyte-specific, have been shown to bind this element in a sequence-specific manner.

Many investigators now favour a multi-hit theory for *in vitro* transformation, in which Tax supplies some of the early hits. Discovering how Tax specifically affects cellular gene transcription may shed light on some of the early events of transformation initiated by Tax. Because transformed cells have an increased expression of interleukin 2 receptor (IL-2R) on their surfaces, a highly favoured hypothesis of early events required for T-cell transformation was the existence of an interleukin 2 (IL-2) autocrine loop. Many investigators suggested that an IL-2 autocrine loop was responsible for early T-cell deregulation and was a necessary step in the evolution of the transformed phenotype. Akagi & Shimotohno (1993) have, however, recently reported that stimulation of Tax-transduced T-cells can occur in an IL-2-independent manner, suggesting IL-2 may not be specifically required for transformation. Primary human T-cells transfected with Tax exhibited markedly higher proliferation in response to anti-CD3 when compared to control cells, while both responded similarly to exogenous IL-2. Furthermore, the Tax-transfected cells continued to grow after anti-CD3 stimulation in the absence of IL-2. These data suggest that stimulation through the T-cell receptor (TCR) may supply signals paramount in T-cell transformation which are not supplied by the IL-2 signal pathway. Existence of an IL-2 autocrine loop may play an essential role in T-cell transformation. However, it now appears that the IL-2 autocrine loop may also simply be a result of the transformation process rather than an active component.

The majority of lymphocytes in the circulation are normally quiescent and require specific mitogenic stimulation to begin dividing. *In vivo*, these cells can become activated and subsequently are able to function as effector cells, and then return to a quiescent state. Activated lymphocytes survive for only a short time; HTLV-I leukaemic cells, however, are able to proliferate and remain activated indefinitely *in vivo*. These observations led to the hypothesis that HTLV-I is able to activate lymphocytes and maintain this state. The discovery that Tax could *trans*-activate the viral LTR raised the possibility that it could also influence cellular gene transcription. Early studies indicated that HTLV-I-infected cells showed unusually high levels of cellular gene products, such as the IL-2 receptor alpha chain (IL-2Rα), which is only transiently expressed on activated T-cells in the periphery (Cross *et al.*, 1987; Depper *et al.*, 1984). Investigators went on to determine that Tax was indeed able to *trans*-activate cellular genes. Subsequently, a wide range of cellular genes have been shown to be *trans*-activated by Tax. Some examples of these are the IL-2Rα, IL-2, granulocyte-macrophage colony-stimulating factor (GM-CSF), and c-*fos* (Depper, 1984; Fujii, Sassone-Corsi & Verma, 1988; Hattori *et al.*, 1991; Inoue *et al.*, 1986*a,b*; Miyatake *et al.*, 1988*a,b*; Nimer *et al.*, 1989). Tax has also been shown to activate a variety of immediate-early response genes such as ERG-1, AP-1, Fra-1, c-*jun*, Jun B, and Jun D (Fujii *et al.*, 1991; Kelly *et al.*, 1992; Nagata *et al.*, 1989; Sakamoto *et al.*, 1992). In addition to the early response genes, Tax also *trans*-activates TGF-B1, c-*myc*, tumour necrosis factor alpha (TNF-α) and parathyroid hormone-related protein (Albrecht, Shakhov & Jongeneel, 1992; Duyao *et al.*, 1992; Ejima *et al.*, 1993; Kim *et al.*, 1990; Lal *et al.*, 1993; Miyatake *et al.*, 1988*a,b*; Wano *et al.*, 1988). The ability of Tax to *trans*-activate such a large array of genes directly involved in both transcriptional control and cell growth and differentiation has led to the hypothesis that Tax is intimately involved in T-cell transformation. It should be pointed out that ATL cells express little or no viral proteins, including Tax. This observation suggests that, while the role of tax in leukaemogenesis seems vital, its effects are likely to occur at an early stage. In addition, maintenance of the tumourigenic phenotype does not appear to require Tax or any other viral protein.

Many researchers have examined the mechanism of cellular gene activation by HTLV-I Tax. Depper *et al.* (1984) showed that HTLV-I-transformed T-cells constitutively express IL-2Rα, suggesting HTLV-I somehow perturbs normal cellular regulation. These initial observations suggest that understanding the mechanism of IL-2Rα deregulation by HTLV-I would help explain how HTLV-I transforms T-cells. Nuclear run-on experiments have shown that IL-2Rα expression normally occurs only after antigenic stimulation via the TCR (Leonard *et al.*, 1985; Maruyama *et al.*, 1987). Transfection assays have suggested that Tax is able to induce constitutive expression of this gene in a mitogen stimulation-like pathway. Transfection of constructs containing the 5' flanking region of the IL-2Rα gene has shown that Tax is able to activate gene expression specifically. Induction of IL-2Rα expression via Tax utilises the same regulatory region necessary for normal mitogen stimulation, again suggesting that Tax interferes with normal T-cell signal pathways (Bohnlein *et al.*, 1988; Lowenthal *et al.*, 1988).

Further analysis of the 5' IL-2Rα promotor has shed some light on how Tax specifically affects transcription. Using deletional analysis, Ballard *et al.* (1988) located a 12-bp element required for Tax *trans*-activation. This 12-bp element shared homology to the NF-κB binding site and was sufficient to confer Tax inducibility on a heterologous promoter (Ruben *et al.*, 1988). Not surprisingly, binding of the kappa enhancer is normally induced by mitogenic stimulation with agents such as phorbol-12-myristate-13-acetate (PMA), further supporting the hypothesis that Tax is able to induce the kappa enhancer by mimicking T-cell activation.

NF-κB is a cellular transcription factor which consists of a p50 and p65 subunit (Kieran *et al.*, 1990). Hirai *et al.* (1992) showed that Tax specifically binds p105, the p50 precursor. Using specific antibodies to both Tax and p105, Hirai *et al.*, were able to precipitate a Tax-p105 complex from both nuclear and cytoplasmic fractions. Because p105 masks its nuclear localization signal (NLS) until cleaved to p50, it is normally found only in the cytoplasm, suggesting Tax may bind p105 and carry it into the nucleus.

Recently, Crenon *et al.* (1993) reported that the c-rel proto-oncogene may also be involved in binding the IL-2Rα kappa enhancer in HTLV-I-infected cells. Because Tax does not directly bind DNA, and therefore, the kappa enhancer, kappa activation via Tax is likely to be caused by the alteration of normal cellular protein binding. The p50–p65 heterodimer, also known as NF-κB, and Rel have recently been proposed to make up a family of transcriptional activators that are normally present in the cell cytoplasm prior to T-cell activation. Crenon *et al.* (1993) demonstrated that, while Tax was able to activate the κB motif in HeLa and Jurkat cells, it had no detectable activity in undifferentiated embryocarcinoma F9 cells. Following induced differentiation, however, F9 cells become responsive to Tax, suggesting that a specific factor(s) found in more mature cells is required for Tax responsiveness. Interestingly, only Rel homodimers and rel-p65 heterodimers were able to activate the κB motif. In DNA affinity precipitation assays, Tax induced the binding of complexes containing Rel, while exhibiting no effect on complexes in which Rel was absent. These experiments suggest that Rel is a functionally active factor capable of gene activation in the presence of Tax in differentiated cells.

Both Crenon's and Hirai's experiments suggest possible mechanisms for the activation of the IL-2Rα kappa enhancer in HTLV-I-infected cells. These experiments suggest that Tax binds the p50 precursor, p105. Binding either sequesters p105 from p65, and thereby inhibits the formation of NF-κB, or facilitates translocation of p105 into the nucleus, where it may bind DNA either alone or as a homodimer or heterodimer. This would leave p65 and Rel free to dimerize, and perhaps further enhance cell activation by binding the kappa enhancer. Alternatively, facilitation by Tax of the movement of p105 to the nucleus may functionally exclude Rel, allowing more p50–p65 heterodimers to form. Binding of Tax to p105 may also enhance cleavage, thereby supplying more p50 to the transcription complex or may allow p105 to directly bind DNA. This model does not address the role of I-κB in transcriptional regulation. I-κB is known to bind both p65 and Rel, effectively sequestering them in the cytoplasm by masking their NLS. Because p50 homodimers alone cannot activate the kappa enhancer, the question that arises is whether Tax can affect the release of p65 or Rel from I-κB, allowing them to enter the nucleus and dimerize with each other or p50. Tax seems to be able to somehow upset the normal complement and/or relative levels of transcription factors known to be involved in lymphocyte activation, and this may be involved in the early events required for T-cell leukaemogenesis.

Similar to what was found with the IL-2Rα gene, Tax is also able to specifically activate GM-CSF, resulting in constitutive expression in HTLV-transformed cells. GM-CSF, like IL-2Rα, is normally expressed by T-cells in response to antigen, mitogen or lectin stimulation. Again, Tax activation utilizes DNA elements important in T-cell activation via the mitogen stimulation pathway. Using transient transfection assays, Miyatake *et al.* (1988*a*,*b*) showed that the 5′ flanking region of GM-CSF was able to drive expression of a heterologous gene when co-transfected with Tax. Deletional analysis indicated that two copies of the repeated sequence, CATTA(A/T), located between nt -48 and -36, were necessary to confer Tax responsiveness (Nimer, 1991). Interestingly, Tax is unable to induce GM-CSF in non-T-cells, suggesting that factors specific to differentiated T-cells are required for Tax function (Nimer *et al.*, 1989). One constant that has emerged from these studies is the repeated utilization of factors normally involved in T-cell activation via the mitogen pathway. Apparently Tax has evolved the ability to simulate T-cell activation by directly affecting cellular gene transcription in T-cells.

In addition to inducing cytokines and various cytokine receptors, Tax has also been shown to alter the regulation of cellular transcription factors. Collectively, these factors are often referred to as immediately early serum response genes, which are intimately involved in cell growth and differentiation. c-*fos* is a proto-oncogene that is transiently expressed after stimulation with lectins, suggesting involvement in T-cell activation and proliferation (Persson *et al.*, 1984; Reed *et al.*, 1986). Fujii *et al.* (1988) showed that Tax induces c-*fos* expression in transient transfection assays. Endogenous c-*fos* expression was also increased in these assays, and was also observed in HTLV-I-infected cells. Normally, c-*fos* promoter activity is serum-dependent, but transfection of a *tax* construct renders the promoter serum-independent. Deletional analysis of the c-*fos* promoter indicated that elements located between nt -362 and 276 were required for *tax* activation. Within this fragment are at least two elements which seem to be required for *tax* trans-activation. These elements consist of a putative cAMP responsive element, similar to that found in the HTLV-I LTR 21-bp repeat, and an element responsive to v-*sis*-conditioned medium (Fujii *et al.*, 1988). Tax may therefore interact with one or more of these elements through cellular proteins, resulting in c-*fos* expression. c-*fos* expression may then activate AP-1 or NF-κB, resulting in the activation of IL-2, the IL-2Rα, and potentially, a variety of other cellular proteins involved in T-cell activation and proliferation (Sassone-Corsi *et al.*, 1988).

In addition to positively up-regulating cellular genes, Tax is also able to negatively affect gene expression. The human β-polymerase gene (huβ-pol) is a cellular enzyme involved in repair of host cellular DNA. Jeang *et al.* (1990) reported down-regulation of the expression of huβ-pol promoter-driven constructs in the presence of Tax, and also observed a decrease in huβ-pol mRNA synthesis in HTLV-I-transformed T-lymphocytes. This could be a possible mechanism to increase the mutation frequency in HTLV-infected cells, and potentially be important in the progression to ATL.

A number of experiments have shown that, although Tax plays an important part in transformation, it alone is necessary but not sufficient. Grassmann *et al.* (1989) have shown that T-cells stably transfected with *tax* constructs alone do not attain IL-2 independence, the hallmark of HTLV-transformed T-cells *in vitro*. Furthermore, while these cells display some traits of a transformed phenotype in regards to proliferation and alteration in surface markers, they lack other important characteristics. Data derived from a transgenic mouse model also suggest that Tax alone is insufficient for leukaemogenesis. Green *et al.* (1989) successfully bred transgenic mice expressing detectable levels of Tax. While these mice were shown to develop neurofibromatosis, none developed the characteristic lymphomas of HTLV-I-infected patients. These studies have strengthened the argument that Tax may be involved in early events of leukaemogenesis, but it alone is insufficient.

HTLV-I virions specifically activate resting T-cells through a mitogenic-like pathway (Gazzolo & Duc Dodon, 1987; Zack *et al.*, 1988). Patient sera directed against HTLV-I antigens and monoclonal antibodies to gp46 can specifically block the proliferative effects of virions. Furthermore, UV-irradiated virions are also able to cause T-cell proliferation, suggesting that this effect is independent of T-cell infection. These data suggest that HTLV-I virions or a component thereof are able to specifically bind a cell surface protein, resulting in T-cell proliferation. In *in vivo* infected cells, this stimulation may work in parallel with the intracellular effects of Tax, and perhaps Rex, ultimately resulting in leukaemogenesis.

Kimata, Palker, Ratner (1993) recently proposed that HTLV-I-induced T-cell proliferation may be mediated through the LFA-3/CD2 pathway. These observations suggest a model in which the HTLV-I virion or some component of it binds the T-cell, possibly through CD2, concurrently with the binding of CD3, resulting in mitogenic stimulation and T-cell proliferation. The stimulated T-cell may then be further activated by Tax, possibly resulting in a transformed phenotype.

Novel HTLV-I transcripts have recently been described, and these may somehow be involved in T-cell transformation. The *tax/rex* region, previously referred to as the X region, is located 3′ of the *env* gene, and 5′ to the 3′ LTR. This region contains four open reading frames, two of which encode Tax and Rex (Fig. 1). Ciminale *et al.* (1992) recently described two novel mRNAs encoded with this same region. These mRNAs encode proteins in transfection assays; however, proteins have not been described in HTLV-I-infected cells. It should therefore be noted that information concerning the putative localization and functions of these new proteins is based solely on *in vitro* transfection assays and awaits confirmation in HTLV-I-infected cells. The first protein, Tof, consists of 241 amino acids and encodes a protein of 30 kDa, which is specifically targeted to the nucleolus (Ciminale *et al.*, 1992). Rof, the second protein, is 152 amino acids and encodes a protein predicted to be 17 kDa but which migrates at 27 kDa on a denaturing gel when isolated from an *in vitro* translation assay (Ciminale *et al.*, 1992). Koralnik, Fullen & Franchini (1993) have, however, shown that *rof* encodes a 12 kDa protein when the cDNA is transfected into cells. These proteins are present at only 5% of the amounts of Env, therefore making detection difficult. Neither Tof nor Rof were able to *trans*-activate the HTLV-I LTR in transfection experiments. Furthermore, neither protein inhibited the functions of Tax or Rex. Tof contains a serine-rich domain similar to the activation domains of transcriptional activators such as oct-1, oct-2 and pit-1. Although Ciminale *et al.* (1992) were unable to demonstrate viral *trans*-activation by Tof, it is still possible that it may play a role in the activation of cellular genes in HTLV-I-infected cells.

Rex is the second HTLV-I regulatory gene, and is expressed from the same doubly spliced mRNA as Tax. Rex is a nuclear phosphoprotein involved in the post-transcriptional regulation of viral gene expression. Specifically, Rex is required for expression of the *gag, pol* and *env* genes. Rex is able to bind the unspliced and singly spliced viral RNAs through the Rex-responsive element (RxRe), resulting in translocation to the cytoplasm (Kalland *et al.*, 1991; Kim *et al.*, 1991). Rex therefore effectively reduces the amount of doubly spliced messages encoding the regulatory genes in favour of the unspliced and singly spliced mRNAs encoding the structural proteins necessary for particle formation. Precisely how Rex mediates RNA migration to the cytoplasm is not completely understood. It has been hypothesized that Rex binding either inhibits spliceosome function or activates a nuclear transport pathway (Bogerd & Greene, 1993; Chang & Sharp, 1989; Malim *et al.*, 1989).

In addition to promoting cell compartmental distribution, some investigators have proposed that Rex may also affect cellular mRNAs. Using pulse-chase experiments, White *et al.* (1991) reported that wild-type Rex was able to stabilise the IL-2Rα message. These data suggest that like Tax, Rex has the ability to alter the expression of cellular genes involved in T-cell activation and proliferation.

Viral gene expression in ATL cells is extremely low. Southern blot analysis indicates that in many but not all cases, large deletions occur in the proviral DNA (Hall *et al.*, 1991). A common feature in these deletions is the retention of the *tax/rex* region encoding Tax, Rex and the two additional open reading frames (Korber *et al.*, 1991). The retention and yet apparent lack of expression of the *tax/rex* region in ATL cells strongly suggests that the effects of this region occur early in tumourigenesis. Whether the expression of Tax and/or Rex is required at undetectable levels to maintain the leukaemic state remains to be determined.

A variety of viral components have been discussed in this review, and these can be incorporated into a model for HTLV-I T-cell leukaemogenesis (Fig. 2). In this model, HTLV-I virions encounter quiescent T-cells, result-

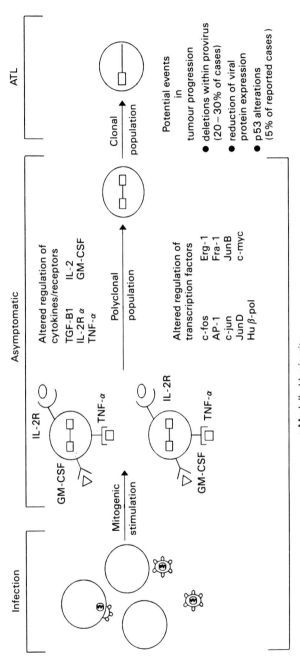

Fig. 2. Model for HTLV-I T-cell tumourigenesis. Initial T-cell activation may be induced by virion T-cell interactions. Activation of the T-cell would then allow the HTLV-I genome to complete reverse transcription, resulting in a productive infection. Expression of both Tax and Rex would then interact with intracellular signals produced by the activated T-cell. Both T-cell activation and the effects of Tax and Rex can alter the production of cellular proteins, as well as the regulation of an array of cellular transcription factors. Infected T-cells that proceed to a leukaemic state may then come under selective pressures by the host immune system and result in deletion or suppression of immunoreactive viral genes. Successful evasion of the host immune system results in outgrowth of a malignant T-cell clone(s) which becomes the predominant tumour clone *in vivo*.

ing in an initial mitogenic-like stimulation. Upon stimulation, the virion is able to enter the cell and successfully reverse transcribe its RNA genome. The proviral DNA then integrates into the host genome, and a small amount of the *tax/rex* message is transcribed. Tax *trans*-activates the LTR and initially increases its own production which, in turn, further up-regulates the viral LTR. At this point, Rex acts by allowing the translocation of transcripts encoding viral structural proteins, resulting in virion production. Release of virions may then further induce T-cell stimulation. Meanwhile, Tax *trans*-activates a vast number of cellular genes, some of which Rex has the potential to stabilize. In addition, the putative Rof and Tof proteins, with as of yet undescribed functions, may further perturb normal T-cell activation and proliferation. This may be sufficient to induce an altered phenotype, most notably IL-2 independence. At this stage, deletion of the potentially immunogenic viral proteins such as Env and Gag may be necessary to avoid the host immune system. The regulatory genes, *tax* and *rex*, are down-regulated at this point. Because both the *tax* and *rex* genes are retained yet appear to be silent in many of the proviruses tested to date, it raises the question as to their potential function after tumourigenesis. This model is illustrated in Fig. 2.

ANIMAL MODELS

A great deal has been learned from *in vitro* studies of HTLV-I; however, there are also important differences from that observed in human disease. It is still not known why leukaemogenic cells isolated from HTLV-I-infected patients are clonal and lack viral gene expression, while *in vitro*-transformed cells do not manifest either of these characteristics. *In vitro* studies have suggested a number of events which may be involved in leukaemogenesis, but the differences observed between *in vitro*-transformed ATL cells under-lines the importance of the development of a suitable animal model. The long latency in infected humans, as well as the low percentage of those infected (only about 1–4%) progress to disease, make the study of events critical in disease difficult (Kondo *et al.*, 1987). In an attempt to develop an animal model for HTLV, some groups have begun looking at bovine leukaemia virus (BLV). BLV is both structurally and biologically similar to HTLV, and many steps in the progression to disease are similar. Leukaemic cells from BLV-infected animals, like those from ATL patients, are clonal and express no viral proteins (Kettman *et al.*, 1980). BLV, in contrast to HTLV-I, forms tumours of B-cell origin, while HTLV-I affects T-cells. Although the BLV system closely resembles HTLV-I and therefore war-rants further investigation, it is still not an ideal model, due to the different target cell and the expense and difficulty of working with a large animal model. Therefore, some investigators have been interested in developing a small animal model for HTLV-I leukaemogenesis.

HTLV-I has been successfully transmitted to both rats and rabbits (Tateno *et al.*, 1984; Uemura *et al.*, 1987). In the rat system, *in vitro*-transformed cells are injected into the animal. Following injection, most rats produce anti-HTLV-I antibodies, and the viral genome is detectable in many organs. Disease, however, has not been observed beyond the production of specific antibodies, suggesting that this model is unable to support the tumourigenic potential of HTLV-I. The recent development of the severe combined immunodeficient (SCID) mouse engrafted with human tissue (SCID-hu) may offer a unique system in which to study HTLV. SCID mice lack any functional T- or B-cells, and are thus unable to mount a secondary immune response or reject allogenic grafts. These mice are implanted with human fetal liver and thymus, which function as a rudimentary organ supplying mature T-cells to the periphery. Studies using this model have shown that it is able to support HIV replication, and to a limited degree, disease progression (Aldovani *et al.*, 1993; Namikawa *et al.*, 1988). Recently, Feuer *et al.* (1993) showed that HTLV-I may also be studied in the SCID mouse. Specifically, human lymphoblastic lymphomas were established in SCID mice injected with human peripheral blood lymphocytes (PBL) interperitoneally. These mice were injected with PBL from patients with ATL. The tumour cells recovered from the mouse resembled ATL cells, in that they were oligoclonal or monoclonal, and they expressed CD4 and CD25 on their surface. These findings validate the SCID model as a potential system for the study of HTLV-I tumourigenesis.

Since the discovery of HTLV and its association with human disease in the early 1980s, much has been learned about the structure and function of the viral genome. Researchers have defined a number of cellular genes *trans*-activated by Tax. Virions have been shown to stimulate resting T-cells, and common phenotypic changes in ATL cells have been observed. Unfortunately, the lack of an adequate animal model has hampered efforts to understand HTLV-I leukaemogenesis. Recent publications, however, suggest that an animal model may be forthcoming. Hopefully this model will reveal vital steps in HTLV leukaemogenesis not observed in *in vitro* studies. These steps may help us to better understand human disease and may help to develop both therapeutic and preventative approaches to the various diseases caused by HTLV-I and HTLV-II.

REFERENCES

Akagi, T. & Shimotohno, K. (1993). Proliferative response of Tax 1-transduced primary human T cells to anti-CD3 antibody stimulation by an interleukin-2-independent pathway. *Journal of Virology*, **67**(3), 1211–17.

Albrecht, H., Shakhov, A. N. & Jongeneel, C. V. (1992). *Trans*-activation of the tumour necrosis factor alpha promoter by the human T-cell leukemia virus type I Tax protein. *Journal of Virology*, **66**(10), 6191–3.

Aldovani, G. M., Feuer, G., Gao, L., Kristeva, M., Chen, I. S. Y. & Zack, J. A.

(1993). HIV-1 infection of the SCID-hu mouse: an animal model for virus pathogenesis. *Nature,* **363**, 732–6.

Ballard, D. W., Bohnlein, E., Lowenthal, J. W., Wano, Y., Franza, B. R. & Greene, W. C. (1988). HTLV-I tax induces cellular proteins that activate the κB element in the IL-2 receptor α gene. *Science,* **241**, 1652–5.

Bangham, C. R. M., Nightingale, S., Cruickshank, J. K. & Daenke, S. (1983). PCR analysis of DNA from multiple sclerosis patients for the presence of HTLV-I. *Science,* **246**, 821.

Bhagavati, S., Ehrlich, G., Kula, R. W. *et al.* (1988). Detection of human T-cell lymphoma/leukemia virus type I DNA and antigen in spinal fluid and blood of patients with chronic progressive myelopathy. *New England Journal of Medicine,* **318**, 1141–7.

Biggar, R. J., Johnson, B. K., Oster, C. *et al.* (1985). Regional variation in prevalence of antibody against human T-lymphotropic virus types I and III in Kenya, East Africa. *International Journal of Cancer,* **35**, 763–7.

Blattner, W. A., Kalyanaraman, V. S., Robert-Guroff, M. *et al.* (1982). The human type-C retrovirus, HTLV, in blacks from the Caribbean region, and relationship to adult T-cell leukemia/lymphoma. *International Journal of Cancer,* **30**, 257–64.

Blayney, D. W., Blattner, W. A., Robert-Guroff, M. *et al.* (1983). The human T-cell leukemia/lymphoma virus (HTLV) in southeastern United States. *Journal of the American Medical Association,* **250**, 1048–52.

Bogerd, H. & Greene, W. C. (1993). Dominant negative mutants of human T-cell leukemia virus type 1 Rex and human immunodeficiency virus type 1 Rev fail to multimerize *in vivo*. *Journal of Virology,* **67**(5), 2496–501.

Bohnlein, E., Lowenthal, J., Siekevitz, M., Ballard, D., Franza, B. R. & Greene, W. C. (1988). The same inducible nuclear proteins regulate mitogen activation of both the interleukin-2 receptor-alpha gene and type 1 HIV. *Cell,* **53**, 827–36.

Bosselut, R., Duvall, J. F., Gegonne, A. *et al.* (1990). The product of the c-ets-1 proto-oncogene and the related Ets2 protein act as transcriptional activators of the long terminal repeat of human T cell leukemia virus HTLV-I. *EMBO Journal,* **9**(10), 3137–44.

Bouzas, M. B., Muchinik, G., Zapiola, I. *et al.* (1991). Human T-cell lymphotropic virus type II infection in Argentina. *Journal of Infectious Diseases,* **164**, 1026–7.

Brady, J., Jeang, K. T., Durall, J. & Khoury, G. (1987). Identification of the p40x-responsive regulatory sequences within the human T-cell leukemia virus type I long terminal repeat. *Journal of Virology,* **61**, 2175–81.

Bunn, P. A., Jr., Schechter, G. P., Jaffe, E. *et al.* (1983). Clinical course of retrovirus-associated adult T-cell lymphoma in the United States. *New England Journal of Medicine,* **309**, 257–64.

Cann, A. J., Rosenblatt, J. D., Wachsman, W., Shah, N. P. & Chen, I. S. Y. (1985). Identification of the gene responsible for human T-cell leukemia virus transcriptional regulation. *Nature,* **318**, 571–4.

Catovsky, D., Rose, M., Goolden, A. W. G. *et al.* (1982). Adult T-cell lymphoma-leukaemia in blacks from the West Indies. *Lancet,* **i**, 639–47.

Chang, D. D. & Sharp, P. A. (1989). Regulation by HIV Rev depends upon recognition of RNA splice sites. *Cell,* **59**, 789–95.

Chen, I. S. Y., Haislip, A. M., Myers, L. W., Ellison, G. W. & Merrill, J. E. (1990). Failure to detect HTLV-related sequences in multiple sclerosis blood. *Archives of Neurology,* **47**, 1064–5.

Chen, I. S. Y., Quan, S. G. & Golde, D. W. (1983). Human T-cell leukemia virus type II transforms normal human lymphocytes. *Proceedings of the National Academy of Sciences, USA,* **80**, 7006–9.

Chen, I. S. Y., Slamon, D. J., Rosenblatt, J. D., Shah, N. P., Quan, S. G. & Wachsman, W. (1985). The x gene is essential for HTLV replication. *Science*, **229**, 54–8.

Ciminale, V., Pavlakis, G. N., Derse, D., Cunningham, C. P. & Felber, B. K. (1992). Complex splicing in the human T-cell leukemia virus (HTLV) family of retroviruses: novel mRNAs and proteins produced by HTLV type I. *Journal of Virology*, **66**(3), 1737–45.

Crenon, I., Beraud, C., Simard, P., Montagne, J., Veschambre, P. & Jalinot, P. (1993). The transcriptionally active factors mediating the effect of the HTLV-1 Tax transactivator on the IL-2Rα κB enhancer include the product of the c-*rel* proto-oncogene. *Oncogene*, **8**, 867–75.

Cross, S. L., Feinberg, M. B., Wolf, J. B., Holbrook, N. J., Wong-Staal, F., Leonard, W. J. (1987). Regulation of the human interleukin-2 receptor α chain promoter: activation of a nonfunctional promoter by the transactivator gene of HTLV-I. *Cell*, **49**, 47–56.

Depper, J. M., Leonard, W. J., Kronke, M., Waldmann, T. A. & Greene, W. C. (1984). Augmented T cell growth factor receptor expression in HTLV-I infected human leukemic T cells. *Journal of Immunology*, **133**, 1691–5.

Duyao, M. P., Kessler, D. J., Spicer, D. B. *et al*. (1992). Transactivation of the c-myc promoter by human T cell leukemia virus type 1 tax is mediated by NF-κB. *Journal of Biological Chemistry*, **267**(23), 16288–91.

Ehrlich, G. D., Glaser, J. B., LaVigne, K. *et al*. (1989). Prevalence of human T-cell leukemia/lymphoma virus (HTLV) type II infection among high risk individuals: type specific identification of HTLVs by polymerase chain reaction. *Blood*, **74**, 1658–64.

Ejima, E., Rosenblatt, J. D., Massari, M. *et al*. (1993). Cell type-specific transactivation of the parathyroid hormone-related protein gene promoter by the human T-cell leukemia virus type-I (HTLV-I) Tax and HTLV-II Tax proteins. *Blood*, **81**(4), 1017–24.

Elovaara, I., Koenig, S., Brewah, A. Y., Woods, R. M., Lehky, T. & Jacobson, S. (1993). High human T-cell lymphotropic virus type I (HTLV-I)-specific cursor cytotoxic T-lymphocyte frequencies in patients with HTLV-I associated neurological diseases. *Journal of Experimental Medicine*, **177**, 1567–73.

Feigal, E., Murphy, E., Vranizan, K. *et al*. (1991). Human T-cell lymphotropic virus types I and II in intravenous drug users in San Francisco: risk factors associated with seropositivity. *Journal of Infectious Diseases*, **164**, 36–42.

Felber, B. K., Paskalis, H., Kleinman-Ewing, C., Wong-Staal, F. & Pavlakis, G. N. (1985). The pX protein of HTLV-I is a transcriptional activator of its long terminal repeats. *Science*, **229**, 675–9.

Feuer, G., Zack, J. A., Harrington, W. J. *et al*. (1993). Establishment of human T-cell leukemia virus type I (HTLV-I) T-cell lymphomas in SCID mice. *Blood*, **82**, 722–31.

Fujii, M., Niki, T., Mori, T. *et al*. Seiki, M. (1991). HTLV-I Tax induces expression of various immediate early serum responsive genes. *Oncogene*, **6**, 1023–9.

Fujii, M., Sassone-Corsi, P. & Verma, I. M. (1988). c-*fos* promoter *trans*-activation by the tax-1 protein of human T-cell leukemia virus type 1. *Proceedings of the National Academy of Sciences, USA*, **85**, 8526–30.

Fujisawa, J., Seiki, M., Kiyokawa, T. & Yoshida, M. (1985). Functional activation of the long terminal repeat of human T-cell leukemia virus type I by a trans-acting factor. *Proceedings of the National Academy of Sciences, USA*, **82**, 2277–2281.

Fujisawa, J., Toita, M. & Yoshida, M. (1989). A unique enhancer element for the transactivator (p40tax) of human T-cell leukemia virus type I that is distinct from cyclic AMP- and 12-o-tetradecanoylphorbol-13-acetate-responsive elements. *Journal of Virology*, **63**(8), 3234–9.

Gazzolo, L. & Duc Dodon, M. (1987). Direct activation of resting T lymphocytes by human T-lymphotropic virus type I. *Nature*, **326**, 714–17.

Gessain, A., Barin, F., Vernant, J. C. *et al.* (1985). Antibodies to human T-lymphotropic virus type I in patients with tropical spastic paraparesis. *Lancet*, **ii**, 407–9.

Gessain, A., Soal, F., Gout, O. *et al.* (1990). High human T-cell lymphotropic virus type I proviral DNA load with polyclonal integration in peripheral blood mononuclear cells of French West Indian, Guianese, and African patients with tropical spastic paraparesis. *Blood*, **75**, 428–33.

Giltin, S. D., Bosselut, R., Gegonne, A., Ghysdael, J. & Brady, J. N. (1991). Sequence-specific interaction of the Ets 1 protein with the long terminal repeat of the human T-lymphotropic virus type I. *Journal of Virology*, **65**(10), 5513–23.

Grassmann, R., Dengler, C., Muller-Fleckenstein, I. *et al.* (1989). Transformation to continuous growth of primary human T lymphocytes by human T-cell leukemia virus type I X-region genes transduced by a *Herpesvirus saimiri* vector. *Proceedings of the National Academy of Sciences, USA*, **86**, 3351–55.

Green, J. E., Begley, G., Wagner, D. K., Waldmann, T. A. & Jay, G. (1989). Transactivation of granulocyte-macrophage colony-stimulating factor and T-lymphotropic virus type I tax gene. *Molecular and Cellular Biology*, **9**, 4731–7.

Greenberg, S. J., Ehrlich, G. D., Abbott, M. A., Hurwitz, B. J., Waldmann, T. A. & Poiesz, B. J. (1989*a*). Detection of sequences homologous to human retroviral DNA in multiple sclerosis by gene amplification. *Proceedings of the National Academy of Sciences, USA*, **86**, 2878–82.

Greenberg, S. J., Jacobson, S., Waldmann, T. A. & McFarlin, D. E. (1989*b*). Molecular analysis of HTLV-I retroviral integration and T-cell receptor arrangement indicates that T-cells in tropical spastic paraparesis are polyclonal. *Journal of Infectious Diseases*, **159**, 741–4.

Hall, W. W., Liu, C. R., Schneenwind, O., Takakashi, H., Kaplan, M. H., Roupe, G. & Vohlne, A. (1991). Deleted HTLV-I provirus in blood and cutaneous lesions of patients with mycosis fungoides. *Science*, **253**, 317–20.

Hattori, M., Uchiyama, T., Poibaba, T., Takatsuki, K. & Uchino, H. (1991). Surface phenotype of Japanese adult T-cell leukemia cells characterized by monoclonal antibodies. *Blood*, **58**, 645–7.

Hidaka, M., Inoue, J., Yosida, M. & Seiki, M. (1988). Post-transcriptional regulator (rex) of HTLV-I initiates expression of viral structural proteins but suppresses expression of regulatory proteins. *EMBO Journal*, **7**(2), 519–23.

Hirai, H., Fujisawa, J., Suzuki, T. *et al.* (1992). Transcriptional activator Tax of HTLV-I binds to the NF-κB precursor p105. *Oncogene*, **7**, 1737–42.

Hjelle, B., Appenzeller, O., Mills, R. *et al.* (1992). Chronic neurodegenerative disease associated with HTLV-II infection. *Lancet*, **339**, 645–6.

Hjelle, B., Scalf, R. & Swenson, S. (1990). High frequency of human T-cell leukemia/lymphoma virus type II infection in New Mexico blood donors: Determination by sequence specific oligonucleotide hybridization. *Blood*, **76**, 450–4.

Hoshino, H., Tanaka, H., Shimotohno, K. *et al.* (1984). Immortalization of peripheral blood lymphocytes of cats by human T-cell leukaemia virus. *International Journal of Cancer*, **34**, 513–17.

Ijichi, S., Eiraku, N., Osame, M. *et al.* (1989). Activated T-lymphocytes in cerebrospinal fluid of patients with HTLV-I associated myelopathy (HAM/TSP). *Journal of Neuroimmunology*, **25**, 251–4.

Inoue, J., Seiki, M., Taniguchi, T., Tsuru, S. & Yoshida, M. (1986*a*). Induction of interleukin-2 receptor gene by p40xI encoded by human T-cell leukemia virus type I. *EMBO Journal*, **5**, 2883–8.

Inoue, J., Seiki, M. & Yoshida, M. (1986*b*). The second pX product p27xIII of HTLV-I is required for *gag* gene expression. *FEBS Letters*, **209**, 187–90.

Inoue, J., Yoshida, M. & Seiki, M. (1987). Transcriptional (p40x) and post-transcriptional (p27xIII) regulators are required for the expression and replication of human T-cell leukemia virus type I genes. *Proceedings of the National Academy of Sciences, USA*, **84**, 3653–7.

Jacobson, S., Lehky, T., Nishimura, M., Robinson, S., McFarlin, D. E. & Dhibjal-but, S. (1993). Isolation of HTLV-II from a patient with chronic, progressive neurological disease clinically indistinguishable from HTLV-I associated myelo-pathy tropical spastic paraparesis. *Annals of Neurology*, **33**, 392–6.

Jacobson, S., Raine, C. S., Mingioli, E. S. & McFarlin, D. E. (1988*b*). Isolation of an HTLV-1-like retrovirus from patients with tropical spastic paraparesis. *Nature*, **331**, 540–3.

Jacobson, S., Zaninovic, V., Mora, O. *et al.* (1988*a*). Immunological findings in neurological diseases associated with antibodies to HTLV-I: activated lympho-cytes in tropical spastic paraparesis. *Annals of Neurology*, **23**, 196–200.

Jason, J. M., McDougal, J. S., Cabradilla, C., Kalyanaraman, V. S. & Evatt, B. L. (1985). Human T-cell leukemia virus (HTLV-I) p24 antibody in New York City blood product recipients. *American Journal of Hematology*, **20**, 129–37.

Jeang, K.-T., Widen, S. G., Semmes, O. J. & Wilson, S. H. (1990). HTLV-I *trans*-activator protein, tax, is a *trans*-repressor of human beta-polymerase gene. *Science*, **247**, 1082–4.

Kalland, K. H., Langhoff, E., Bos, J. H., Gottlinger, H. & Haseltine, W. A. (1991). Rex-dependent nucleolar accumulation of HTLV-I mRNAs. *New Biologist*, **3**, 389–97.

Kannagi, M., Harada, S., Maruyama, I. *et al.* (1991). Predominant recognition of human T cell leukemia virus type I (HTLV-I) pX gene products by human CD8+ cytotoxic T-cells directed against HTLV-I-infected cells. *International Immunology*, **3**, 761–7.

Kelly, K., Davis, P., Mitsuya, H. *et al.* (1992). A high proportion of early response genes are constitutively activated in T cells by HTLV-I. *Oncogene*, **7**, 1463–70.

Kettman, R., Cleuter, Y., Mammerickx, M. *et al.* (1980). Genomic integration of bovine leukemia virus: Comparison of persistent lymphocytes with lymph node tumour form of enzootic bovine leukosis. *Proceedings of the National Academy of Sciences, USA*, **77**, 2577–80.

Khabbaz, R. F., Hartel, D., Lairmore, M. *et al.* (1991). HTLV-II infection in a cohort of New York IVDU: an old infection? *Journal of Infectious Diseases*, **163**, 252–6.

Kieran, M., Blank, V., Logeat, F. *et al.* (1990). The DNA binding subunit of NF-kappa B is identical to factor KBF1 and homologous to the rel oncogene product. *Cell*, **62**(5), 1007–8.

Kim, B. S., Kehrl, J. H., Burton, J. *et al.* (1990). Transactivation of the transforming growth factor B1 (TGF-B1) gene by human T lymphotrophic virus type I Tax: a potential mechanism for the increased production of TGF-B1 in adult T cell leukemia. *Journal of Experimental Medicine*, **172**, 121–9.

Kim, J. H., Kaufman, P. A., Hanly, S. M., Rimsky, L. T. & Greene, W. C. (1991). Rex transregulation of human T-cell leukemia virus type II gene expression. *Journal of Virology*, **65**, 405–14.

Kimata, J. T., Palker, T. J. & Ratner, L. (1993). The mitogenic activity of human T-cell leukemia virus type I is T-cell associated and requires the CD2/LFA-3 activation pathway. *Journal of Virology*, **67**(6), 3134–41.

Kira, J. I., Koyanagi, Y., Hamakado, T., Itoyama, Y., Yamamoto, N. & Goto, I. (1991*a*). HTLV-II in patients with HTLV-I-associated myelopathy. *Lancet*, **338**, 64–5.

Kira, J., Koyanagi, Y., Yamada, T. *et al.* (1991*b*). Increased HTLV-I proviral DNA in HTLV-I-associated myelopathy: a quantitative polymerase chain reaction study. *Annals of Neurology*, **29**, 194–201.

Kitada, H., Chen I. S. Y., Shah, N. P., Cann, A. J., Shimotohno, K. & Fan, H. (1987). U3 sequences form the HTLV-I and -II LTRs confer pX protein responsiveness to a murine leukemia virus LTr. *Science*, **235**, 901–4.

Kiyokawa, T., Yamaguchi, K., Nishimura, Y., Yoshiki, K. & Takatsuki, K. (1991). Lack of anti-HTLV-II seropositivity in HTLV-I-associated myelopathy and adult T-cell leukaemia. *Lancet*, **338**, 451.

Kondo, T., Kono, H., Nonaka, H. *et al.* (1987). Risk of adult T-cell leukemia/lymphoma in HTLV-I carriers. *Lancet*, **ii**, 159.

Koralnik, I. J., Fullen, J. & Franchini, G. (1993). The p12I, p13II, p30II proteins encoded by human T-cell leukemia/lymphotropic virus type I open reading frames I and II are localized in three different cellular compartments. *Journal of Virology*, **67**(4), 2360–6.

Korber, B., Okayama, A., Donnelly, R., Tachibana, N. & Essex, M. (1991). Polymerase chain reaction analysis of defective human T-cell leukemia virus type I proviral genomes in leukemic cells of patients with adult T-cell leukemia. *Journal of Virology*, **65**, 5471–6.

Kwok, S., Kellogg, D., Ehrlich, G., Poiesz, B., Bhagavati, S. & Sninsky, J. J. (1988). Characterization of a sequence of human T cell leukemia virus type I from a patient with chronic progressive myelopathy. *Journal of Infectious Diseases*, **158**, 1193–7.

Lal, R. B., Rudolph, D. L., Folks, T. M. & Hooper, W. C. (1993). Over expression of insulin-like growth factor receptor type-I in T-cell lines infected with human T-lymphotropic virus types-I and -II. *Leukemia Research*, **17**(1), 31–5.

Lee, H., Swanson, P., Shorty, V. S., Zack, J. A., Rosenblatt, J. D. & Chen, I. S. Y. (1989). High rate of HTLV-II infection in seropositive IV drug abusers from New Orleans. *Science*, **244**, 471–5.

Lee, T. H., Coligan, J. E., Sodroski, J. G. *et al.* (1984). Antigens encoded by the 3′-terminal region of human T-cell leukemia virus: evidence for a functional gene. *Science*, **226**, 57–61.

Leonard, W. J., Kronke, M., Peffer, N. J., Depper, J. M. & Greene, W. C. (1985). Interleukin-2 receptor gene expression in normal human T lymphocytes. *Proceedings of the National Academy of Sciences, USA*, **82**, 6281–5.

Lowenthal, J. W., Bohnlein, E., Ballard, D. W. & Greene, W. C. (1988). Regulation of interleukin 2 receptor alpha subunit (Tac or CD25 antigen) gene expression: binding of inducible nuclear proteins to discrete promoter sequences correlates with transcriptional activation. *Proceedings of the National Academy of Sciences, USA*, **88**, 4468–72.

Malim, M. H., Hauber, J., Lee, S. Y., Maizel, J. V. & Cullen, B. R. (1989). The HIV-1 rev trans-activator acts through a structured target sequence to activate nuclear export of unspliced viral mRNA. *Nature*, **338**, 254–7.

Maruyama, M., Shibuya, H., Harada, H. *et al.* (1987). Evidence for aberrant activation of the interleukin-2 autocrine loop by HTLV-I-encoded p40^{xI} and T3/Ti complex triggering. *Cell*, **48**, 343–50.

Merelli, E., Sola, P., Marasca, R., Salati, R. & Torelli, G. (1993). Failure to detect genomic material of HTLV-I or HTLV-II in mononuclear cells of Italian patients with multiple sclerosis and chronic progressive myelopathy. *European Neurology*, **33**, 23–6.

Merino, F., Robert-Guroff, M., Clark, J., Biondo-Bracho, M., Blattner, W. A. & Gallo, R. C. (1984). Natural antibodies to human T-cell leukemia/lymphoma virus in healthy Venezuelan populations. *International Journal of Cancer*, **34**, 501–6.

Miwa, M., Shimotohno, K., Hoshino, H., Fujino, M. & Sugimura, T. (1984). Detection of pX proteins in human T-cell leukemia virus (HTLV)-infected cells by using antibody against peptide deduced from sequences of X-IV DNA of HTLV-I and Xc DNA of HTLV-II proviruses. *Gann*, **75**, 752–5.

Miyamoto, K., Nakamura, T., Nagataki, S., Tomita, N. & Kitajima, K. (1992). Absence of HTLV-II co-infection in HTLV-I-associated myelopathy patients. *Japanese Journal of Cancer Research*, **83**, 415–17.

Miyatake, S., Seiki, M., Malefijt, R. D. *et al.* (1988*a*). Activation of T cell-derived lymphokine genes in T cells and fibroblasts: effects of human T cell leukemia virus type I p40x protein and bovine papilloma virus encoded E2 protein. *Nucleic Acids Research*, **16**(14), 6547–66.

Miyatake, S., Seiki, M., Yoshida, M., Arai, K. (1988*b*). T-cell activation signals and human T-cell leukemia virus type I-encoded p40x protein activate the mouse granulocyte-macrophage colony-stimulating factor gene through a common DNA element. *Molecular and Cellular Biology*, **8**(12), 5581–7.

Miyoshi, I., Taguchi, H., Fujishita, M. *et al.* (1982). Transformation of monkey lymphocytes with adult T-cell leukemia virus. *Lancet*, **i**, 1016.

Miyoshi, I., Kubonishi, I., Yoshimoto, S. *et al.* (1983*a*). Type C virus particles in a cord T-cell line derived by co-cultivating normal human cord leukocytes and human leukaemic T cells. *Nature*, **294**, 770–1.

Miyoshi, I., Yoshimoto, S., Taguchi, H. *et al.* (1983*b*). Transformation of rabbit lymphocytes with T-cell leukemia virus. *Gann*, **74**, 1–4.

Montagne, J., Beraud, C., Crenon, I. *et al.* (1990). Tax 1 induction of the HTLV-I 21 bp enhancer requires cooperation between two cellular DNA-binding proteins. *EMBO Journal*, **9**(3), 957–64.

Muchardt, C., Seeler, J. S., Nirula, A., Gong, S. & Gaynor, R. (1992*a*). Transcription factor AP-2 activates gene expression of HTLV-I. *EMBO Journal*, **11**(7), 2573–81.

Muchardt, C., Seeler, J. S. & Gaynor, R. G. (1992*b*). Regulation of HTLV-I gene expression by Tax and AP-2. *New Biologist*, **4**(5), 541–50.

Murphy, E. L., Engstrom, J. W., Miller, K., Sacher, R. A., Busch, M. P. & Hollingsworth, C. G. (1993). HTLV-II associated myelopathy in a 43-year-old woman. *Lancet*, **341**, 757–8.

Murphy, E. L., Hanchard, B., Figueroa, J. P. *et al.* (1989). Modeling the risk of adult T-cell leukemia/lymphoma in persons infected with human T-lymphotropic virus type I. *International Journal of Cancer*, **43**, 250–3.

Nagai, H., Kinoshita, T., Imamura, J. *et al.* (1991). Genetic alteration of p53 in some patients with adult T-cell leukemia. *Japanese Journal of Cancer Research*, **82**, 1421–7.

Nagata, K., Ohtani, K., Nakamura, M. & Sugamura, K. (1989). Activation of endogenous c-fos proto-oncogene expression by human T-cell leukemia virus type-I-encoded p40x protein in the human T-cell line, Jurkat. *Journal of Virology*, **63**(8), 3220–6.

Namikawa, R., Kaneshima, H., Lieberman, M., Weissman, I. L. & McCune, J. M. (1988). Infection of SCID-hu mouse by HIV-1. *Science*, **242**, 1684–6.

Nimer, S. (1991). Tax responsiveness of the GM-CSF promoter is mediated by mitogen-inducible sequences other than κB. *New Biologist*, 3(10), 997–1004.

Nimer, S. D., Gasson, J. C., Hu, K., Smalberg, I., Chen, I. S. Y. & Rosenblatt, J. D. (1989). Activation of the GM-CSF promoter by HTLV-I and -II tax proteins. *Oncogene*, 4, 671–6.

Nishimura, M., Adachi, A., Maeda, M. *et al.* (1988). Analysis of the provirus genome integrated in T cell lines established from the cerebrospinal fluid of patients with human T lymphotropic virus type-I-associated myelopathy (HAM). *Journal of Neuroimmunology*, 20, 33–7.

Osame, M., Igata, A., Matsumoto, M., Usuku, K., Izumo, S. & Kosaka, K. (1987a). HTLV-I associated myelopathy: A report of 85 cases. *Annals of Neurology*, 22, 116.

Osame, M., Matsumoto, M., Usuku, K. *et al.* (1987b). Chronic progressive myelopathy associated with elevated antibodies to human T-lymphotropic virus type I and adult T-cell leukemia-like cells. *Annals of Neurology*, 21, 117–22.

Persson, H., Hennighausen, L., Taub, R., De Grado, W. & Leder, P. (1984). Antibodies to human c-myc oncogene product: Evidence of an evolutionarily conserved protein induced during cell proliferation. *Science*, 225, 687–93.

Poiesz, B. J., Ruscetti, F. W., Gazdar, A. F., Bunn, P. A., Minna, J. D. & Gallo, R. C. (1980). Detection and isolation of type C retrovirus particles from fresh and cultured lymphocytes of a patient with cutaneous T-cell lymphoma. *Proceedings of the National Academy of Sciences, USA*, 77, 7415–19.

Popovic, M., Lange-Wantzin, G., Sarin, P. S., Mann, D. & Gallo, R. C. (1983). Transformation of human umbilical cord blood T cells by human T-cell leukemia/lymphoma virus. *Proceedings of the National Academy of Sciences, USA*, 80, 5402–6.

Reddy, E. P., Sandberg-Wollheim, M., Mettus, R. V., Ray, P. E., DeFreitas, E. & Koprowski, H. (1989). Amplification and molecular cloning of HTLV-I sequences from DNA of multiple sclerosis patients. *Science*, 243, 529–33.

Reed, J. C., Alpers, J. D., Nowell, P. C. & Hoover, R. G. (1986). Sequential expression of proto-oncogenes during lectin-stimulated mitogenesis of normal human lymphocytes. 3982–6.

Richardson, J. H., Wucherpfenning, K. W., Endo, N., Rudge, P., Dalgleish, A. G. & Hafler, D. A. (1989). PCR analysis of DNA from multiple sclerosis patients for the presence of HTLV-I. *Science*, 246, 821–3.

Robert-Guroff, M., Weiss, S. H., Giron, J. A. *et al.* (1986). Prevalence of antibodies to HTLV-I, -II, and -III in intravenous drug abusers from an AIDS endemic region. *Journal of the American Medical Association*, 255, 3133–7.

Rosenblatt, J. D., Cann, A. J., Slamon, D. J. *et al.* (1988). HTLV-II trans-activation is regulated by two overlapping nonstructural genes. *Science*, 240, 916–19.

Rosenblatt, J. D., Golde, D. W., Wachsman, W. *et al.* (1986). A second HTLV-II isolate associated with atypical hairy-cell leukemia. *New England Journal of Medicine*, 315, 372–5.

Ruben, S., Poteat, H., Tan, T. H. *et al.* (1988). Cellular transcription factors and regulation of IL-2 receptor gene expression by HTLV-I *tax* gene product. *Science*, 241, 89–92.

Sakamoto, K., Nimer, S. D., Rosenblatt, J. D. & Gasson, J. C. (1992). HTLV-I and HTLV-II Tax *trans*-activate the human EGR-1 promoter through different cis-acting sequences. *Oncogene*, 7, 2125–30.

Sakashita, A., Hattori, T., Miller, C. W. *et al.* (1992). Mutation of the p53 gene in adult T-cell leukemia. *Blood*, 79, 477–80.

Sandler, S. G. (1986). HTLV-I and -II: new risks for recipients of blood transfusions? *Journal of the American Medical Association,* **256**, 2245–6.

Sarin, P. S., Rodgers-Johnson, P., Sun, D. K. *et al.* (1989). Comparison of a human T-cell lymphotropic virus type I strain from cerebrospinal fluid of a Jamaican patient with tropical spastic paraparesis with a prototype human T-cell lymphotropic virus type I. *Proceedings of the National Academy of Sciences, USA,* **86**, 2021–5.

Sassone-Corsi, P., Lamph, W. W., Kamps, M. & Verma, I. M. (1988). c-fos-associated cellular p39 is related to nuclear transcription factor, AP-1. *Cell,* **54**, 553–60.

Saxinger, W., Blattner, W. A., Levine, P. H. *et al.* (1984). Human T-cell leukemia virus (HTLV-I) antibodies in Africa. *Science,* **225**, 1473–6.

Saxon, A., Stevens, R. H. & Golde, D. W. (1978). T-lymphocyte variant of hairy-cell leukemia. *Annals of Internal Medicine,* **88**, 323–6.

Seiki, M., Eddy, R., Shows, T. B. & Yoshida, M. (1984). Nonspecific integration of the HTLV provirus into adult T-cell leukaemia cells. *Nature,* **309**, 640–2.

Seiki, M., Hattori, S. & Yoshida, M. (1982). Human adult T-cell leukemia virus: molecular cloning of the provirus DNA and the unique terminal structure. *Proceedings of the National Academy of Sciences, USA,* **79**, 6899–902.

Shimotohno, K., Miwa, M., Salomon, D. J. *et al.* (1985). Identification of new gene products coded from X regions of human T-cell leukemia viruses. *Proceedings of the National Academy of Sciences, USA,* **82**, 302–6.

Shimotohno, K., Takano, M., Terunsho, T. & Miwa, M. (1986). Requirement of multiple copies of a 21-nucleotide sequence in the U3 regions of human T-cell leukemia virus type I and type II long terminal repeats for transactivation of transcription. *Proceedings of the National Academy of Sciences, USA,* **83**, 8112–16.

Slamon, D. J., Shimotohno, K., Cline, M. J., Golde, D. W. & Chen, I. S. Y. (1984). Identification of the putative transforming protein of the human T-cell leukemia viruses HTLV-I and HTLV-II. *Science,* **226**, 61–5.

Slamon, D. J., Press, M. F., Souza, L. M. *et al.* (1985). Studies of the putative transforming protein of the type I human T-cell leukemia virus. *Science,* **228**, 1427–30.

Sodroski, J., Rosen, C., Goh, W. C. & Haseltine, W. (1985). A transcriptional activator protein encoded by the x-lor region of the human T-cell leukemia virus. *Science,* **228**, 1430–4.

Su, I.-J., Chan, H.-L., Kuo, T.-T. *et al.* (1985). Adult T-cell leukemia/lymphoma in Taiwan. *Cancer,* **56**, 2217–20.

Sugito, S., Yamato, K., Sameshima, Y., Yokota, J., Yano, S. & Miyoshi, I. (1991). Adult T-cell leukemia: Structures and expression of the p53 gene. *International Journal of Cancer,* **49**, 880–5.

Suzuki, T., Fujisawa, J., Toita, M. & Yoshida, M. (1993). The trans-activator Tax of human T-cell leukemia virus type I (HTLV) interacts with cAMP-responsive element (CRE) binding and CRE modulator proteins that bind to the 21-base-pair enhancer of HTLV-1. *Proceedings of the National Academy of Sciences, USA,* **90**, 610–4.

Tateno, M., Kondo, N., Itoh, T., Chubachi, T., Togashi, T. & Yoshiki, T. (1984). Rat lymphoid lines with human T-cell leukemia virus production. I. Biological and serological characterization. *Journal of Experimental Medicine,* **159**, 1105–16.

Tedder, R. S., Shanson, D. C., Jeffries, D. J. *et al.* (1984). Low prevalence in the UK of HTLV-I and HTLV-II infection in subjects with AIDS, with extended lymphadenopathy, and at risk of AIDS. *Lancet,* **ii**, 125–8.

Uemura, Y., Kotani, S., Yoshimoto, S. *et al.* (1987). Mother-to-offspring transmission of human T cell leukemia virus type I in rabbits. *Blood,* **69**, 1255–8.

Vernant, J. C., Maurs, L., Gessain, A. *et al.* (1987). Endemic tropical spastic paraparesis associated with human T-lymphotropic virus type I: a clinical and seroepidemiological study of 25 cases. *Annals of Neurology,* **21**, 123–30.

Wano, Y., Feinberg, M., Hosking, J. B., Begerd, H. & Greene, W. C. (1988). Stable expression of the tax gene of type I human T-cell leukemia virus in human T cells activates specific cellular genes involved in growth. *Proceedings of the National Academy of Sciences, USA,* **85**, 9733–7.

Weiss, R., Teich, N., Varmus, H., Coffin, J. (eds) (1984). *RNA Tumour Viruses.* Cold Spring Harbor NY: Cold Spring Harbor Laboratory.

White, K. N., Nosaka, T., Kanamori, H., Hatanaka, M. & Honjo, T. (1991). The nucleolar localisation signal of the HTLV-I protein p27rex is important for stabilisation of IL-2 receptor α subunit mRNA by p27rex. *Biochemical and Biophysical Research Communications,* **175**(1), 98–103.

Yamato, K., Oka, T., Hirai, M., Iwahara, Y., Sugito, S., Tsuchida, N. & Miyoshi, I. (1993). Aberrant expression of the p53 tumour suppressor gene in adult T-cell leukemia and HTLV-I infected cells. *Japanese Journal of Cancer Research,* **84**, 4–8.

Yoshida, M., Miyoshi, I. & Hinuma, Y. (1982). Isolation and characterization of retrovirus from cell lines of human adult T-cell leukemia and its implication in the disease. *Proceedings of the National Academy of Science, USA,* **79**, 2031–5.

Yoshida, M., Osame, M., Usuku, K., Matsumoto, M. & Igata, A. (1987). Viruses detected in HTLV-I-associated myelopathy and adult T-cell leukaemia are identical on DNA blotting. *Lancet,* **i**, 1085–6.

Yoshida, M., Osame, M., Kawai, H. *et al.* (1989). Increased replication of HTLV-I in HTLV-I associated myelopathy. *Annals of Neurology,* **26**, 331–5.

Yoshimura, T., Fujisawa, J. & Yoshida, M. (1990). Multiple cDNA clones encoding nuclear proteins that bind to the tax-dependent enhancer of HTLV-I: All contain a leucine zipper structure and basic amino acid domain. *EMBO Journal,* **9**(8), 2537–42.

Zack, J. A., Cann, A. J., Lugo, J. P. & Chen, I. S. Y. (1988). AIDS virus production from infected peripheral blood T cells following HTLV-I-induced mitogenic stimulation. *Science,* **240**, 1026–9.

Zella, D., Mori, L., Sala, M. *et al.* (1990). HTLV-II infection in Italian drug abusers. *Lancet,* **336**, 575–6.

BOVINE LEUKAEMIA VIRUS: BIOLOGY AND MODE OF TRANSFORMATION

A. BURNY[1,2], L. WILLEMS[1], I. CALLEBAUT[1,4],
E. ADAM[2,3], I. CLUDTS[2], F. DEQUIEDT[1],
L. DROOGMANS[2], C. GRIMONPONT[1],
P. KERKHOFS[3], M. MAMMERICKX[3],
D. PORTETELLE[1], A. VAN DEN BROEKE[2],
AND R. KETTMANN[1]

[1]*Faculty of Agronomy, B5030 Gembloux, Belgium*
[2]*Department of Molecular Biology, University of Brussels, B1640*
Rhode-St-Genèse, Belgium
[3]*National Institute for Veterinary Research, B1180 Brussels,*
Belgium
[4]*Department of Biological Macromolecules, University of Paris 6,*
F75252 Paris cedex 05, France

INTRODUCTION

Bovine leukaemia (lymphoma, lymphosarcoma) is a contagious disease induced by bovine leukaemia virus (BLV), a retrovirus exogenous to the bovine species. It is a chronic disease, evolving over extended periods (1–8 years), with tumours developing in only a small number of infected animals. The same virus infects sheep where it induces tumours with very high frequency (for reviews, see Burny *et al.*, 1980, 1984, 1985, 1990; Burny & Mammerickx, 1987; Ghysdael *et al.*, 1984; Kettmann *et al.*, 1993).

GENOME AND GENE PRODUCTS

The BLV proviral genomic structure has been described in detail by Rice *et al.* (1984, 1985); Rice, Stephens & Gilden (1987*a*) and Sagata *et al.* (1985*a*) and is illustrated in Fig. 1. The precursors of these structural proteins have also been identified both *in vivo* and by *in vitro* synthesis (Ghysdael, Kettmann & Burny, 1979; Bruck *et al.*, 1984*a*; Yoshinaka *et al.*, 1986). It is apparent that: (1) the gag polyprotein contains the viral structural proteins p15 (109 aa), p24 (215 aa) and p12 (69 aa); (2) a protease, p14 (195 aa), is coded by an open reading frame overlapping the *gag* gene to the left and the *pol* gene to the right; *gag, pro* and *pol* genes are in three different reading frames; (3) the *pol* gene yielding a translated sequence of 852 aa represents the reverse transcriptase and integrase; (4) the *env* gene codes for a 72 000 mol. wt. *env* precursor (Pr 72^{env}: 515 aa including the signal peptide) and is

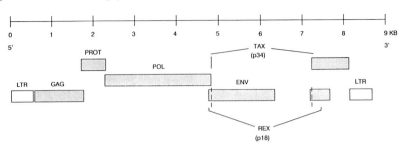

Fig. 1. Genomic structure of BLV provirus.

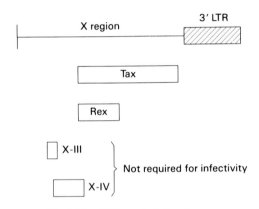

Fig. 2. The BLV proviral X region.

cleaved to form the two glycosylated envelope proteins gp51 (268 aa) and gp30 (214 aa); (5) two overlapping open reading frames (ORFs), located between *env* and the 3' LTR (long terminal repeat) code for the TAX (p34; 309 aa) and REX (p18; 156 aa) proteins.

The complete nucleotide sequence of the 8714 base pair BLV provirus revealed the presence of a region of 1817 base pairs (called the X region) between the *env* gene and the 3'LTR. This region, also found in the HTLV-I and -II genomes (Seiki *et al.*, 1983; Shimotohno *et al.*, 1985), contains several ORFs (Fig. 2). A subgenomic mRNA of 2.1 kilobases (kb) containing two overlapping ORFs has been identified in BLV-infected cells. This mRNA is generated by a complex splicing mechanism resulting in juxtaposition of the 5' end of the *env* gene and two overlapping X-ORFs (Mamoun *et al.*, 1985; Rice *et al.*, 1987*b*). Translation of this mRNA can yield at least two distinct proteins depending upon which initiation codon is used (Sagata *et al.*, 1985*b*,*c*; Rice *et al.*, 1987*b*). The BLV long open reading frame (LOR) in the X region encodes a protein (apparent mol. wt. of 34 000) referred to here as p34^{tax} (TAX) that is predominantly located in the nucleus of infected cells (Sagata *et al.*, 1985*c*). The short open reading frame (SOR) product was identified as a nuclear phosphoprotein, referred to here as p18^{rex} (REX), with an apparent mol. wt. of 18 000 (Rice *et al.*, 1987*b*). Both proteins are recognized by sera from cattle or sheep in the tumour phase of infection (Yoshinaka & Oroszlan, 1985; Willems *et al.*, 1988; Powers *et al.*, 1991). In addition to *tax* and *rex*, BLV and HTLV may have the potential to express other small ORFs in the X region (Rice *et al.*, 1984, 1987*b*; Sagata *et al.*, 1985*a*; Seiki *et al.*, 1983; Shimotohno *et al.*, 1985). These ORFs have been termed XBL-III and XBL-IV for BLV and pXI and pXII for HTLV. Proteins encoded by these genes would have to be translated from mRNAs that are spliced differently from the mRNA for the TAX and REX proteins. Recently, complex splicing was shown for HTLV and BLV (Ciminale *et al.*, 1992; Berneman *et al.*, 1992; Koralnik *et al.*, 1992; Alexandersen *et al.*, 1993). Four alternatively spliced mRNAs, encoding potential new regulatory proteins, have been recently identified in BLV-infected cattle. Interestingly, among them, the presence of the GIV mRNA seems correlated to the development of persistent lymphocytosis (Alexandersen *et al.*, 1993) but is not required for *in vivo* infection (Willems *et al.*, 1992*b*).

Because of its genome structure, proviral nucleotide sequence, and the size and amino acid sequence of the structural and nonstructural viral proteins, BLV is an obvious relative of the human T lymphotropic viruses (HTLV-I and -II) and the simian T lymphotropic virus (STLV-I) (Weiss *et al.*, 1984; Gallo & Wong-Staal, 1990). These four viruses replicate via *trans*-activation of the enhancer–promoter regions of their own LTRs. Their TAX protein probably plays a major role in transformation of the target cell either via transactivation of cellular genes or via interference with the transcription machinery through protein–protein interactions. The diseases induced have similar pathologies, characterized by the absence of chronic viremia, a long latency period and a lack of preferred proviral integration sites in the tumours.

Oroszlan *et al.* (1982) have shown that the major core proteins of BLV and HTLV-I share significant homology in both their N- and C-terminal regions. In preliminary studies, anti-HTLV-I sera failed to cross-react with BLV p24 in a competitive RIA; however, hyperimmune goat anti-HTLV-I serum did react with BLV p24 upon denaturation (Oroszlan *et al.*, 1984). Similarly, using a competitive ELISA, Onuma *et al.* (1987) observed competition between an anti-BLV p24 monoclonal antibody and some human anti-HTLV-I sera only after denaturation of the antigen. Using sera from selected infected hosts and non-denatured virion proteins, Zandomeni *et al.* (1991) showed that BLV, HTLV-I, HTLV-II and STLV-I share antigenic determinant(s) that distinguish them from other retroviruses. Antibodies against these determinants were found only in infected hosts with high antibody titres against the major *gag* proteins.

TRANSMISSION OF BLV

BLV is found worldwide, and in temperate climates the virus spreads mainly through iatrogenic transfer of infected lymphocytes. In warm climates and areas heavily populated by haematophageous insects, there are indications of insect-borne propagation of the virus. Cases of natural infection are documented in cattle, sheep, capybara and water buffaloes. Infection can be experimentally transmitted to goats, pigs, rabbits, rhesus monkeys, chimpanzees and buffaloes. There is considerable documentation to support the fact that horizontal transmission is most common (Burny *et al.*, 1980; Burny & Mammerickx, 1987). Since infected cells are the best potential vehicles for infectious BLV particles, the concentration of BLV-positive cells in the blood or other body fluids plays a major role in the success or failure of BLV transmission. As few as 926 lymphocytes (more than 50% containing proviral information) from a cow with PL have been shown to transmit infection to serologically negative sheep (Mammerickx *et al.*, 1987).

Recently, Willems *et al.* (1992c) successfully infected sheep through the intradermal inoculation of proviral DNA (100 μg) mixed with a cationic lipid solution. Seroconversion occurred one to two months after infection. Thus, a cloned BLV provirus is capable of inducing infection in the absence of other cellular factors, underlining the fact that the provirus encodes all of the genes required for virus infectivity.

VIRUS-CELL INTERPLAY

The BLV envelope is composed of the membrane glycoproteins, gp51 and gp30, and is directly involved in infectivity events such as receptor recognition and membrane fusion.

Studies have been performed, on the one hand, to demonstrate the importance of the gp51–gp30 complex in the virus–host interplay and to identify, at the molecular level, the specific regions involved in protective antigenic properties, membrane fusion and receptor recognition activities, and on the other hand to identify the BLV receptor on specific target cells.

The immunological importance of gp51

All available data indicate that the external envelope glycoprotein gp51 (apparent mol. wt. of 51 000) may be a critical viral target for the immune system of naturally or experimentally infected animals (Bruck *et al.*, 1984b; Burny *et al.*, 1990). The gp51 is not only responsible for binding to a cell receptor and determining the tropism of BLV, but is the first antigen to be recognized by antibodies in the sera of newly infected hosts. These polyclonal antibodies neutralize virus pseudotypes, block induction of syncytia and exhibit complement-dependent cytolytic activity.

Moreover, lambs or calves immunized passively with purified immuno-globulins derived from the serum of infected sheep, the colostrum of infected sheep or cows, can successfully resist an infectious challenge provided they have sufficiently high anti-gp51 antibody titres.

Surveys of published vaccination trials reviewed by Portetelle *et al.* (1993) reveal that most promising results have been obtained using gp51 in its native configuration.

In order to characterize regions of gp51 that potentially could act as an effective BLV subunit vaccine or as an efficient diagnostic probe, mono-clonal antibodies, synthetic peptides and more recently molecular modeling have been used to precisely identify crucial epitopes involved in the biological activities of BLV gp51.

Mouse monoclonal antibodies against gp51 have identified eight different epitopes referred to as A, B, B', C, D, D', E, F, G, H (Bruck *et al.*, 1982*a,b*). Three of these (F, G and H) have been shown to inhibit the BLV-induced early polykaryocytosis (EPK) and to neutralize vesicular stomatitis virus (VSV)-BLV pseudotypes. For vaccination trials, recombinant *env* vaccinia virus appears to efficiently produce the gp51 in its native confor-mation and therefore to properly present the three neutralizing epitopes F, G and H, whose reactivities depend on conformation and glycosylation. Vaccination trials in sheep using BLV *env*-vaccinia virus recombinants confer some protection against infection. However, protection seems not always to correlate with high levels of neutralizing antibodies (Portetelle *et al.*, 1993).

A number of synthetic peptides of the gp51 have also been obtained and characterized for their immunogenic properties, yielding information about the molecular aspects of the immune response required for protection (Fig. 3). Antipeptide antibodies were demonstrated to neutralize VSV–BLV pseudotypes and inhibit BLV-induced syncytia formation. Four (aa 39–40, 78–92, 144–157, 177–192) out of a total of 31 anti-peptides antisera have neutralizing activities. On the other hand, inhibition of syncytia formation is associated with the antibodies that are directed against peptides 64–73 (belonging to F phenotype), 98–117 (belonging to H phenotype) and 177–192.

Some helper T cells determinants on gp51 have been defined. For example, peptides 98–117, 169–188, 177–192 appear to be either involved in the stimulation of the helper T cells or constitute the T helper epitope (Callebaut *et al.*, 1993*a*). The first two peptides have been demonstrated to stimulate T helper cells. Similarly, peptides 121–140 and 131–150 have been shown to stimulate cytotoxic T cells (Gatei *et al.*, 1993).

Sequence analysis based on 2D hydrophobic cluster analysis (HCA) shows that retroviral envelope glycoproteins are organized, like influenza A hemagglutinin, into a stem and head (Fig. 3(*a*)). HCA gives an atomic-scale model for the bovine leukaemia virus (BLV) head. This head probably does

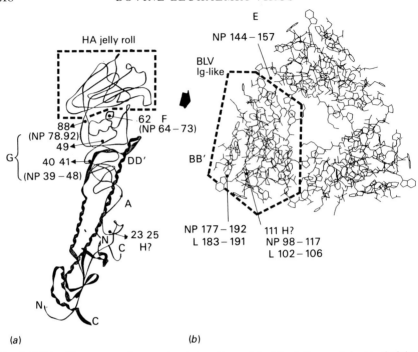

Fig. 3. BLV-gp51 representation constructed on the model of the Influenza A hemagglutinin HA1 (small ribbon)/HA2 (wide ribbon) complex: (*a*) monomeric model highlighting regions from the stem and heads; (*b*) BLV-gp51 head trimer. According to our modeling studies which predict another 3D organization for the BLV-gp51 head domain (immunoglobulin fold instead of jelly roll): stem would contain the sites A, D, D′; small head the sites F, G and peptides 39–48, 64–73 and 78–92 eliciting neutralizing properties (NP); large globular head the sites B, B′, E and peptides 98–117, 177–192 and 144–157 eliciting neutralizing antibodies. Unknown position for site H (?); Loops (L) are also indicated.

not possess a 'jelly-roll' fold like the hemagglutinin head, but rather a combination of 'Greek key' motifs corresponding to the overall topology of constant immunoglobulin domains. Like haemagglutinin, the envelope head may also exhibit trimeric organization (Fig. 3(*b*)).

Neutralizing epitopes known so far are concentrated within an exposed region on the independently modelled head tip, thereby defining a potential receptor binding site, which occupies a similar position to that of haemagglutinin. This study suggests general features for the fold of retroviral *env* heads and the structure of their receptor binding sites (Callebaut *et al.*, in preparation).

The BLV-cellular receptor

Using a lambda gt11 cDNA library derived from BLV permissive bovine cells (MDBK cell line), a cDNA (BLVRcp1) was isolated and characterized. BLVRcp1 encodes a polypeptide that represents part of the putative BLV receptor and selectively binds the envelope glycoprotein gp51 (Ban *et al.*, 1993). BLVRcp1 codes for a protein fragment of 729 amino acids; the missing piece of this protein has recently been identified to generate an ORF that encodes a putative receptor of 849 aa. This ORF contains a signal sequence, an extracellular domain including the BLV gp51 binding region, a transmembrane region and a putative intracellular cytoplasmic domain. This molecule appears to be of the EGF–receptor structural type.

No homology with sequences in the nucleotide or amino acid data bases could be found. Southern blot analysis revealed that the putative BLV receptor is widely distributed among mammals, which suggests that it plays an important role in the physiology of many cell types. Elucidating the mechanism by which the virus is confined to a few cell types *in vivo* awaits more detailed analysis.

The biological importance of gp30

The gp30 (apparent mol. wt. of 30 000) anchors the envelope both to the membrane of infected cells and in virions, and contains, like other enveloped viruses, a hydrophobic N-terminal segment (fusion peptide) involved in membrane fusion.

At the molecular level, Brasseur *et al.* (1988); Brasseur (1991) have suggested that the fusion peptides may act as 'sided insertional helices' and insert obliquely into the membrane, thus leading to its destabilization. Moreover, the replacement of the BLV fusion peptide with its simian immunodeficiency virus (SIV) counterpart does not modify the fusogenic capacity of vaccinia virus BLV-env recombinants, suggesting that oblique insertion is critical and supersedes amino acid sequence for enveloped viruses fusion efficiency (Voneche *et al.*, 1992).

More recent theoretical analysis suggests that these fusion peptides are divided into two regions: one is highly hydrophobic and inserts into the lipid bilayer in a preferential α helical oblique orientation and the other, which is particularly rich in small residues such as alanine, glycine, serine and threonine, confers mobility to the polypeptide chain (Callebaut *et al.*, 1993*b*). Obliquity and mobility are thus probably responsible for destabilization of the target membrane.

BLV-target cell(s) in vivo

Most of the available data indicate that the natural target cell for BLV belongs to the B cell lineage. Cows with PL have an elevated number of surface immunoglobulin (sIg)-bearing B-lymphocytes (Muscoplat *et al.*, 1974). Nylon-adherent bovine cells highly enriched for sIg-positive B-lymphocytes have been found to be the main producers of BLV in

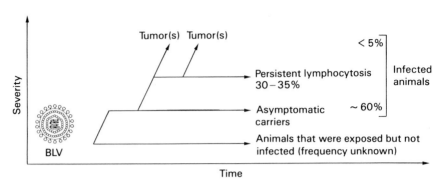

Fig. 4. BLV-induced pathogenesis.

short-term cultures (Paul *et al.*, 1977). BLV-infected B-cells from cows with PL and the majority of B-cells from BLV-infected sheep with haematological disorders, express high levels of surface IgM, CD11c and CD5 which reflects the activated state of these B cells (Depelchin *et al.*, 1989; Letesson *et al.*, 1990; Matheise *et al.*, 1992). Seventeen cases of BLV-induced bovine lymphoid tumours were examined by Heeney and Valli (1990). The conclusion derived from this study was that BLV-induced tumours are composed of mature B-cells.

PATHOGENESIS INDUCED BY BLV

BLV particles were first observed by Miller *et al.* (1969) in short-term cultures of PBL from BLV-infected animals in PL. Numerous attempts to detect the virus in body fluids from PL animals and in tumours thought to be of the EBL type failed until it was discovered that viremia can only be monitored during the first 10–12 days post-infection prior to the appearance of permanent anti-viral neutralizing antibodies. The continued presence of anti-BLV antibodies in infected animals indicates the existence of a constant antigenic stimulation, which is mediated either via viral proteins and/or virus particles produced by B-lymphocytes and possibly other cell types. Considering the long-term progression of BLV infection, cattle fall into three major groups (Fig. 4). The first and largest of these groups (about 60%) includes those animals that develop a persistent infection and humoral immune response but are normal in every other respect. The second group, representing perhaps 30 to 35% of all BLV-infected cattle, develop PL. This lymphocytosis is due to expansion of the B-lymphocyte population. Some of these B cells carry the BLV provirus (that population is polyclonal as far as the integration of the provirus is concerned, Kettmann *et al.*, 1980*b*) but others do not, and it has been suggested that the latter population is due to immune expansion in response to the BLV infection. Expression of a spliced mRNA encoding a potential new BLV regulatory protein seems correlated to development of PL (Alexandersen *et al.*, 1993). A third, and much smaller group (less than 5%), includes those animals that develop lympho-sarcoma or localized lymphoid tumours. These tumours appear monoclonal with respect to the integration site of the provirus (Kettmann *et al.*, 1980*b*).

BLV is not only the aetiological agent of bovine leukosis but also of the rare natural occurrence of ovine leucosis. Although less than 5% of BLV-infected cattle go on to develop monoclonal tumours, it has been shown that nearly all experimentally BLV-infected sheep progress to, and die in, the tumour phase of the disease (Mammerickx *et al.*, 1976). Infection, once established, is lifelong. The progression of infection is linked to the infectious dose that seems to determine the length of the latency period before the onset of the neoplastic phase (Mammerickx *et al.*, 1987, 1988).

Furthermore, when BLV is injected into rabbits, it induces profound perturbations in the white cell compartment (neutropenia, lymphopenia) together with diarrhoea, and wasting syndrome which is followed by death at a time when the antibody titre has dropped to zero (Burny *et al.*, 1985; Altanerova, Ban & Altaner, 1989; Wyatt *et al.*, 1989).

VIRUS EXPRESSION

In vivo studies

BLV expression *in vivo* is thought to be transcriptionally blocked since neither virus particles nor viral proteins or RNA have been readily detected in freshly isolated PBL or tumour cells (Baliga & Ferrer, 1977; Kettmann *et al.*, 1980*a*, 1982). Even transcriptionally competent proviruses are silent in BLV-infected tumour cells (Van den Broeke *et al.*, 1988). This phenomenon was initially attributed to the presence of a blocking factor in the plasma of BLV-infected cattle (Gupta & Ferrer, 1982). It seems likely that host factors suppress and/or stimulate BLV expression, with their relative concentration in the animal acting as an important determinant of susceptibility or resistance to the development of PL and lymphosarcoma in infected cattle (Zandomeni *et al.*, 1992). Using *in situ* hybridization, the presence of viral RNA has only been detected in 1/2 000 to fewer than 1/500 000 of freshly isolated PBL from clinically normal BLV-infected sheep (Lagarias & Radke, 1989). Thirty-eight BLV-induced tumours (16 bovine, 22 ovine) were examined by the same technique, and in about 35% of the tumours, an average of 1/5 000 cells contained viral RNA (Fig. 5) (Van den Broeke, personal cummunication). However, it cannot presently be excluded that these positive cells are non-neoplastic BLV-infected PBL that have infil-trated the tumour. Using the sensitive polymerase chain reaction after reverse transcription (RT–PCR), the presence of the *tax/rex* 2.1 kb mRNA was found in 3/12 normal seropositive animals, in 6/6 animals with PL, and in 5/8 tumours but not in BLV seronegative animals (Jensen, Rovnak &

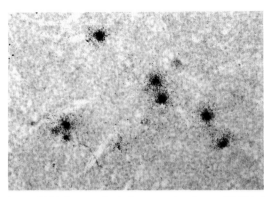

Fig. 5. Hybridization to RNA in a cytological preparation from a BLV-induced tumour with ^{35}S-labelled anti-sense RNA corresponding to the 3' region of the BLV genome. (Magnification, ×100).

Cockerell, 1991). The fact that some BLV-induced tumours were negative for the 2.1 kb mRNA species, suggests that expression of p34[tax] is not necessary to maintain the neoplastic state as previously proposed (Kettmann *et al.*, 1982, 1985).

It has recently been reported that BLV transcripts can be amplified by RT–PCR from total or cytoplasmic-enriched bovine lymphocyte RNA from all BLV-infected cells but only a few aleukemic animals. Using primer pairs that differentiate the spliced transcripts (genomic, envelope and *tax/rex*), it appears that a trend toward the exclusion of both genomic and envelope RNAs, with retention of the *tax/rex* message, occurs as the leukemia/lymphosarcoma develops (Haas, Divers & Casey, 1992). The constant presence of *tax/rex* mRNA in the cytoplasm of cells from cows at all stages of infection led these authors to propose that persistent low-level expression of either of these gene products may be important to the pathogenesis of BLV infection.

Taken altogether, the minimal *in vivo* expression of structural genes could be viewed as a unique strategy designed to evade immunosurveillance and permit the persistent and low level infection that is characteristic of BLV. This could also explain the extremely low genetic variability of BLV compared to the HIV/SIV lentivirus group (Willems *et al.*, 1993*a*).

Direct inoculation of a cloned BLV provirus in sheep has recently permitted the study of viral infectivity of genetic mutants *in vivo* (Willems *et al.*, 1993*b*). Three BLV variants cloned from BLV-induced tumours and 12 *in vitro*-modified proviruses were isolated and analysed for viral expression in cell culture. These proviruses were then inoculated in sheep to assess viral infectivity *in vivo* (refer to 'Transmission of BLV'). Out of three variants cloned from BLV-induced tumours (344, 395 and 1345), one (344) was found to be infectious *in vivo*. This particular provirus was used to engineer

12 BLV mutants. A hybrid between the 5' region of the complete but non-infectious provirus 395 and the 3' end of mutant 344 was found to be infectious *in vivo*, suggesting that the *tax/rex* sequences were defective in virus 395. As expected, several regions of the BLV genome appear to be essential for viral infection; the *pro*, the *pol* and the *env* genes. Even discrete modifications in the fusion peptide located at the NH_2 end of the transmembrane gp30 glycoprotein, that do not impair its fusogenic capacity, destroy infectious potential. In contrast, mutations and deletions in the X3–X4 region present between the *env* gene and the 3' *tax/rex* region do not interfere with viral infection *in vivo*. The function of this region is unknown, and could provide a target for the introduction of foreign sequences.

In vitro studies

Transactivation

The transactivator protein p34[tax].

Transcriptional initiation signals for retroviruses lie within the LTR sequences (Fig. 1). To study these signals, various cell lines have been transfected with recombinant plasmids carrying the bacterial chloramphenicol acetyl-transferase (CAT) gene coupled to the BLV LTR. Transient expression of CAT activity was found only in the established BLV-producer cell line derived from foetal lamb kidney cells (FLK; Van der Maaten and Miller, 1976) and in BLV-infected bat lung cells (Derse, Caradona & Casey, 1985; Rosen *et al.*, 1985). Enhancer activity of the BLV LTR sequences was apparent only in virus-producing cells (Derse & Casey, 1986; Rosen *et al.*, 1986*a*). It has been suggested that the major determinant for efficient functioning of BLV transcriptional elements is the presence of BLV-induced (viral or cellular) transacting factors because CAT gene expression directed by a plasmid containing the BLV LTR was only observed in BLV-infected cells (Rosen *et al.*, 1985).

Transient expression of a cDNA encompassing the X region ORFs (see Genome and gene products pp. 213-16) demonstrated that a product encoded by this region can activate, in *trans*, gene expression directed by the BLV LTR (Rosen *et al.*, 1986*b*; Willems *et al.*, 1987*a*). Infection of fibroblasts with recombinant murine retroviruses exclusively expressing p34, provided further evidence that the X-LOR gene product (p34[tax], TAX) specifically transactivates the BLV LTR (Willems *et al.*, 1987*b*). Similar observations were obtained using other recombinant plasmids expressing the same product (Derse, 1987). Synthesis of functional p34[tax] in a recombinant baculovirus system demonstrated that the p34[tax] product is phosphorylated in insect SF9 cells and requires mammalian cellular factors for transactivation (Chen *et al.*, 1989).

Attempts to identify the region(s) responsible for the functional activity of the transactivator protein by mutational analysis (amino acid deletions or

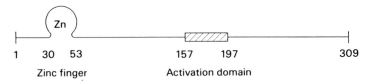

Fig. 6. Localization of the zinc finger and an activation domain on BLV TAX.

insertions) have been unsuccessful because most of the modifications abrogated transactivation (Willems *et al.*, 1989; Sakurai *et al.*, 1991), and suggests that BLV p34^{tax} is under heavy evolutionary constraints. By constructing hybrids between a well-defined region of TAX and the specific DNA binding portion of the yeast GAL4 transactivator protein (amino acids 1 to 147), the TAX domain between amino acids 157–197 was identified as an activating region (Willems, Kettmann & Burny, 1991). This segment is approximately located in the middle of p34^{tax} and is globally neutral (a net charge of zero) (Fig. 6). This putative TAX domain contains 24% of leucine residues, which are potentially involved in heterologous protein interactions.

The *cis*-acting response elements and cellular activators.

Sequences upstream of the TATA element in the BLV LTR were analysed for their ability to respond to p34^{tax} using the CAT transient expression assay (Gorman, Moffat & Howard, 1982). Deletion experiments suggested there were multiple cis-acting elements present that respond to p34^{tax} activation (Derse & Casey, 1986; Derse, 1987; Katoh, Yoshinaka & Ikawa, 1989). *Trans*-activation of the BLV LTR by p34^{tax} requires the presence of a 75-base pair element upstream of the viral promoter in the LTR U3 region. This fragment was identified as a cell-specific 'enhancer' because its activity is restricted to cells productively infected with BLV and contains one or more elements responsive to p34. Nucleotide sequence analysis shows that this *cis*-acting element contains two 21 base pair repeats that are centred around position 148 and located 123 base pairs upstream of the RNA start site. A third, related element was found at position-48 of the RNA cap site and is apparently poorly active in the absence of the upstream sequences (Fig. 7). These three repeats are very similar, with respect to their nucleotide sequence and spatial array, to elements found in the LTRs of HTLV-I and II.

All of these elements contain the common 8-base pair core sequence, TGACGTCA, which has been proposed to be a cyclic AMP (cAMP)-responsive element (CRE) (Montminy *et al.*, 1986) or a binding site for the cellular transcription factor, ATF (Lee *et al.*, 1987). The mutually exclusive

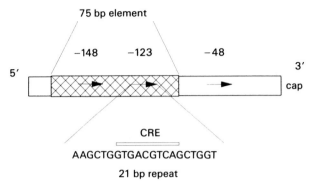

Fig. 7. The U3 region of the BLV LTR.

trans-activation of the BLV and HTLV promoters by their cognate TAX proteins suggests that the shared CRE represents the functional centre for a larger response element whose specificity is determined by the flanking nucleotides (Derse, 1987). Adenovirus-5 E3 and E4, c-fos and somatostatin regulatory regions containing a CRE/ATF element have been reported to exhibit responsiveness to $p34^{tax}$ in a CAT transient expression assay (Katoh *et al.*, 1989).

A lambda gt11 cDNA library, constructed with mRNA isolated from a BLV-induced tumour, was screened with an oligonucleotide corresponding to the 21-base pair *tax* activation response element, and permitted the identification of the bovine CREB2 protein (Willems *et al.*, 1992*a*). In the absence of $p34^{tax}$, both the CREB2 protein and the cyclic AMP-dependent protein kinase A can activate the BLV LTR. Thus, CREB2 might be a key element for regulating BLV expression, and play a role in the de-repression of proviral expression in the absence of TAX and regulation of TAX-induced trans-activation during the replicative cycle.

Trans-regulation of mRNA processing: the role of $p18^{rex}$.

The BLV 2.1 kb mRNA encoding $p34^{tax}$ is translated, in an alternative reading frame, to produce the nuclear phosphoprotein $p18^{rex}$. Co-transfection experiments using mutated BLV proviruses in combination with BLV-X region expression plasmids indicate that p18 is required to regulate, in *trans*, the accumulation of full-length (genomic) and single-spliced (*env*) viral mRNAs. In contrast, synthesis of the double-spliced X mRNA is not influenced by p18 expression. As a consequence, immediately after its induction viral transcription would predominantly yield the doubly-spliced X mRNA. Translation of this message and the subsequent accumulation of p34^{tax} and p18^{rex} would result in both the amplification of viral transcription and a switch to the events that permit expression of the structural genes. Large regional deletions and substitutions of proviral sequences has located the elements essential for p18 regulation in the 3' LTR. Furthermore, sequences within a 250-nucleotide region between the AATAAA signal and poly (A) site were found to be essential for efficient mRNA 3'-end processing and responsiveness to p18^{rex} regulation (Derse, 1988).

Transformation

The BLV genome is devoid of a typical oncogene (Deschamps, Kettmann & Burny, 1981). All BLV-induced tumours are clonal and contain minimally part of a provirus that is randomly integrated in the host genome (Kettmann *et al.*, 1980*b*, 1982, 1983; Grégoire *et al.*, 1984). All deleted proviral copies in tumour cells have been shown to preserve the X region, stressing its probable important role in tumorigenesis. Considering that the presence of virus is mandatory during the initial steps of the process and apparently dispensable later on (Kettmann *et al.*, 1982, 1985; Van den Broecke *et al.*, 1988), we can speculate that transient expression of viral functions can lead to the permanent expression of critical cellular genes. Two possibilities must be considered: (1) either the transformation process rests entirely upon regulatory mechanisms without DNA alterations (transactivation of host functions; possible mechanisms by which a gene may be switched on by a transient signal and remain on after that signal has disappeared); or (2) alterations in the cellular genome make the cell more susceptible to progression towards the neoplastic state (through mutations, deletions, amplifications, translocations, visible chromosomal abnormalities, etc). Aneuploidy and structural modifications have been reported in BLV-induced lymphosarcoma and are observed in cultured tumour cells from cow and sheep in the tumour phase where karyotypic analysis shows that profound rearrangement of one X chromosome is a common event (Popescu, personal communication). The major question that arises is whether chromosomal abnormalities are a primary event in tumourigenesis involving oncogenes, growth factor receptor genes, differentiation genes or tumour suppressor genes, or a secondary step in the transformation process and reflect an adaptation of the cancer cells to tissue culture conditions or to abnormalities in metabolic pathways.

It is generally accepted that BLV p34^{tax} acts as a positive regulator of viral transcription. A current hypothesis for the induction of leukaemogenesis suggests that the transactivation of cellular gene(s) is the first step in transformation. The identity of these genes and their mode of action remains to be determined. Transcriptional activation of the interleukin-2 (IL-2) gene and its receptor (IL-2R) permits the preferential proliferation of the HTLV-infected T-cells by an autocrine mechanism (Inoue *et al.*, 1986), although subsequent steps are required to stabilize this activation state.

In the BLV system, p34^{tax} has been shown to immortalize primary rat embryo fibroblasts (REF) in culture. Furthermore, TAX, in co-operation with the Ha-ras oncogene, can induce foci *in vitro* that will develop into tumours *in vivo* (Willems *et al.*, 1990). Thus, BLV TAX can be classified as a member of the immortalizing oncogene subgroup, with other proteins such as adenovirus E1A, SV40 large T, mutated p53 and *myc* gene products. It has recently been reported that TAX, mutated in its putative zinc finger structure (Fig. 6), may completely lose its *trans*-activating activity but maintain its transforming capacity (Willems *et al.*, 1992*b*). Thus, *trans*-activation of the BLV LTR and target cell transformation appear to be independent functions of TAX involving different functional domains of the protein.

CONCLUSIONS

Bovine leukaemia virus has probably evolved over a very long period. The host–virus relationship is characterized by minimal pathogenicity. Only a small percentage of infected cattle ultimately develop the neoplastic disease. In other hosts such as sheep or rabbits, viral propagation leads to tumour development (in sheep) or induces a wasting syndrome (in rabbits). Host–virus interplay is obviously very complex. A set of conditions, occurring with very low frequency and exclusively in the B cell compartment, transforms a normal cell into a tumorous entity, leading to the development of a clonal neoplasia. It is believed that TAX plays a major role in such an event, via direct interactions with transcription factors, or more indirectly via co-operation with oncogenic products like the RAS proteins. Other viral products like REX could interfere with the nuclear splicing machinery ultimately modifying the relative concentrations of mRNA populations and altering otherwise finely tuned physiological equilibria. Proteins coded by other open reading frames of the virus genome seem to play an elusive role in the infection cycle and are challenging objects with respect to why and how they were conserved in evolution.

The BLV external glycoprotein is the smallest retroviral envelope glyco-protein known today. Molecular modelling based on the hydrophobic cluster analysis (HCA) suggests that its folding pattern is of the immuno-globulin type. Such theoretical considerations are strongly reinforced by the observation that immunologically significant epitopes are localized in loop regions that connect the strands of the Greek–Key fold. Further refinements of these ongoing studies are still necessary. The approach appears, however, fruitful enough to suggest that the combination of predictive and immunological methods is a safe path in the approach of the tertiary structure of retroviral envelope proteins.

Analysis of retroviral systems has a bearing on basic biological phenomena such as cancer induction, immune deficiency and virus–cell fusion. BLV contributes its share to every one of these major issues.

ACKNOWLEDGEMENTS

The work performed in the authors' laboratory was helped financially by the Fonds Cancérologique de la Caisse Générale d'Epargne et de Retraite, the Belgian Service de Programmation de la Politique Scientifique (SPPS) and the Belgian National Fund for Scientific Research (FNRS). I. C. is Research Assistant, R. K. is Research Director and L. W. is Research Associate of the FNRS. L. D., C. G. and F. D. are Fellows of the FNRS-Télévie. E. A. is a fellow from IRSIA. The authors warmly thank Y. Cleuter, R. Martin and M. Nuttinck for excellent technical assistance. The authors also thank Y. Braet and M. Prévot for preparing all documents.

REFERENCES

Alexandersen, S., Carpenter, S., Christensen, J. (1993). Identification of alternatively splicing mRNAs encoding potential new regulatory proteins in cattle infected with bovine leukemia virus. *Journal of Virology*, **67**, 39–52.

Altanerova, V., Ban, J. & Altaner, C. (1989). Induction of immune deficiency syndrome in rabbits by bovine leukemia virus. *AIDS*, **3**, 755–8.

Baliga, V. & Ferrer, J. (1977). Expression of the bovine leukemia virus and its internal antigen in blood lymphocytes. *Proceedings of the Society of Experimental Biology and Medicine*, **156**, 388–91.

Ban, J., Portetelle, D. & Altaner, C. (1993). Isolation and characterization of a 2.3-kp cDNA fragment encoding the binding domain of the bovine leukemia virus cell receptor. *Journal of Virology*, **67**, 1050–7.

Berneman, Z., Gartenhaus, R. & Reitz, M. (1992). Expression of alternatively spliced human T-lymphotropic virus type I pX mRNA in infected cell lines and in primary uncultured cells from patients with adult T-cell leukemia/lymphoma and healthy carriers. *Proceedings of the National Academy of Sciences, USA*, **89**, 3005–9.

Brasseur, R. (1991). Differentiation of lipid-association helices by use of three-dimensional molecular hydrophobicity potential calculations. *Journal of Biological Chemistry*, **266**, 16120.

Brasseur, R., Cornet, B., Burny, A., Vandenbranden, M. & Ruysschaert, J. M. (1988). Mode of insertion into a lipid membrane of the N-terminal HIV peptide segment. *AIDS Research in Human Retroviruses*, **4**, 83–90.

Bruck, C., Mathot, S., Portetelle, D. *et al.* (1982*a*). Monoclonal antibodies define eight independent antigenic regions on the bovine leukemia virus (BLV) envelope glycoprotein gp51. *Virology*, **122**, 342–52.

Bruck, C., Portetelle, D., Burny, A. & Zavada, J. (1982*b*). Topographical analysis by monoclonal antibodies of BLV-gp51 epitopes involved in viral functions. *Virology*, **122**, 353–62.

Bruck, C., Rensonnet, N. & Portetelle, D. (1984*a*). Biologically active epitopes of bovine leukemia virus glycoprotein gp51: their dependence on protein glycosylation and genetic variability. *Virology,* **136**, 20–31.

Bruck, C., Portetelle, D., Mammerickx, M., Mathot, S. & Burny, A. (1984*b*). Epitopes of BLV glycoprotein gp51 recognized by sera infected cattle and sheep. *Leukemia Research,* **8**, 315–21.

Burny, A. & Mammerickx, M. (1987). '*Enzootic Bovine Leukosis and Bovine Leukemia Virus*', Boston: Nijhoff Publ.

Burny, A., Bruck, C. & Chantrenne, H. (1980). Bovine leukemia virus: molecular biology and epidemiology. In *Viral Oncology*, ed. G. Klein, pp. 231–289. New York: Raven Press.

Burny, A., Bruck, C. & Cleuter, Y. *et al*. (1985). Bovine leukemia virus, a versatile agent with various pathogenic effects in various animal species. *Cancer Research,* **45**, 4578–82.

Burny, A., Bruck, C. & Cleuter, Y. (1984). Leukemogenesis by bovine leukemia virus. In *Mechanism of Viral Leukemogenesis*, ed. J. Goldman and O. Jarrett, pp. 229–260. London: Churchill Livingstone.

Burny, A., Cleuter, Y. & Kettmann, R. *et al*. (1990). Bovine leukemia: facts and hypotheses derived from the study of an infectious cancer. In *Retroviruses Biology and Human Disease*, ed. R. Gallo and F. Wong-Staal, pp. 9–32. New York: Marcel Dekker.

Callebaut, I., Voneche, V. & Mager, A. *et al*. (1993*a*). Mapping of B-neutralizing and T-helper cell epitopes on the bovine leukemia virus external glycoprotein gp51. *Journal of Virology*, in press.

Callebaut, I., Tasso, A., Brasseur, R., Burny, A., Portetelle, D. & Mornon, J. P. (1993*b*). Common prevalence of alanine and glycine in mobile reactive centre loops of serpins and viral fusion peptides. Do prions possess a fusogenic peptide? *Journal of Computer-Aided Molecular Design*, in press.

Chen, G., Willems, L. & Portetelle, D. *et al*. (1989). Synthesis of functional bovine leukemia virus (BLV) p34^{tax} protein by recombinant baculoviruses. *Virology,* **173**, 343–7.

Ciminale, V., Pavlakis, G., Derse, D., Cunningham, C. & Felber, B. (1992). Complex splicing in the human T-cell leukemia virus (HTLV) family of retroviruses: novel mRNAs and proteins produced by HTLV type I. *Journal of Virology,* **66**, 1737–45.

Depelchin, A., Letesson, J., Lostrie, N., Mammerickx, M., Portetelle, D. & Burny, A. (1989). Bovine leukemia virus (BLV) infected B cells express a marker similar to the CD5 T-cell marker. *Immunology Letters,* **20**, 69–76.

Derse, D. (1987). Bovine leukemia virus transcription is controlled by a virus-encoded transacting factor and by *cis*-acting response elements. *Journal of Virology,* **61**, 2462–71.

Derse, D. (1988). *Trans*-acting regulation of bovine leukemia virus mRNA processing. *Journal of Virology,* **62**, 1115–19.

Derse, D. & Casey, J. (1986). Two elements in the bovine leukemia virus long terminal repeat that regulate gene expression. *Science,* **231**, 1437–40.

Derse, D., Caradona, S. & Casey, J. (1985). Bovine leukemia virus long terminal repeat: a cell type specific promoter. *Science,* **227**, 317–20.

Deschamps, J., Kettmann, R. & Burny, A. (1981). Experiments with cloned complete tumour-derived bovine leukemia virus information prove that the virus is totally exogenous to its animal species. *Journal of Virology,* **40**, 605–9.

Gallo, R. & Wong-Staal (1990). *Retrovirus Biology and Human Disease*. pp. 409, New York: Marcel Dekker.

Gatei, M., Good, M., Daniel, R. & Lavin, M. (1993). T-cell responses to highly conserved CD4 and CD8 epitopes on the outer membrane protein of bovine leukemia virus: relevance to vaccine development. *Journal of Virology, 67*, 1796–802.

Ghysdael, J., Bruck, C., Kettmann, R. & Burny, A. (1984). Bovine leukemia virus. *Current Topics in Microbiology and Immunology, 112*, 1–19.

Ghysdael, J., Kettmann, R. & Burny, A. (1979). Translation of BLV virion RNAs in heterologous protein synthesizing systems. *Journal of Virology, 29*, 1087–98.

Gorman, C., Moffat, L. & Howard, B. (1982). Recombinant genomes which express chloramphenicol acetyltransferase in mammalian cells. *Molecular and Cellular Biology, 2*, 1044–51.

Gregoire, D., Couez, D., Deschamps, J. *et al.* (1984). Different bovine leukemia virus-induced tumours harbor the provirus in different chromosomes. *Journal of Virology, 50*, 275–9.

Gupta, P. & Ferrer, J. (1982). Expression of bovine leukemia virus genome is blocked by a non immunoglobulin protein in plasma from infected cattle. *Science, 215*, 405–7.

Haas, L., Divers, T. & Casey, J. (1992). Bovine leukemia virus gene expression *in vivo*. *Journal of Virology, 66*, 6223–5.

Heeney, J. & Valli, V. (1990). Transformed phenotype of enzootic bovine lymphoma reflects differentiation-linked leukemogenesis. *Laboratory Investigations, 62*, 339–45.

Inoue, J., Seiki, M., Taniguchi, T., Tsuru, S. & Yoshida, M. (1986). Induction of interleukin 2 receptor gene expression by p40^x encoded by human T-cell leukemia virus type 1. *EMBO Journal, 5*, 2883–8.

Jensen, W., Rovnak, J. & Cockerell, G. (1991). *In vivo* transcription of the bovine leukemia virus tax/rex region normal and neoplastic lymphocytes of cattle and sheep. *Journal of Virology, 65*, 2484–90.

Katoh, I., Yoshinaka, Y. & Ikawa, Y. (1989). Bovine leukemia virus trans-activator p38^{tax} activates heterologous promoter with a common sequence known as a cAMP-responsive element or the binding site of a cellular transcription factor ATF. *EMBO Journal, 8*, 497–503.

Kettmann, R., Marbaix, G., Cleuter, Y., Portetelle, D., Mammerickx, M. & Burny, A. (1980a). Genomic integration of bovine leukemia provirus and lack of viral RNA expression in the target cells of cattle with different response to BLV infection. *Leukemia Research, 4*, 509–19.

Kettmann, R., Cleuter, Y., Mammerickx, M. *et al.* (1980b). Genomic integration of bovine leukemia provirus: comparison between persistent lymphocytosis and lymph node tumour form of enzootic bovine leukosis. *Proceedings of the National Academy of Sciences, USA, 77*, 2577–81.

Kettmann, R., Deschamps, J., Cleuter, Y., Couez, D., Burny, A. & Marbaix, G. (1982). Leukemogenesis by bovine leukemia virus: provirus DNA integration and lack of RNA expression of viral long terminal repeat and 3′ proximate cellular sequences. *Proceedings of the National Academy of Sciences, USA, 79*, 2465–9.

Kettmann, R., Deschamps, J., Couez, D., Claustriaux, J-J., Palm, R. & Burny, A. (1983). Chromosome integration domain for bovine leukemia provirus in tumours. *Journal of Virology, 47*, 146–50.

Kettmann, R., Cleuter, Y., Gregoire, D. & Burny, A. (1985). Role of the 3′ long open reading frame region of bovine leukemia virus in the maintenance of cell transformation. *Journal of Virology, 54*, 899–901.

Kettmann, R., Burny, A., Callebaut, I. *et al.* (1993). Bovine leukemia virus. In *The Retroviridae*, ed. J. A. Levy, vol. 3, Plenum Press, in press.

Koralnik, I., Gessain, A., Klotman, M., Lo Monico, A., Berneman, Z. & Franchini,

G. (1992). Protein isoforms encoded by the pX region of human T-cell leukemia/ lymphotropic virus type I. *Proceedings of the National Academy of Sciences, USA,* **89**, 8813–17.

Lagarias, D. & Radke, K. (1989). Transcriptional activation of bovine leukemia virus in blood cells from experimentally infected asymptomatic sheep with latent infection. *Journal of Virology,* **63**, 2099–107.

Lee, K., Hai, T., Siva Raman, L., Thimmappaya, B., Hurst, H., Jones, N. & Green, N. (1987). A cellular protein, activating factor activates transcription of multiple E1A-inducible adenovirus early promoters. *Proceedings of the National Academy of Sciences, USA,* **84**, 8355–9.

Letesson, J., Mager, A., Mammerickx, M., Burny, A. & Depelchin, A. (1990). B cells from bovine leukemia virus (BLV)-infected sheep with hematological disorders express the CD5 T-cell marker. *Leukemia,* **4**, 377–9.

Mammerickx, M., Dekegel, D., Burny, A. & Portetelle, D. (1976). Study on the oral transmission of bovine leukosis to sheep. *Veterinary Microbiology,* **1**, 347–50.

Mammerickx, M., Portetelle, D., De Clercq, K. & Burny, A. (1987). Experimental transmission of enzootic bovine leukosis to cattle, sheep and goats: infectious doses of blood and incubation period of the disease. *Leukemia Research,* **11**, 353–8.

Mammerickx, M., Palm, R., Portetelle, D. & Burny, A. (1988). Experimental transmission of enzootic bovine leukosis in sheep: latency period of the tumoral disease. *Leukemia,* **2**, 103–7.

Mamoun, R., Astier-Gin, R., Kettmann, R., Deschamps, J., Reybeyrotte, N. & Guillemain, B. (1985). The pX region of the bovine leukemia virus is transcribed as a 2.1 kb mRNA. *Journal of Virology,* **54**, 625–9.

Matheise, J., Delcommenne, M., Mager, A., Didembourg, C. & Letesson, J. (1992). CD5$^+$ B cells from bovine leukemia virus infected cows are activated cycling cells responsive to interleukin 2. *Leukemia,* **6**, 304–9.

Miller, J., Miller, L., Olson, C. & Gillette, K. (1969). Virus-like particles in phytohemagglutinin-stimulated lymphocyte cultures with reference to bovine lymphosarcoma. *Journal of the National Institute,* **43**, 1297–305.

Montminy, M., Sevarino, K., Wagner, J., Mandel, G. & Goodman, R. (1986). Identification of a cyclic-AMP-responsive element within the rat somatostatin gene. *Proceedings of the National Academy of Sciences, USA,* **83**, 6682–6.

Muscoplat, C., Johnson, D. & Pomeroy, K. *et al.* (1974). Lymphocyte surface immunoglobulin: frequency in normal and lymphocytotic cattle. *American Journal of Veterinary Research,* **35**, 593–5.

Onuma, M., Tsukiyama, K, Ohya, K., Morishima, Y. & Ohno, R. (1987). Detection of cross-reactive antibody to BLV p24 in sera of human patients infected with HTLV. *Microbiological Immunology,* **31**, 131–7.

Oroszlan, S., Sarngadharan, M., Copeland, T., Kalyanaraman, V., Gilden, R. & Gallo, R. (1982). Primary structure analysis of the major internal protein p24 of human type C T-cell leukemia virus. *Proceedings of the National Academy of Sciences, USA,* **79**, 1291–5.

Oroszlan, S., Copeland, T., Kalyanaraman, V., Sarngadharan, M., Schultz, A. & Gallo, R. (1984). Chemical analysis of human T-cell leukemia virus structural proteins. In *Human T-Cell Leukemia/Lymphoma Virus,* ed. R. Gallo, M. Essex and L. Gross, pp. 101–110, Cold Spring Harbor Laboratory.

Paul, P., Pomeroy, K., Castro Johnson, D., Muscoplat, C. & Sorensen, D. (1977). Detection of bovine leukemia virus in B-lymphocytes by the syncytia induction assay, *Journal of National Cancer Institute,* **59**, 1269–71.

Portetelle, D., Callebaut, I., Bex, F. & Burny, A. (1993). Vaccination against animal retroviruses. In *Progress in Vaccinology. Vol 4: Veterinary Vaccines*, ed. R. Pandey, S. Hoglund and G. Prasad, pp. 87–138, Berlin: Springer Verlag.

Powers, M., Grossman, D., Kidd, L. & Radke, K. (1991). Episodic occurrence of antibodies against the bovine leukemia virus rex protein during the course of infection in sheep, *Journal of Virology*, **65**, 4959–65.

Rice, N., Stephens, R., Couez, D. *et al.* (1984). The nucleotide sequence of the *env* gene and post-*env* region of bovine leukemia virus, *Virology*, **138**, 82–93.

Rice, N., Stephens, R., Burny, A. & Gilden, R. (1985). The *gag* and *pol* genes of bovine leukemia virus: nucleotide sequence and analysis, *Virology*, **142**, 357–77.

Rice, N., Stephens, R. & Gilden, R. (1987a). Sequence analysis of the bovine leukemia virus genome. In *Enzootic Bovine Leukosis and Bovine Leukemia Virus*, ed. A. Burny & M. Mammerickx, pp. 115–144, Boston: Nijhoff Publ.

Rice, N., Simek, S., Dubois, G., Showalter, S., Gilden, R. & Stephens, R. (1987b). Expression of the bovine leukemia virus X region in virus-infected cells. *Journal of Virology*, **61**, 1577–85.

Rosen, C., Sodroski, J., Kettmann, R., Burny, A. & Haseltine, W. (1985). *Trans*-activation of the bovine leukemia virus long terminal repeat in BLV-infected cells. *Science*, **227**, 320–2.

Rosen, C., Sodroski, J., Kettmann, R. & Haseltine, W. (1986a). Activation of enhancer sequences in type II human T-cell leukemia virus and bovine leukemia virus long terminal repeats by virus-associated *trans*-acting regulatory factors, *Journal of Virology*, **57**, 738–44.

Rosen, C., Sodroski, J., Willems, L. *et al.* (1986b). The region of bovine leukemia virus genome encodes a trans-activator protein. *EMBO Journal*, **5**, 2585–9.

Sagata, N., Yasunaga, J., Tsuzuku-Kawamura, J., Ohishi, K., Ogawa, Y. & Ikawa, Y., (1985a). Complete nucleotide sequence of the genome of bovine leukemia virus: its evolutionary relationship to other retroviruses. *Proceedings of the National Academy of Sciences, USA*, **82**, 677–82.

Sagata, N., Yasunaga, T. & Ikawa, Y. (1985b). Two distinct polypeptides may be translated from a single spliced mRNA of the X genes of human T-cell leukemia and bovine leukemia viruses. *FEBS Letters*, **192**, 37–42.

Sagata, N., Tsuzuku-Kawamura, J., Nagayoshi-Aida, M., Shimuzu, F., Imagawa, K. & Ikawa, Y. (1985c). Identification and some biochemical properties of the major XBL gene product of bovine leukemia virus. *Proceedings of National Academy of Sciences, USA*, **82**, 7879–83.

Sakurai, M., Taneda, A., Nagoya, H. & Sekikawa, K. (1991). Construction and functional characterization of mutants of the bovine leukemia virus trans-activator protein $p34^{tax}$. *Journal of General Virology*, **72**, 2527–31.

Seiki, M., Hattori, Y., Hirayama, Y. & Yoshida, M. (1983). Human adult T-cell leukemia virus: complete nucleotide sequences of the provirus genome integrated in leukemia cell DNA. *Proceedings of the National Academy of Sciences, USA*, **80**, 3618–22.

Shimotohno, K., Takahashi, Y., Shimizu, N. (1985). Complete nucleotide sequence of an infectious clone of human T-cell leukemia virus type II: an open reading frame for the protease gene. *Proceedings of the National Academy of Sciences, USA*, **82**, 3103–7.

Van den Broeke, A., Cleuter, Y., Chen, G. *et al.* (1988). Even transcriptionally competent proviruses are silent in bovine leukemia virus-induced sheep tumour cells. *Proceedings of the National Academy of Sciences, USA*, **85**, 9263–7.

Van der Maaten, M. & Miller, J. (1976). Replication of bovine leukemia virus in monolayer cell culture, *Bibliographica Haematologia*, **43**, 360–2.

Voneche, V., Portetelle, D. & Kettmann, R. *et al.* (1992). Fusogenic segments of bovine leukemia virus and simian immunodeficiency virus are interchangeable and mediate fusion by means of oblique insertion in the lipid bilayer of their target cells. *Proceedings of the National Academy of Sciences, USA*, **89**, 3810–14.

Weiss, R., Teich, N., Varmus, M. & Coffin, J. (1984). *RNA Tumor Viruses*, vol. 1, pp. 1292, NY: Cold Spring Harbor Laboratory, Cold Spring Harbor.

Willems, L., Bruck, C., Portetelle, D., Burny, A. & Kettmann, R. (1987*a*). Expression of a cDNA clone corresponding to the long open reading frame (XBL-I) of the bovine leukemia virus. *Virology*, **160**, 55–9.

Willems, L., Gegonne, A., Chen, G., Burny, A., Kettmann, R. & Ghysdael, J. (1987*b*). The bovine leukemia virus p34 is a transactivator protein. *EMBO Journal*, **6**, 3385–9.

Willems, L., Chen, G., Burny, A. & Kettmann, R. (1988). Expression in bacteria of β-galactosidase fusion proteins carrying antigenic determinants of two X gene products of bovine leukemia virus. *Leukemia*, **2**, 1–5.

Willems, L., Chen, G., Portetelle, D., Mamoun, R., Burny, A. & Kettmann, R. (1989). Structural and functional characterization of mutants of the bovine leukemia virus transactivator protein p34. *Virology*, **171**, 615–18.

Willems, L., Heremans, H., Chen, G. *et al.* (1990). Cooperation between bovine leukemia virus transactivator protein and Ha-ras oncogene in cellular transformation. *EMBO Journal*, **9**, 1577–81.

Willems, L., Kettmann, R. & Burny, A. (1991). The amino acid (157–197) peptide segment of bovine leukemia virus p34[tax] encompass a leucine-rich globally neutral activation domain. *Oncogene*, **6**, 159–63.

Willems, L., Kettmann, R., Chen, G., Portetelle, D., Burny, A. & Derse, D. (1992*a*). A cyclic AMP-responsive DNA-binding protein (CREB2) is a cellular transactivator of the bovine leukemia virus long terminal repeat. *Journal of Virology*, **66**, 766–72.

Willems, L., Grimonpont, C., Heremans, H. *et al.* (1992*b*). Mutations in the bovine leukemia virus TAX protein can abrogate the long terminal repeat-directed transactivating activity without concomitant loss of transforming potential. *Proceedings of the National Academy of Sciences, USA*, **89**, 3957–61.

Willems, L., Portetelle, D., Kerkhofs, P. *et al.* (1992*c*). *In vivo* transfection of bovine leukemia provirus into sheep. *Virology*, **189**, 775–7.

Willems, L., Thienpont, E., Kerkhofs, P., Burny, A., Mammerickx, M. & Kettmann, R. (1993*a*). Bovine leukemia virus, an animal model for the study of intrastrain variability. *Journal of Virology*, **67**, 1086–9.

Willems, L., Kettmann, R., Dequiedt, F. *et al.* (1993*b*). *In vivo*, infection of sheep by bovine leukemia virus mutants. *Journal of Virology*, **67**, 4078–85.

Wyatt, C., Wingett, D. & White, J. *et al.* (1989). Persistent infection of rabbits with bovine leukemia virus associated with development of immune dysfunction. *Journal of Virology*, **63**, 4498–506.

Yoshinaka, Y. & Oroszlan, S. (1985). Bovine leukemia virus post-envelope gene coded protein: evidence for expression in natural infection. *Biochemical-Biophysical Research Communications*, **131**, 347–54.

Yoshinaka, Y., Katoh, I., Copeland, T., Smythers, G. & Oroszlan, S. (1986). Bovine leukemia virus protease: purification, chemical analysis, and *in vitro* processing of gag polyproteins. *Journal of Virology*, **57**, 826–32.

Zandomeni, R., Carrera-Zandomeni, M., Esteban, E. & Ferrer, J. (1991). The *trans*-activating C-type retroviruses share a distinct epitope(s) that induces antibodies in certain infected hosts. *Journal of General Virology,* **72**, 2113–19.

Zandomeni, R., Carrera-Zandomeni, M., Esteban, E., Donawich, W. & Ferrer, J. (1992). Induction and inhibition of bovine leukemia virus expression in naturally infected cells. *Journal of General Virology,* **73**, 1915–24.

TRANSMISSION AND CONTROL OF FELINE LEUKAEMIA VIRUS INFECTIONS

O. JARRETT

University of Glasgow, Department of Veterinary Pathology
Bearsden, Glasgow G61 1QH, UK.

INTRODUCTION

Feline leukaemia virus (FeLV) infection of the domestic cat is often considered as a model system in which to study the molecular events in leukaemogenesis, and indeed the study of FeLV has been most rewarding for this purpose. However, because FeLV is an important pathogen in a popular companion animal, a great deal of research has centred on understanding the natural history of the virus and devising methods to control the infection. Consequently, probably more is known about these aspects in FeLV infection than in infection with any other oncogenic retrovirus.

FeLV was discovered during experiments to investigate the cause of lymphosarcoma in pet cats (Jarrett *et al.*, 1964). This condition, and other tumours of the haemopoietic system, had long been recognized as by far the most common cancers of the cat. The occurrence of multiple cases of these diseases in individual households of cats over relatively short periods of time suggested that they might be caused by an infectious agent, with a virus similar to those known to cause analogous diseases in laboratory mice and domestic poultry as the most likely candidate. Once the virus was isolated, diagnostic methods were developed so that the epidemiology of the infection could be defined (Hardy *et al.*, 1969). It became clear that the virus was ubiquitous among domestic cats throughout the world. For example, the prevalence in healthy cats in the UK has been estimated recently as 5% (Hosie, Robertson & Jarrett, 1989). Transmission was shown to occur vertically and horizontally by the spread of virus from persistently infected cats (Hardy *et al.*, 1975a; Jarrett *et al.*, 1973). However, many exposed cats were found to recover from the infection and become immune (Hardy *et al.*, 1976). These last two findings laid the groundwork for the development of measures for control of the infection, first by management practices and subsequently by vaccination, which have been very successful in reducing the incidence of the infection and its associated diseases in many populations of cats (Hardy *et al.*, 1975a, b; Weijer & Daams, 1978).

FeLV subgroups

The source of FeLV infection is the persistently viraemic cat. FeLVs may be isolated readily from the plasma of these cats and grown in cultures of feline embryonic fibroblasts (Jarrett, Laird & Hay, 1970). In these cells, three subgroups of the virus can be distinguished by interference assay (Sarma & Log, 1971, 1973). This assay is based on the observation that cells infected with virus of one subgroup are resistant to superinfection with viruses of the same subgroup while remaining susceptible to viruses of other subgroups. Subgroup is determined by elements in the envelope surface glycoprotein, gp70. The three subgroups, A, B or C, have an unusual distribution in that all isolates contain FeLV-A while approximately half contain FeLV-B in addition (Jarrett et al., 1978). FeLV-C is uncommon, being detected in only approximately 1–2% of isolates.

FeLV-A is the archetypal FeLV which is readily transmissible among cats in nature. The other two subgroups are ultimately derived from FeLV-A, either by recombination with endogenous FeLV-related env genes in the case of FeLV-B (Stewart et al., 1986), or by mutation within the env gene in the case of FeLV-C (Rigby et al., 1992). Generation of these novel viruses is important in the pathogenesis of FeLV-related diseases since the occurrence of FeLV-B is associated with an increased risk of infection and of tumour production, and of FeLV-C with fatal anaemia.

In the persistently infected cat, virus is found free in the plasma in relatively high concentrations. This virus originates in the bone marrow in which essentially all nucleated cells are infected and release virus during mitosis (Rojko et al., 1979). The virus is also produced by cells of most epithelial surfaces including those of the respiratory tract, intestine, bladder, mammary gland and salivary gland. Natural infection occurs mainly by transfer of contaminated saliva between cats in the course of mutual licking and grooming (Francis, Essex & Hardy, 1977), across the placenta to the developing foetus (Hoover, Rojko & Quackenbush, 1983), or in the milk (Pacitti, 1989).

Age-related susceptibility of cats to FeLV

The susceptibility of cats to FeLV is very dependent upon age (Hoover et al., 1976). The foetus and the newborn kitten are most susceptible. Up to the age of approximately 14 weeks, kittens are readily infected with FeLV either in nature or experimentally. However, after that age, kittens rapidly become resistant to infection until by 6 months of age only around 15% of cats can be infected with a dose of virus which would infect all newborn kittens. The age resistance to FeLV exhibited by kittens over 4 months of age poses particular problems in determining the efficacy of vaccines, as

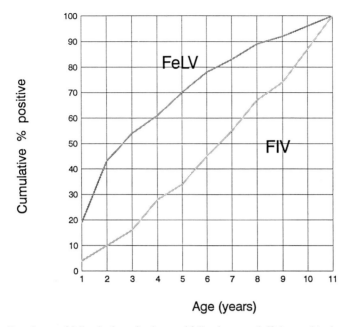

Age (years)

Fig. 1. Prevalence of feline leukaemia virus and feline immunodefficiency virus in pet cats.

described below. The influence of age on the prevalence of FeLV-related diseases in the field is clearly demonstrated by considering the cumulative prevalence of FeLV infection in sick pet cats, shown in Fig. 1. The age at which 50% of sick FeLV-positive cats occur is approximately 2.5 years compared to 6.5 years for sick cats infected with feline immunodeficiency virus (FIV) (Hosie et al., 1989). The most plausible explanation for this difference is that most FeLV infections in the field occur at an early age and the latent period to the development of disease in infected cats is quite short, while infection with FIV is not age-restricted and the latent period to the occurrence of AIDS is long.

Transmissibility of each FeLV subgroup is different

The situation described above relates only to infection with FeLV-A. The efficiency of transmission of the other two subgroups is much less. Even newborn kittens are relatively resistant to FeLV-B and only about 25% of kittens infected as neonates by inoculation with this subgroup become persistently viraemic. In the case of FeLV-C infection, newborn kittens are completely susceptible but become resistant after approximately 4 weeks of age. The resistance of kittens to infection with subgroups B and C can be overcome by inoculation with mixtures of A and B or C (Jarrett & Russell, 1978; Jarrett et al., 1984). In the case of inoculation of a mixture of FeLV-A

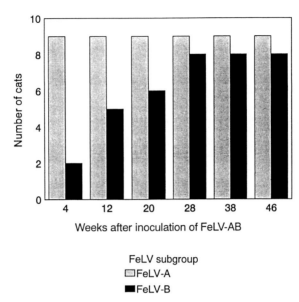

Fig. 2. Feline leukaemia virus subgroups isolated from cats infected with a mixture of subgroups A and B.

and B into neonatal kittens, FeLV-A grows rapidly and appears in the plasma after 3–4 weeks of inoculation while FeLV-B grows more slowly, as shown in Fig. 2. From the Figure, it is also seen that not all cats inoculated with the mixture become viraemic with FeLV-B. It seems that, in these circumstances, FeLV-A enhances the growth of viruses of the other two subgroups, probably by the formation of phenotypic mixtures which permit the spread of FeLV-B or C into normally resistant cells through their FeLV-A envelopes. Virus does not appear to be spread very effectively, if at all, by contact from cats which are viraemic with either FeLV-B or -C alone (Sarma *et al.*, 1978; Jarrett, 1981). However FeLV-B can be transmitted naturally from cats that are persistently viraemic with both FeLV-A and B (Jarrett & Russell, 1978). As in the case of experimental infections with mixtures, the growth of FeLV-B in the recipient cats is restricted to the extent that in a proportion of these cats, FeLV-B is never found in the plasma and presumably dies out. Hence the survival of FeLV-B in the cat population almost certainly requires the constant generation of new viruses through recombination between FeLV-A and endogenous FeLV *env* genes (Stewart *et al.*, 1986).

FeLV-C is associated with anaemia

While FeLV-C is isolated from only a very small proportion of all isolates, viruses of this subgroup are found in almost 40% of cats with a severe, non-regenerative anaemia, analogous to pure red cell aplasia in man (Onions *et al.*, 1982). Four independent isolates of FeLV-C induced an identical disease following inoculation of young kittens (Mackey *et al.*, 1975; Onions *et al.*, 1982). Examination of the nucleotide sequence of the *env* genes of several FeLV-C isolates indicated that each had been generated as a mutant of FeLV-A (Rigby *et al.*, 1992; Brojatsch *et al.*, 1992). These mutations clustered in variable region 1 of the *env* gene suggesting that this region is responsible for subgroup specificity of the virus and therefore presumably is involved in binding to cellular receptors. The variety of amino acid changes among FeLV-C isolates also indicates that each virus had arisen as an individual mutation. Following inoculation into newborn kittens, FeLV-C is acutely pathogenic causing fatal anaemia within a few weeks of inoculation (Onions *et al.*, 1982). Therefore it would seem reasonable that, following the generation of a FeLV-C in a cat viraemic with FeLV-A, only a short period elapses before fatal disease develops. Indeed, experimental inoculation of FeLV-C into adult cats already viraemic with FeLV-A leads to the appearance of FeLV-C in the plasma and then rapidly to fatal disease (Jarrett *et al.* 1984). Therefore the opportunity for spread of FeLV-C in nature is likely to be very limited since the virus is lost with the death of the cat, which may explain the infrequent isolation of this virus in the field.

CONSEQUENCES OF FeLV INFECTION

Persistent viraemia

Following natural infection with FeLV, there are three possible outcomes. The animals may become persistently viraemic, may recover or may develop a latent infection. A summary of the consequences of FeLV infection is given in Fig. 3. Persistently infected animals have a continuous viraemia with titres of approximately 1×10^6 focus forming units per ml of plasma (Jarrett *et al.*, 1982). In addition, virus is excreted from the mouth, urine and in the milk. Viraemic animals appear to be immunotolerant or anergic and make no, or very slight, antibody responses to viral proteins (Hardy *et al.*, 1976). The prognosis for these animals is extremely poor, 85% of them dying within three and half years of infection (McClelland, Hardy & Zuckerman, 1980). Following intranasal instillation of virus into kittens, which simulates natural infection, virus appears in the blood approximately three weeks after infection (Jarrett *et al.*, 1982). From the nasopharynx, virus spreads to regional lymph nodes and then to the bone marrow where the large number of rapidly dividing haemopoietic cells provides an ideal substrate for the rapid growth of virus (Rojko *et al.*, 1979). Subsequently, the virus spreads haematogenously to infect the stem cells of many epithelial tissues.

Recovery from FeLV infection

At any time during this cycle infection may be terminated, presumably by an immune response to the virus, as indicated in Fig. 3. Virus neutralizing

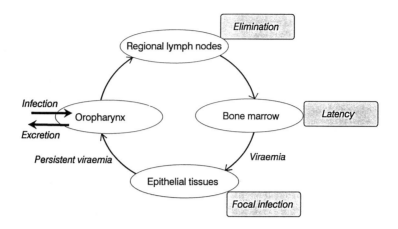

The consequences of the immune response to FeLV are indicated in the shaded boxes.

Fig. 3. Spread of feline leukaemia virus in the infected cat.

antibodies are found in the sera of recovered cats (Sarma *et al.*, 1974; Hardy *et al.*, 1976) and protect cats from subsequent challenge. It is likely that cytotoxic T cell responses are also involved in recovery but methods have yet to be devised to investigate their possible role. If the immune response occurs before viral growth is firmly established in the bone marrow, then the virus may be eliminated from the animal. However, it is likely that in most infections the cessation of productive infection takes place after the bone marrow has become involved since a transient viraemia often occurs prior to recovery (Jarrett *et al.*, 1982). In this case the immune response appears to contain infection within the marrow, and possibly other tissues, producing what has been termed a 'latent' infection (Post & Warren, 1980; Rojko *et al.*, 1982). Shortly after natural infection by contact, a high proportion of apparently recovered cats appears to have a latent infection, but this proportion declines with time until by three years after infection only about 10% of cats still carry virus in their bone marrow (Pacitti & Jarrett, 1985). These findings lead to the suspicion that most 'recovered' cats may not really recover from FeLV infection at all but may harbour small foci of covert infection for the rest of their lives. Supportive evidence for this proposition comes from the observation that virus neutralizing antibody levels to FeLV in recovered cats do not decline over a period of several years suggesting that there is continuous antigenic challenge from within these animals (Pacitti & Jarrett, 1985). This non-productive persistent state may be significant in pathogenesis and diagnosis of infection in a small proportion of cats.

However, it is very doubtful if these animals have a major impact on the transmissibility of FeLV in the field.

CONTROL OF FeLV INFECTIONS

Control by management

The control of FeLV infections in pet cats relies on appropriate management combined with vaccination. Previously, many infections occurred in multiple cat households, often containing pedigree cats used for breeding. Before the availability of vaccines, control measures were designed to separate viraemic from non-viraemic cats in these households with the aim of removing the source of infection. Simple test-and-removal programmes based on rapid diagnostic methods have been highly successful in reducing or eliminating FeLV infections in cat households (Hardy et al., 1975b; Weijer and Daams, 1978).

Control by vaccination

The observations that FeLV was horizontally transmitted and that recovered cats mounted an immune response which protected them from further challenge suggested that vaccination might be possible. The feasibility of vaccination was demonstrated (Jarrett et al., 1975), and commercial vaccines based on inactivated viral products became available. One of the major difficulties in developing FeLV vaccines has been the establishment of a single reliable method for demonstrating vaccine efficacy. A problem in assessing efficacy of FeLV vaccines is that kittens over 14 weeks of age rapidly become resistant to FeLV, as indicated above. Hence it is often difficult to achieve persistent viraemia in a high proportion of cats following artificial or natural challenge of kittens. Because of this problem it has been difficult to compare the efficacy of the several vaccines which are available. A common way to circumvent these difficulties has been to administer corticosteroid at the same time as an oronasal challenge in order to reduce the susceptibility of the kittens. The rationale for this process is that corticosteroid increases the susceptibility of macrophages to the virus and overcomes age-related resistance (Hoover et al., 1981). A second method has been to use intraperitoneal challenge (Marciani et al., 1991). Both of these methods, of course, suffer from the objection that they do not accurately simulate the events during natural infection. Nevertheless, there are data that current vaccines tested by these methods do afford a reasonable measure of protection both on experimental challenge and in the field. The third method of challenge is by administration of a relatively high dose of virus oronasally to 14 week old kittens. In this way, a high proportion of unvaccinated kittens becomes persistently infected (Tartaglia et al., 1993). However, a potential disadvantage of this method is that only a short period of two weeks is available between vaccination and challenge in which time the immune response may not have developed sufficiently in vaccinated animals to protect the kittens from challenge.

Another problem in the development of FeLV vaccines has been that the available immunogens have not been particularly potent. The perceived wisdom is that attenuated viruses would not be acceptable as a vaccine for FeLV. This is probably true since lifetime studies would be required to ensure safety of the product. In the absence of a modified live virus vaccine, inactivated or subunit vaccines have been used. All of the currently available vaccines developed in the USA are based on chemically inactivated FeLV or FeLV products. Efficacies of 80–100% have been claimed for these products. However, there has been controversy about the efficacy of some of these vaccines under natural conditions of challenge. More recently a subunit vaccine incorporating the protein moiety of the surface glycoprotein of the virus, has been produced in Europe and has been shown to afford good protection (Marciani *et al.*, 1991). This product is expressed in *E. coli* and therefore is neither glycosylated nor in a native conformation. Nevertheless, it does appear to protect cats from infection.

Other experimental vaccines have been reported which have yet to be developed for commercial use and of these, two are of considerable interest. The first comprises gp70 derived from whole virus incorporated into immune stimulating complexes (ISCOMs) (Osterhaus *et al.*, 1985). This is the only product which has consistently induced detectable neutralizing antibodies in a large proportion of vaccinated animals and therefore might be expected to give high levels of protection on challenge. The second vaccine is a canarypox FeLV *env–gag* recombinant virus (Tartaglia *et al.*, 1993). In contrast to the ISCOM vaccine, this vaccine induced no neutralising antibodies before challenge. Nevertheless, it gave excellent protection against oronasal instillation of virus. This finding prompted the suggestion that cell mediated immunity might have been the determinant of protection in this case, and that perhaps epitopes within the gag proteins were responsible. However, the absence of neutralizing antibodies in vaccinated cats prior to challenge does not necessarily imply either that the vaccine did not prime for a neutralizing antibody response or that an immune mechanism other than antibody is necessary for protection. It is quite clear that antibody alone can protect cats from challenge with FeLV. Thus, kittens which naturally acquire maternal antibodies (Haley *et al.*, 1985, Jarrett, Russell & Stewart, 1977) or cats to whom antibody is passively administered are resistant to challenge (Hoover *et al.*, 1977). Of course, these findings do not rule out a major role for cytotoxic T cell responses in immunity to FeLV.

The success of management practices in the control of FeLV infections is well recognized. In future it will be important to investigate by field studies how well the infection can be controlled by vaccination alone. On a more basic level, it will be of considerable veterinary and comparative interest to determine the immune mechanisms leading to both natural and vaccinal immunity and to define the immunogens and epitopes responsible. In this way even more effective FeLV vaccines may be produced, and valuable lessons will be learned for the development of vaccines against other retroviral infections of animals and man.

REFERENCES

Brojatsch, J., Kristal, B. S., Viglianti, G. A., Khiroya, R., Hoover, E. A. & Mullins, J. I. (1992). Feline leukemia virus subgroup C phenotype evolves through distinct alterations near the N terminus of the envelope surface glycoprotein. *Proceedings of the National Academy of Sciences, USA*, **89**, 8457–61.

Francis, D. P., Essex, M. & Hardy, W. D. J. (1977). Excretion of feline leukaemia virus by naturally infected pet cats. *Nature*, **269**, 252–4.

Haley, P. J., Hoover, E. A., Quackenbush, S. L., Gasper, P. W. & Macy, D. W. (1985). Influence of antibody infusion on the pathogenesis of experimental feline leukemia virus infection. *Journal of the National Cancer Institute*, **74**, 821–5.

Hardy, W. D., Geering, G., Old, L. J. & deHarven, E. (1969). Feline leukemia virus: occurrence of viral antigen in the tissues of cats with lymphosarcoma and other diseases. *Science*, **166**, 1019–21.

Hardy, W. D. J., Hess, P. W., Essex, M., Cotter, S., McClelland, A. J. & MacEwen, G. (1975a). Horizontal transmission of feline leukemia virus in cats. *Bibliotheca Haematologica*, **40**, 67–74.

Hardy, W. D. J., Hess, P. W., MacEwen, E. G. et al. (1976). Biology of feline leukemia virus in the natural environment. *Cancer Research*, **36**, 582–8.

Hardy, W. D. J., McClelland, A. J., Zuckerman, E. E. et al. (1975b). Prevention of the contagious spread of the feline leukemia virus between pet cats. *Bibliotheca Haematologica*, **43**, 511–14.

Hoover, E. A., Olsen, R. G., Hardy, W. D. J., Schaller, J. P. & Mathes, L. E. (1976). Feline leukemia virus infection: age-related variation in response of cats to experimental infection *Journal of the National Cancer Institute*, **57**, 365–9.

Hoover, E. A., Rojko, J. L. & Quackenbush, S. L. (1983). Congenital feline leukemia virus infection. *Leukaemia Reviews International*, **1**, 7–8.

Hoover, E. A., Rojko, J. L., Wilson, P. L. & Olsen, R. G. (1981). Determinants of susceptibility and resistance to feline leukemia virus: I. Role of macrophages. *Journal of the National Cancer Institute*, **67**, 889–98.

Hoover, E. A., Schaller, J. P., Mathes, L. E. & Olsen, R. G. (1977). Passive immunity to feline leukemia: evaluation of immunity from dams naturally infected and experimentally vaccinated. *Infections and Immunology*, **16**, 54–9.

Hosie, M. J., Robertson, C. & Jarrett, O. (1989). Prevalence of feline leukemia virus and antibodies to feline immunodeficiency virus in cats in the United Kingdom. *Veterinary Record*, **125**, 293–7.

Jarrett, O. (1981). Natural occurrence of subgroups of feline leukemia virus. *Cold Spring Harbor Conferences on Cell Proliferation*, **8**, 603–11.

Jarrett, O., Golder, M. C. & Stewart, M. F. (1982). Detection of transient and persistent feline leukemia virus infections. *Veterinary Record*, **110**, 225–8.

Jarrett, O., Golder, M. C., Toth, S., Onions, D. E. & Stewart, M. F. (1984). Interaction between feline leukemia virus subgroups in the pathogenesis of erythroid hypoplasia. *International Journal of Cancer*, **34**, 283–8.

Jarrett, O., Hardy, W. D. J., Golder, M. C. & Hay, D. (1978). The frequency of occurrence of feline leukemia virus subgroups in cats. *International Journal of Cancer*, **21**, 334–7.

Jarrett, O., Laird, H. M. & Hay, D. (1970). Growth of feline leukemia virus in human, canine and porcine cells. *Bibliotheca Haematologica*, 387–92.

Jarrett, O. & Russell, P. H. (1978). Differential growth and transmission in cats of feline leukemia viruses of subgroups A and B. *International Journal of Cancer*, **21**, 466–72.

Jarrett, O., Russell, P. H. & Stewart, M. F. (1977). Protection of kittens from feline leukemia virus infection by maternally-derived antibody. *Veterinary Record*, **101**, 304–5.

Jarrett, W., Jarrett, O., Mackey, L., Laird, H. M., Hardy, W. J. & Essex, M. (1973). Horizontal transmission of leukemia virus and leukaemia in the cat. *Journal of the National Cancer Institute*, **51**, 833–41.

Jarrett, W., Jarrett, O., Mackey, L., Laird, H. M., Hood, C. & Hay, D. (1975). Vaccination against feline leukemia virus using a cell membrane antigen system. *International Journal of Cancer*, **16**, 134–41.

Jarrett, W. F. H., Crawford, E., Martin, W. B. & Davie, F. (1964). Leukemia in the cat. A virus-like particle associated with leukemia (lymphosarcoma). *Nature*, **202**, 567.

Mackey, L., Jarrett, W., Jarrett, O. & Laird, H. M. (1975). Anemia associated with feline leukemia virus infection in cats. *Journal of the National Cancer Institute*, **54**, 209–17.

Marciani, D. J., Kensil, C. R., Beltz, G. A., Chung-ho, H., Cronier, J. & Aubert, A. (1991). Genetically engineered subunit vaccine against feline leukemia virus: protective immune response in cats. *Vaccine*, **9**, 89–96.

McClelland, A. J., Hardy, W. D. & Zuckerman, E. E. (1980). *Feline Leukaemia Virus*, ed. W. D. Hardy, M. Essex & A. J. McClelland, pp. 121–126, New York: Elsevier/North-Holland.

Onions, D., Jarrett, O., Testa, N., Frassoni, F. & Toth, S. (1982). Selective effect of feline leukemia virus on early erythroid precursors. *Nature*, **296**, 156–8.

Osterhaus, A., Weijer, K., Uytdehaag, F., Jarrett, O., Sundquist, B. & Morein, B. (1985). Induction of protective immune response in cats by vaccination with feline leukemia virus iscom. *Journal of Immunology*, **135**, 591–6.

Pacitti, A. (1989). *Current Veterinary Therapy*, ed. R. W. Kirk, pp. 526–529. Philadelphia: W. B. Saunders Company.

Pacitti, A. M. & Jarrett, O. (1985). Duration of the latent state in feline leukemia virus infections. *Veterinary Record*, **117**, 472–4.

Post, J. E. & Warren, L. (1980). *Feline Leukaemia Virus*, ed. W. D. J. Hardy, M. Essex & A. J. McClelland, pp. 151–155, Amsterdam: Elsevier.

Rigby, M. A., Rojko, J. L., Stewart, M. A. *et al.* (1992). Partial dissociation of subgroup C phenotype and in vivo behaviour in feline leukemia viruses with chimeric envelope genes. *Journal of General Virology*, **73**, 2839–47.

Rojko, J. L., Hoover, E. A., Mathes, L. E., Olsen, R. G. & Schaller, J. P. (1979). Pathogenesis of experimental feline leukemia virus infection. *Journal of the National Cancer Institute*, **63**, 759–68.

Rojko, J. L., Hoover, E. A., Quackenbush, S. L. & Olsen, R. G. (1982). Reactivation of latent feline leukaemia virus infection. *Nature*, **298**, 385–8.

Sarma, P. S. & Log, T. (1971). Viral interference in feline leukaemia–sarcoma complex. *Virology*, **44**, 352–8.

Sarma, P. S. & Log, T. (1973). Subgroup classification of feline leukemia and sarcoma viruses by viral interference and neutralization tests. *Virology*, **54**, 160–9.

Sarma, P. S., Log, T., Skuntz, S., Krishnan, S. & Burkley, K. (1978). Experimental horizontal transmission of feline leukemia viruses of subgroups A, B and C. *Journal of the National Cancer Institute*, **60**, 871–4.

Sarma, P. S., Sharar, A., Walters, V. & Gardner, M. (1974). A survey of cats and humans for prevalence of feline leukemia-sarcoma virus neutralizing serum antibodies. *Proceedings of the Society of Experiments in Biology and Medicine*, **145**, 560–4.

Stewart, M. A., Warnock, M., Wheeler, A. *et al.* (1986). Nucleotide sequences of a feline leukemia virus subgroup A envelope gene and long terminal repeat and evidence for the recombinational origin of subgroup B viruses. *Journal of Virology*, **58**, 825–34.

Tartaglia, J., Jarrett, O., Neil, J. C., Desmettre, P. & Paoletti, E. (1993). Protection of cats against feline leukemia virus by vaccination with a canarypox recombinant, ALVAC-FL. *Journal of Virology*, **67**, 2370–5.

Weijer, K. & Daams, J. H. (1978). The control of lymphosarcoma/leukaemia and feline leukaemia virus. *Journal of Small Animal Practice*, **19**, 631–7.

PROGRESSION OF RETROVIRUS INDUCED RODENT T CELL LYMPHOMAS, AND REGULATION OF T CELL GROWTH; AN INSERTIONAL MUTAGENESIS BASED GENETIC STRATEGY

PHILIP N. TSICHLIS, ANTONIOS MAKRIS, CHRISTOS PATRIOTIS, CYRIL B. GILKS,[1] ALFONSO BELLACOSA[2] and SUSAN E. BEAR

Fox Chase Cancer Center, Philadelphia, PA 19111, USA.
[1]Present address: Department of Pathology, University of British Columbia, 2211 Wesbrook Mall, Room G227, Vancouver, BC, V6T-2B5, Canada.
[2]Present address: Institute of Human Genetics, Catholic University Medical School, L.go F. Vito 00168 Roma, Italy.

INTRODUCTION

Preneoplastic and neoplastic cells evolve continuously. Preneoplastic cells evolve into cells with a full neoplastic phenotype while neoplastic cells progress to acquire enhanced malignant potential. The process of continuous selection operating on preneoplastic and neoplastic cells is called tumour progression (Foulds, 1969; Yeatman & Nicholson, 1993).

Tumour cells are genetically unstable (Cheng & Loeb, 1993). Owing to their instability, they become genetically heterogeneous despite their origin from a single cell clone (Nicholson, 1991; Nowell, 1990; Cavenee, 1989). Tumour progression selects from this heterogeneous population, cells that exhibit a growth advantage, or enhanced ability to metastasize into distant sites. The enhanced metastatic potential of selected cell clones depends on a series of attributes (Liotta, 1986; Aanavoorian *et al.*, 1993) including: a) the ability to invade locally and enter the lymphatic or blood vessels, b) the ability to exit into the extravascular space by adhering to high endothelial venules in specific organs, and by dissolving the basal membrane, and c) the ability to grow into the environment of their new site. These processes may require the activity of genes involved in the expression of specialized cellular functions. Thus, lymphoid cell tumours may metastasize by utilising mechanisms that normal lymphocytes employ to regulate homing (Pabst & Binns, 1989; Holzmann & Weissman, 1989; Pals *et al.*, 1989) into different lymphoid organs.

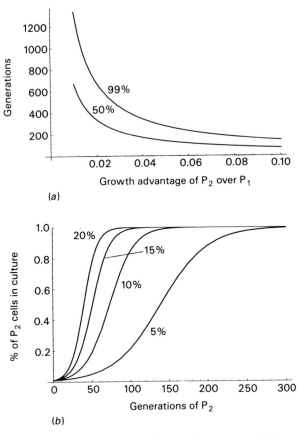

Fig. 1. Outcome of the growth selection operating on cultures composed of two populations of cells (P1 and P2) that differ in growth rates. (*a*) The two curves show, as a function of the growth advantage of P2 over P1, the number of generations required for P2 to reach 50 or 99% of the total number of cells in the culture. (*b*) Dynamic evolution of a culture composed of two populations of cells (P1 and P2) of which the faster growing (P2) exhibits a 5, 10, 15 or 20% growth advantage over the slow growing one and represents 1% of the population at time 0.

GROWTH SELECTION

The selection of cell clones from a heterogeneous population growing in a given site depends on their growth advantage relative to other clones (Fig. 1). Fig. 1(*a*) shows the outcome of selection in cultures composed of two populations of cells (P1 and P2) of which the one (P2), representing 1% of the total population, exhibits a growth advantage over the other (P1). The two curves show quantitatively how the magnitude of the growth advantage of P2 over P1 influences the number of generations required for the P2 cells to reach 50% or 99% of the total cell number. Fig. 1(*b*) shows the dynamic evolution of a culture composed of two populations of cells (P1 and P2) of which the one (P2), representing 1% of the overall population at the beginning of the experiment, exhibits a 5, 10, 15 or 20% growth advantage over the other (P1). These results demonstrate that even small differences in the rate of growth between populations of cells lead to rapid changes in the overall composition of the culture.

Pertinent to the understanding of the metastatic phenotype is the fact that cell growth may not be an absolute but a conditional attribute, which depends on combinations of complex sets of factors. One of these factors, the cellular microenvironment, may promote or inhibit the growth of a given cell clone. During development, cells migrating into new sites are selected positively or negatively by spatially determined processes which either stimulate or inhibit cellular proliferation and differentiation (Oppenheimer & Lefevre, 1989) or, alternatively, instruct the cells to undergo programmed cell death (PCD) (Ellis, Yuan & Horvitz, 1991). Similarly to the growth of differentiating normal cells, the growth of tumour cells also depends on microenvironmental influences. In retrovirus-induced neoplasms, the activation of genes by provirus insertion may deliver spatially determined conditional proliferative signals to the affected cell clones. In the course of our studies on Moloney murine leukaemia virus (MoMuLV)-induced T cell lymphomas in rats, we have identified several loci which appear to contribute in the induction of these tumours (Lazo & Tsichlis, 1990; Tsichlis & Lazo, 1991). One of these (*Mlvi-5*) was the target of provirus integration in 8.3% of the tumours (unpublished). Rearrangement of this locus in at least one T cell lymphoma line (2769) appeared to have diametrically opposite effects on cell growth in different microenvironments. Provirus insertion in *Mlvi-5* was positively selected in the primary tumour-bearing animal. However, examination of the spleens of nude mice transplanted with 2769 cells revealed that cells carrying a provirus in the *Mlvi-5* locus were counterselected (Fig. 2).

The interplay between genes activated by provirus integration in individual metastatic tumours and the cellular microenvironment may provide clues towards understanding the biochemical basis of tumour metastasis.

GENETIC INSTABILITY AND PROGRESSION OF RETROVIRUS-INDUCED NEOPLASMS. SPONTANEOUS INSERTIONAL MUTAGENESIS

The genetic instability of tumour cells gives rise to genetic heterogeneity among clonally derived cell populations and fuels tumour progression (Nicholson, 1991; Nowell, 1990; Cavenee, 1989). In retrovirus-induced lymphoid cell tumours, a significant factor contributing to the genetic instability and heterogeneity among clonally-derived cells is the virus (Bear *et al.*, 1989; Patriotis *et al.*, 1993; Gilks *et al.*, 1993).

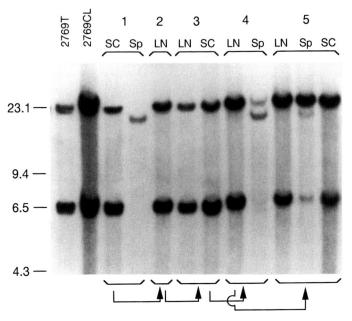

Fig. 2. Counterselection of transplanted 2769 rat thymoma cells, carrying a provirus in the *Mlvi*-5 locus, in the spleen of nude mice. An apparently homogeneous population of 2769 cells (~10^7 cells) harbouring a provirus in the *Mlvi*-5 locus (2769T) were transplanted IP in a Balb/c nude mouse. Southern blot analysis of *Bam*HI digested genomic DNA from tumour cells growing subcutaneously (SC) or in the spleen of the recipient nude mouse and hybridization to an *Mlvi*-5 probe revealed lack of splenic infiltration by the tumour cells. The band detected in the spleen represents the unrearranged mouse *Mlvi*-5 locus. Following additional passages of the tumour cells in nude mice, according to the scheme shown by arrows at the bottom of the Figure, rat thymoma cells began to infiltrate the nude mouse spleen. However, the majority of the infiltrating cells had lost the rearranged *Mlvi*-5 allele. This demonstrates that serial transplantation selects for cells that can infiltrate the nude mouse spleen and that the selected cells may have deleted the rearranged *Mlvi*-5 locus. Alternatively, the selected cells may represent a rare population in which the *Mlvi*-5 locus was in the germline configuration. These cells could have been selected because of their ability to infiltrate the nude mouse spleen.

In the course of our studies of MoMuLV-induced T cell lymphomas in rats we observed that the pattern of provirus integration in individual tumours was unstable (Bear *et al.*, 1989; Patriotis *et al.*, 1993; Gilks *et al.*, 1993). Occasionally, a given integrated provirus detected by Southern blot analysis of tumour cell DNA was lost. More commonly, however, we observed a stepwise increase in the number of integrated proviruses during progression (Fig. 3). Both the loss as well as the gain of a provirus could result from the rearrangement of proviruses integrated in the genome of the tumour cells during earlier steps in oncogenesis. Alternatively, they could result from the net loss or gain of integrated proviruses during progression.

Given that insertion mutations caused by provirus integration play a major role in retrovirus oncogenesis, we asked whether the instability of the

T S   1 2 3

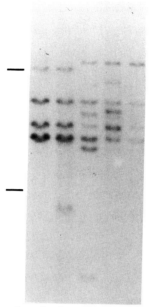

Fig. 3. Instability of the pattern of provirus integration in a MoMuLV induced T-cell lymphoma during progression. Southern blot analysis of *Eco*RI digested genomic DNA from the thymus and spleen of the tumour bearing rat 2769 and hybridization to a MoMuLV LTR probe. Given that *Eco*RI does not cleave the MoMuLV genome each integrated provirus is likely to be defined by a single band. Tumour cells derived from the thymus of the primary tumour bearing animal were passaged IP in three nude mice and cell lines were established in culture from the tumour cells isolated from each of them. Southern blot analysis of genomic DNA from these three cell lines and hybridization to the MoMuLV LTR probe shows that the pattern of provirus insertion in tumour cells is unstable and that it is characterized by both gains and losses of integrated proviruses. Please note that, overall, the gains are more common than the losses.

pattern of provirus insertion contributed to tumour progression. Our initial hypothesis, based on the apparent selection of cells that lost or gained integrated proviruses during progression, was that the genetic changes associated with the changing patterns of retrovirus integration enhanced the growth potential of the affected cell clones. To address this hypothesis, a rat T cell lymphoma line (4437) containing three integrated proviruses was subcloned in microtitre wells to generate 57 single cell clones. DNA from all the clones was subjected to Southern blot analysis and hybridization to a MoMuLV LTR probe. The results (Fig. 4) showed that approximately 95% of the clones were identical to each other in that they contained the same three integrated proviruses as the original mass culture. The low percentage of cells with new provirus insertions suggests that growth selection, if not promoted by the genetic changes associated with provirus integration, should give rise to cell clones that carry the same integrated proviruses as the

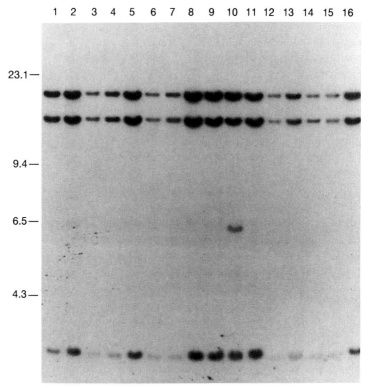

Fig. 4. The majority of the MoMuLV-induced rat thymoma cells in an apparently homogeneous culture are identical with regard to the pattern of provirus insertion. Cells from the apparently homogeneous MoMuLV-induced rat thymoma line 4437 were subcloned into 57 single cell subclones, in microtitre plates. Genomic DNA from all the subclones was digested with *Eco*RI and following Southern blotting it was hybridized to a MoMuLV LTR probe. This shows a representative sample of the results. Lane 1 represents the original mass culture. Lanes 2–16 represent individual single cell clones.

original culture. However, our work has shown that the selected cells, in the majority of cases contain new integrated proviruses (Bear *et al.*, 1989; Patriotis *et al.*, 1993; Gilks *et al.*, 1993). Based on these findings, we tentatively concluded that the majority of genetic events associated with the changing patterns of provirus integration contributed to the growth selection associated with tumour progression.

Net gain of integrated proviruses during progression

To investigate the validity of this conclusion, we proceeded to determine the biological significance of the genetic changes associated with the net gain of integrated proviruses, the most common manifestation of instability in the pattern of provirus insertion. To this end, three randomly selected proviruses detected late in oncogenesis were cloned and two of them were analysed to determine whether they identify loci targeted by provirus insertion in multiple tumours. This was done by hybridizing genomic DNA probes derived from the sequences flanking these proviruses to 130 DNA samples from 40 MoMuLV induced primary T cell lymphomas and their passages in nude mice and in culture. These experiments revealed that both new proviruses identified loci of common integration, targeted by the provirus in the late stages of oncogenesis (*Tpl*-1 and *Tpl*-2, Tumour progression loci 1–2) (Bear *et al.*, 1989; Patriotis *et al.*, 1993; Bellacosa *et al.*, 1994).

Further studies revealed that the *Tpl*-1 locus identified a target of provirus insertion located within 1 kb 5' of the first exon of the *Ets*-1 protooncogene (Bellacosa *et al.*, 1994), while the *Tpl*-2 locus identified a gene encoding a serine/threonine protein kinase (Patriotis *et al.*, 1993; Makris *et al.*, 1993). The Tpl-2 kinase was shown to be identical or closely related to the kinase encoded by an oncogene (Cot) which was identified by its accidental activation during transfection of genomic DNA from a human thyroid carcinoma cell line into SHOK cells (Miyoshi *et al.*, 1991), and the protein kinase Est-1 which was isolated by expression cDNA cloning from a Ewing's sarcoma cDNA library (Chan *et al.*, 1993). Please note that Cot should not be confused with another serine/threonine protein kinase, also named Cot which is required for hyphal elongation in *Neurospora crassa* (Yarden *et al.*, 1992).

The Tpl-*1 locus*

Provirus integration in the *Tpl*-1/*Ets*-1 locus is associated with a modest enhancement of the steady state levels of *Ets*-1 mRNA in the tumour cells (Bellacosa *et al.*, 1994). The sites of transcriptional initiation, and the relative ratio of full length mRNA transcripts and transcripts arising by differential splicing of exon VII or differential polyadenylation are not affected by provirus insertion (Bellacosa *et al.*, 1994). The subtlety of the effects of provirus integration on the expression of *Ets*-1, combined with the strong selection operating on cells with a provirus in the *Tpl*-1/*Ets*-1 locus (Bear *et al.*, 1989), led us to suggest that provirus insertion may affect the fine regulation of the gene during cell cycle progression (Bellacosa *et al.*, 1994).

The Tpl-2 *locus*

Tpl-2 encodes a 55 kd serine/threonine protein kinase which is expressed primarily in spleen, thymus and lung (Makris *et al.*, 1993). *Tpl*-2 rearrangements were detected in approximately 22.5% of all the tumours. Since these rearrangements were detected in cell lines maintained in culture as well as in primary tumours, it was concluded that the growth selection of cells with

Tpl-2 rearrangements operates not only in culture but also *in vivo* (Patriotis *et al.*, 1993).

Tpl-2 spans a 35 kb genomic DNA region and contains eight exons (Makris *et al.*, 1993). During tumour progression provirus insertion occurs in the last intron and in the same transcriptional orientation with the *Tpl*-2 gene. This leads to the enhanced expression of a truncated RNA transcript in which the 3′ exon is replaced by intron and proviral LTR sequences. This transcript encodes a protein in which the 43 C-terminal amino acids encoded by the eighth exon are replaced by seven amino acids encoded by the last intron (Patriotis *et al.*, 1993, Makris *et al.*, 1993) (Fig. 5).

Tpl-2 is transcribed from two alternating promoters P1 and P2 (Fig. 5). The RNA transcripts originating in the two promoters harbour different 5′ untranslated regions derived from the alternate noncoding exons 1A and 1B (Makris *et al.*, 1993). Transcripts with 5′ untranslated regions derived from exon 1B are translated more efficiently than transcripts with 5′ untranslated regions derived from exon 1A. Utilization of the P2 promoter, giving rise to exon 1B containing RNA transcripts, was detected primarily in tumour cells (Makris *et al.*, 1993) and in normal T cells during the early steps of T cell activation.

Exposure of normal rat spleen cells to concanavalin A (ConA) induces the expression of *Tpl*-2 within the first 60 minutes from the time of exposure suggesting that *Tpl*-2 may be involved normally in the transition from a quiescent to the G1 phase of the cell cycle (Patriotis *et al.*, 1993). Enhanced *Tpl*-2 expression during ConA stimulation is also associated with a shift in promoter usage from promoter P1 to promoter P2.

Provirus insertion in *Tpl*-2 elicits a complex set of events that contribute to its role in tumour progression. First, the steady state level of *Tpl*-2 mRNA is enhanced perhaps because of a combination of enhancer insertion and mRNA stabilization due to truncation and deletion of potentially destabilizing sequences in the 3′ untranslated region of the gene. Secondly, the kinase encoded by the activated *Tpl*-2 locus carries a deletion of its C-terminal tail (Patriotis *et al.*, 1993; Makris *et al.*, 1993). This deletion renders the kinase constitutively active and enhances its transforming potential (Patriotis *et al.*, unpublished observations). Finally, perhaps due to the physiological state of activation of the tumour cells, *Tpl*-2 is transcribed primarily from promoter P2 (Makris *et al.*, 1993). The promoter P2 derived, exon 1B containing RNA transcripts, are translated more efficiently than the transcripts originating in promoter P1.

Tpl-2 was mapped near the centromere on rat chromosome 17 (Yeung *et al.*, 1993), mouse chromosome 18 and human chromosome 10p11 (Justice *et al.*, 1993). The *Tpl*-2 map location in humans defines a region characterized by karyotypic abnormalities associated with adult T cell leukaemia (ATL) and acute non-lymphocytic leukaemia (ANLL) (Shiraishi *et al.*, 1985; Fujita *et al.*, 1986; Bernstein, MacDougall & Pinto, 1984). The same region could

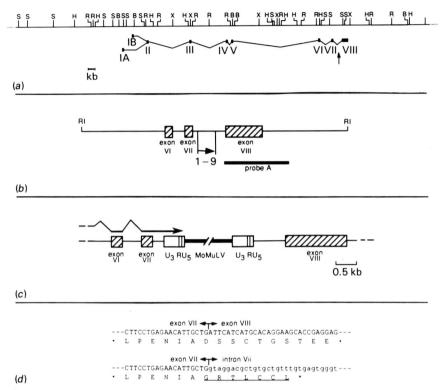

Fig. 5. Genomic organization of the *Tpl-2* gene. Sites and outcome of provirus insertion in the *Tpl-2* locus. (*a*) Restriction endonuclease map and intron/exon structure of the *Tpl-2* locus. The exons are numbered with uppercase Roman numerals and they have been placed directly under their map position in the restriction endonuclease map. The arrow pointing to the last intron shows the region where all known provirus insertions in this locus were detected. S, *Sst*I; H, *Hin*dIII; R, *Eco*RI; X, *Xho*I; B, *Bam*HI. 1(*A*) and 1(*B*) define the two alternate first exons. (*b*) Map of the 3′ end of the *Tpl-2* gene showing the positions of the last three exons relative to the neighbouring *Eco*RI sites, and the region in the last intron where all provirus insertions (1 to 9) were detected. (*c*) mRNA transcripts derived from the rearranged *Tpl-2* locus terminate in the 5′ proviral LTR. (*d*) The truncated mRNA transcripts encode a protein which lacks 43 C terminal amino acids encoded by exon VIII. These 43 amino acids were replaced by a 7-amino-acid peptide encoded by intron vii sequences (underlined).

also be involved in the hereditary syndrome, multiple endocrine neoplasia type II (Matthew *et al.*, 1987; Narod *et al.*, 1992).

Growth selection of cells with a provirus in the Tpl-1 *or* Tpl-2 *locus.*

The apparent selection of cells containing new provirus insertions in both *Tpl-1* and *Tpl-2* late in oncogenesis was confirmed in the experiments shown in Fig. 6 (Bear *et al.*, 1989; Patriotis *et al.*, 1993). Fig. 6(*a*) concentrates on the *in vitro* passage of a group of sublines of a cell line derived from tumour

(a)

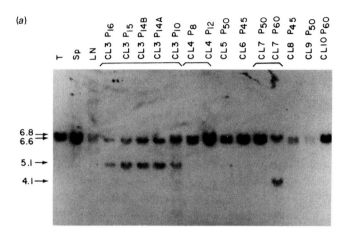

Provirus insertions in the _Tpl_-1 locus in 2772 cells

(b)

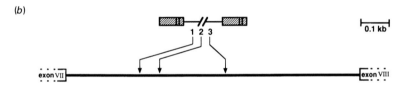

Provirus insertions in the _Tpl_-2 locus in 2769 cells

Fig. 6. Growth selection of cells with a provirus in the *Tpl*-1 or *Tpl*-2 locus. (*a*) Provirus integration occurred independently twice during passage of 2772 tumour sublines in culture, and it was rapidly selected. Southern blot analysis of *Sac*I-digested genomic DNA isolated from the primary tumour 2772 and different passages of eight 2772 tumour sublines hybridized to a *Tpl*-1/*Ets*-1 probe. In primary 2772 tumour cells the probe detects two *Sac*I bands which represent two alleles of a polymorphic *Tpl*-1/*Ets*-1 locus. The 2772 rat therefore was heterozygous for this *Sac*I RFLP. The two events occurred independently because (i) the rearranged band generated by each event was different in size and (ii) a different allele was rearranged by each event. The rearrangement was followed by rapid selection as suggested by the fact that cells carrying the rearrangement dominated the culture within a few passages. (*b*) Sites of provirus integration in the *Tpl*-2 locus in three sublines derived from tumour 2769. Precise mapping of the site of integration was determined by PCR, using a genomic DNA template, cloning and sequencing.

2772. Southern blot analysis of genomic DNA from sequential cell passages and hybridization to a *Tpl*-1 probe showed that provirus insertion in this locus was associated with rapid selection. The same experiment revealed independent *Tpl*-1 provirus insertions in two 2772 sublines. The rapid selection of cells carrying independent provirus insertions in the *Tpl*-1 locus provides strong evidence that this genetic change enhances the growth potential of the tumour cells.

The high frequency of provirus integration in *Tpl*-1 in cells derived from the tumour 2772 (2/10) is in sharp contrast with the low frequency of provirus integration in this locus in primary tumour tissues (1/79) and other tumour cell lines (1/66 DNAs tested). The difference in the frequency of provirus insertion between the cell lines derived from tumour 2772 and all the other thymic lymphomas we tested is statistically significant at the $P = 0.02$ level. This may mean that the effects of provirus insertion in *Tpl*-1 are restricted to cells that exist within a narrow developmental window. Alternatively, the effects of *Tpl*-1 activation on cell growth may depend on interactions between *Tpl*-1 and other genes activated in tumour 2772. In either case, activation of a given gene by provirus insertion may affect cell growth only in a subset of tumour cells. Therefore, a given locus may be associated with tumour progression only in the context of a specific cell phenotype (Bear *et al.*, 1989).

The apparent selection of cells containing a provirus in the *Tpl*-2 locus was confirmed in a similar experiment. During continuous passage in culture, all three sublines derived from tumour 2769 acquired independent provirus insertions in *Tpl*-2 (Patriotis *et al.*, 1993) (Fig. 6(*b*)).

Selection of cells that contain genetically rearranged integrated proviruses

In addition to the net gain of integrated proviruses, tumour progression is often associated with the rearrangement of existing proviruses (Lazo & Tsichlis, 1988; S. Bear *et al.*, unpublished observations). The selection of cells carrying rearranged proviruses suggests that these rearrangements may facilitate the expression of neighbouring genes. This would imply that the genes neighbouring the rearranged proviruses contribute to oncogenesis. Based on these theoretical arguments we cloned a 6.5 kb genomic DNA fragment containing a rearranged provirus from the MoMuLV induced thymoma 6890 and its derivative cell lines. Characterization of the generated genomic clones revealed that the cloned provirus consisted of a solo LTR (S. Bear *et al.*, unpublished observations). Using flanking sequence probes, we showed rearrangements of the DNA region targeted by this virus in 4.3% of the tumours. One of the cell lines carrying a rearrangement was derived from a tumour containing this locus in the germline configuration. On this basis the locus was named *Tpl*-4 (tumour progression locus-4). Further characterization of this locus is in progress.

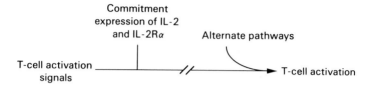

Fig. 7. Diagrammatic representation of the process of T-cell activation. The contribution of *Gfi*-1 was mapped at a time point subsequent to the interaction of IL-2 with the IL-2 receptor. Therefore, it may be involved in the transduction of signals originating in the IL-2 receptor.

GENETIC ANALYSIS OF T CELL ACTIVATION

These data suggested that mutagenesis of retrovirus induced tumours caused by the changing pattern of provirus insertion is a powerful genetic tool that can be used to identify genes whose mutation is associated with selectable phenotypes. One such phenotype is the progression of IL-2 dependent T cell lymphoma lines to IL-2 independence. Given that the interaction between IL-2 and its receptor is a critical event in T cell activation (Fig. 7), the combination of this insertional mutagenesis strategy with the selection for IL-2 independence can be used to identify genes involved in the regulation of the proliferation of T cells following activation (Gilks *et al.*, 1993). To isolate IL-2 independent mutants from IL-2 dependent T cell lymphoma lines undergoing spontaneous insertional mutagenesis, we employed two selection strategies: a) IL-2 was withdrawn from long-term mass cultures of 10^8 tumour cells. After the number of viable cells had decreased by approximately 1000-fold, the remaining viable cells were distributed at a concentration of 10^2–10^3/ml in 0.2 ml microcultures mixed with 10^5 splenocytes as feeder layers. Growth of IL-2 independent mutants was detected within 2–4 weeks. b) When the number of viable cells decreased by approximately 1000-fold following IL-2 withdrawal, the cells were placed in media containing 1/10 the usual concentration of IL-2 (10 units/nl). Following partial recovery, the IL-2 was completely withdrawn. Selection was achieved in the absence of feeder layers. Using a combination of these strategies on two IL-2 dependent rat T cell lymphoma lines (2780 and 5675), we have isolated a total of 33 IL-2 independent mutants. Of these mutants, 30 were derived from 5675 and three from 2780 cells.

Southern blot analysis of genomic DNA from the IL-2 independent lines and hybridization to a MoMuLV LTR probe showed that all the lines derived from tumour 2780 were identical in that all three of them contained a Single 9.2 kb newly integrated provirus (Gilks *et al.*, 1993). The same type of analysis performed in 20 out of the 30 IL-2 independent sublines derived from the tumour cell line 5675 revealed that 16 had acquired new integrated proviruses during transition from the IL-2 dependence to the IL-2 indepen-

Table 1. *Changes in the pattern of provirus integration during the transition of the T cell lymphoma line 5675 from the IL-2 dependent to the IL-2 independent phenotype*

	Number of new provirus insertions.			
	0	1	2	>2
Number of IL-2 independent sublines containing a given number of new integrated proviruses.	4	3	2	11[a]

[a] 8 out of these 11 sublines lost a 23 kb provirus which was present in the parental 5675[d]T cell lymphoma line.

dence phenotype. Five of these sublines contained only 1 or 2 newly integrated proviruses (Table 1).

The Gfi-*1 locus*

The novel integrated provirus detected in one IL-2 independent cell line (2780^{i-5}) selected from spontaneously mutagenised 2780 cells was cloned. A probe derived from the cellular sequences flanking the provirus detected a DNA rearrangement in one additional IL-2 independent cell line out of a total of 24 tested. No rearrangements were detected in IL-2 dependent cell lines. Because of its apparent association with the progression to IL-2 independent growth, the locus targeted by this provirus was named *Gfi*-1 (growth factor independence-1).

A genomic *Gfi*-1 DNA probe detected RNA transcripts in normal adult rat thymus, spleen and testis, in IL-2 independent cell lines carrying a provirus in the *Gfi*-1 locus, and in several additional IL-2 independent cell lines which lacked a *Gfi*-1 provirus. Low levels of expression were detected in IL-2 dependent 2780 cells maintained in IL-2 containing media. Using the same genomic DNA probe we isolated full length cDNA clones from a library constructed from oligo(dT) primed RNA derived from a cell line (2775) lacking a *Gft*-1 provirus. Sequence analysis of this cDNA revealed that *Gfi*-1 encodes a protein with six zinc fingers of the C_2H_2 type (Gilks *et al.*, 1993).

A gene activated by insertional mutagenesis could lead to IL-2 independence by one of three possible mechanisms, as suggested by the simple linear model of T cell activation shown in Fig. 7: (i) Activation of the IL-2 gene either directly or indirectly; (ii) activation of genes expressed during T cell activation as a result of the interaction of IL-2 with its receptor; and (iii) activation of genes which induce IL-2 independent proliferation of T cells by triggering alternate pathways which bypass the need for earlier activation events. Given that *Gfi*-1 does not induce IL-2 expression in IL-2 independent 2780 cells and that it is normally expressed in T cells, we conclude that its activity would be required after the interaction of IL-2 with its receptor.

This was indeed confirmed by Northern blot analysis of polyadenylated RNA from ConA stimulated rat splenocytes and hybridization to the *Gfi*-1 cDNA probe. The level of expression of *Gfi*-1 in unstimulated splenocytes is low. *Gfi*-1 expression started to increase by 12–16 hours, reaching very high levels by 50 hours (Gilks *et al.*, 1993). Precise timing of events during ConA stimulation of bulk splenocytes is not possible as the cells progress asynchronously towards mitosis. However, *Gfi*-1 expression occurs well after IL-2 and IL-2R expression (1–5 hours) and corresponds to the approximate timing of cells entering S-phase. These data suggest that *Gfi*-1 may be active during the transition from the G_1 to the S phase of the cell cycle (Gilks *et al.*, 1993).

The cDNA clone of *Gfi*-1 was introduced into the *Eco*RI site of the MoMuLV based retrovirus vector LXSN and it was transferred into the PA317 amphotropic packaging cell line (Miller & Rosman, 1989). The *Gfi*-1 retrovirus produced by the transfected PA317 cells was used to infect 2780 IL-2 dependent cells. After removal of IL-2, four of six infected cultures gave rise to IL-2 independent lines. These data provided evidence that expression of *Gfi*-1 in IL-2 dependent 2780 cells via retrovirus mediated gene transfer contributes to the emergence of the IL-2 independence phenotype (Gilks *et al.*, 1993).

It should be noted here that both the genes identified using selection for IL-2 independence (*Gfi* loci) as well as the genes identified using selection for enhanced growth in culture (*Tpl* loci) contribute to the growth regulation of T cell lymphomas. Therefore, it is likely that they both contribute to the process of T cell activation.

SPONTANEOUS VS INDUCED INSERTIONAL MUTAGENESIS

The results in the preceding paragraphs confirmed that spontaneous insertional mutagenesis of retrovirus induced neoplasms is a powerful tool that can be used to identify genes associated with selectable phenotypes. Such phenotypes include the rate of proliferation of the tumour cells *in vivo* or in culture, the pattern of tumour metastasis, the progression of growth factor dependent tumour cells to growth factor independence, the development of chemotherapeutic drug resistance, programmed cell death and others.

The identification of genes associated with these phenotypes could also be accomplished by an insertional mutagenesis strategy that depends on superinfection of the tumour cells with a second virus (induced insertional mutagenesis). To avoid superinfection resistance, this approach requires that the *env* gene of the second virus belongs to a different subgroup than the *env* gene of the virus responsible for tumour induction. The induced insertional mutagenesis strategy has the distinct advantage that it permits the measurement of both the background and experimental mutation frequencies. The difference between these mutation frequencies gives an estimate of the probability that the selected phenotypes are due to the genetic changes associated with provirus integration. This is not feasible in individual experiments utilizing the spontaneous insertional mutagenesis strategy. The data presented in this report, however, showed that the majority of genetic events associated with the changing patterns of provirus integration contribute to the emergence of cells expressing the selected phenotype. Therefore, estimating the efficiency of spontaneous insertional mutagenesis in individual experiments may be desirable but not critical. While this disadvantage of the spontaneous insertional mutagenesis strategy is not critical, the advantages it offers make it the method of choice in many applications. a) During *in vivo* studies on tumour progression, the use of spontaneous insertional mutagenesis permits oncogenesis to proceed naturally. Induced insertional mutagenesis on the other hand depends on the superinfection of the tumour cells with another virus which may require an intermediate step of culture *in vitro*. Moreover, superinfection of the tumour cells could give rise to a generalized infection of the tumour bearing animal with a second virus and/or recombinants between the two parental viruses. Both the *in vitro* culture of the tumour cells and the superinfection of the tumour bearing animal with additional viruses could have unpredictable effects on tumour progression. b) Spontaneous insertional mutagenesis avoids superinfection with additional retroviruses which may alter the tumour cell phenotype by direct action. Recent studies have shown that the envelope glycoproteins encoded by the spleen focus forming virus (SFFV) and the murine polytropic retroviruses deliver proliferative signals to the infected cells through their interaction with receptors of the hemopoietin receptor family (Li *et al.*, 1990; Li & Baltimore, 1991; Tsichlis & Bear, 1991). c) Cells undergoing spontaneous insertional mutagenesis frequently carry a single new integrated provirus. On the other hand, the number of new provirus insertions in cells subjected to induced insertional mutagenesis cannot be controlled and it is usually high. Therefore, the analysis of the mutant cells selected following the strategy of spontaneous insertional mutagenesis is usually simpler. d) Spontaneous insertional mutagenesis can be used retrospectively.

CONCLUSIONS

The data discussed in this report indicate that insertional mutagenesis is a powerful tool that can be used to identify genes responsible for tumour progression and other selectable phenotypes. The use of insertional mutagenesis to identify genes contributing to IL-2 independence confirmed its usefulness in studies aiming at the genetic analysis of the cellular responses to external stimuli. Modifications of this strategy based on our developing understanding of signal transduction could lead to highly focused studies that will genetically dissect this process in mammalian cells.

By comparison to other methods that have been used to analyse tumour cells during progression, including the construction and screening of subtraction cDNA libraries (Lee *et al.*, 1991) and differential PCR display (Liang & Pardee, 1992), insertional mutagenesis has the advantage that it identifies the basic genetic changes and not secondary phenotypic changes associated with changing phenotypes. Moreover, this method allows the identification of genes responsible for selectable phenotypes without the influence of any preconceived notions of what their nature might be.

Insertional mutagenesis of retrovirus induced tumour cells can be accomplished by two strategies: spontaneous, which depends on the monitoring of the changing patterns of provirus insertion in retrovirus infected tumour cells, and induced, which depends on superinfection by a second retrovirus. The spontaneous mutagenesis strategy is the method of choice for many applications, because it is simple, it can be used retrospectively, and does not require superinfection with additional viruses that may have unpredictable effects on the tumour cell phenotype.

ACKNOWLEDGEMENTS

We would like to thank Sam Litwin for the generation of the growth curves in Fig. 1. We would also like to thank the members of this laboratory for comments on the manuscript and Pat Bateman for secretarial assistance. Work in this laboratory has been supported by NIH grants CA-38047, CA-51893 and CA-56110, from the ACS grant MV-524 and from the CTR grant #2976. Additional support is provided from NIH grants CA-06927 and RR-05539, and an appropriation from the Commonwealth of Pennsylvania to the Fox Chase Cancer Center. A.M. and C.P. are fellows of the Leukaemia Society of America, Inc. and A.B. was supported by a fellowship from the Greenwald Foundation for Leukaemia and Lymphoma Research.

REFERENCES

Aanavoorian, S., Murphy, A. N., Stetler-Stevenson, W. G. & Liotta, L. A. (1993). Molecular aspects of tumour cell invasion and metastasis. *Cancer*, **71**, 1368–83.

Bear, S. E., Bellacosa, A., Lazo, P. A. *et al.* (1989). Provirus insertion in *Tpl*-1, an *Ets*-1 related oncogene, is associated with tumour progression in MoMuLV induced rat thymic lymphomas. *Proceedings of the National Academy of Sciences, USA*, **86**, 7495–9.

Bellacosa, A., Datta, K., Bear, S. E. *et al.* (1994). Effects of provirus integration in the *Tpl*-1/*Ets*-1 locus in MoMuLV induced rat T cell lymphomas: level of expression, polyadenylation, transcriptional initiation, and differential splicing of the *Ets*-1 mRNA. *Journal of Virology*, in press.

Bernstein, R., MacDougall, L. G. & Pinto, M. R. (1984). Chromosome patterns in 26 South African children with acute nonlymphocytic leukaemia (ANLL). *Cancer Genetics and Cytogenetics*, **11**, 199–214.

Cavenee, W. K. (1989). Tumour progression stage: specific losses of heterozygosity. *International Symposium of the Princess Takamatsu Cancer Research Fund*, **20**, 33–42.

Chan, A. M., Chedid, M., McGovern, E. S., Popescu, N. C., Miki, T. & Aaronson, S. A. (1993). Expression cDNA cloning of a serine kinase transforming gene. *Oncogene*, **8**, 1329–33.

Cheng, K. C. & Loeb, L. A. (1993). Genomic instability and tumour progression: mechanistic considerations. *Advances in Cancer Research*, **60**, 121–56.

Ellis, R. E., Yuan, J. & Horvitz, H. R. (1991). Mechanisms and functions of cell death. *Annual Review of Cell Biology*, **7**, 663–98.

Foulds, L. (1969). *Neoplastic Development*. Academic Press.

Fujita, K., Fukuhara, S., Nasu, K. *et al.* (1986). Recurrent chromosome abnormalities in adult T-cell lymphomas of peripheral T-cell origin. *International Journal of Cancer*, **37**, 517–24.

Gilks, C. B., Bear, S. E., Grimes, H. L. & Tsichlis, P. N. (1993). Progression of interleukin-2 (IL-2) dependent rat T cell lymphoma lines to IL-2 independent growth following activation of a gene (*Gfi-1*) encoding a novel zinc finger protein. *Molecular and Cellular Biology*, **13**, 1759–68.

Holzmann, B. & Weissman, I. L. (1989). Integrin molecules involved in lymphocyte homing to Peyer's patches. *Immunological Reviews*, **108**, 45–61.

Justice, M. J., Gilbert, D. J., Kinzler, K. W. *et al.* (1993). A molecular genetic linkage map of mouse chromosome 18 reveals extensive linkage conservation with human chromosomes 5 and 18. *Genomics*, **13**, 1281–8.

Lazo, P. A. & Tsichlis, P. N. (1988). Recombination between two integrated proviruses one of which was inserted near c-*myc* in a retrovirus induced rat thymoma: implications for tumour progression. *Journal of Virology*, **62**, 788–94.

Lazo, P. A. & Tsichlis, P. N. (1990). Biology and pathogenesis of retroviruses. *Seminars in Oncology*, **17**, 269–94.

Lee, S. W., Tomasetto, C. & Sager, R. (1991). Positive selection of candidate tumour-suppressor genes by subtractive hybridization. *Proceedings of the National Academy of Sciences, USA*, **88**, 2825–9.

Li, J. P., D'Andrea, A. D., Lodish, H. F. & Baltimore, D. (1990). Activation of cell growth by binding of Friend spleen focus-forming virus gp55 glycoprotein to the erythropoietin receptor. *Nature*, **343**, 762–4.

Li, J. P. & Baltimore, D. (1991). Mechanism of leukemogenesis induced by mink cell focus-forming murine leukaemia viruses. *Journal of Virology*, **65**, 2408–14.

Liang, P. & Pardee, A. B. (1992). Differential display of eukaryotic messenger RNA by means of the polymerase chain reaction. *Science*, **257**, 967–71.

Liotta, L. A. (1986). Tumour invasion and metastases – role of the extracellular matrix: Rhoads Award Lecture, *Cancer Research*, **46**, 1–7.

Makris, A., Patriotis, C., Bear, S. E. & Tsichlis, P. N. (1993). Genomic organization and expression of *Tpl-2* in normal cells and Moloney murine leukaemia virus-induced rat T-cell lymphomas: activation by provirus insertion. *Journal of Virology*, **67**, 4283–9.

Matthew, C. G. P., Chin, K. S., Easton, D. F. *et al.* (1987). A linked genetic marker for multiple endocrine neoplasia type 2A on chromosome 10. *Nature*, **328**, 527–8.

Miller, A. D. & Rosman, G. J. (1989). Improved retroviral vectors for gene transfer and expression. *BioTechniques*, **7**, 980–90.

Miyoshi, J., Higashi, T., Mukai, H., Ohuchi, T. & Kakunaga, T. (1991). Structure and transforming potential of the human cot oncogene encoding a putative protein kinase. *Molecular and Cellular Biology*, **11**, 4088–96.

Narod, S. A., Lavoue, M. F., Morgan, K. *et al.* (1992). Genetic analysis of 24 French families with multiple endocrine neoplasia type 2A. *American Journal of Human Genetics*, **51**, 469–77.

Nicholson, G. L. (1991). Gene expression, cellular diversification and tumour progression to the metastatic phenotype. *Bioessays*, **13**, 337–42.

Nowell, P. C. (1990). Cytogenetics of tumour progression. *Cancer*, **65**, 2172–7.

Oppenheimer, S. B. & Lefevre, G. (1989). *Introduction to Embryonic Development*, Chapter 11, pp. 292–311.

Pabst, R. & Binns, R. M. (1989). Heterogeneity of lymphocyte homing physiology: several mechanisms operate in the control of migration to lymphoid and non-lymphoid organs *in vivo*. *Immunological Reviews*, **108**, 83–109.

Pals, S. T., Horst, E., Scheper, R. J. & Meijer, C. J. (1989). Mechanisms of human lymphocyte migration and their role in the pathogenesis of disease. *Immunological Reviews*, **108**, 111–33.

Patriotis, C., Makris, A., Bear, S. E. & Tsichlis, P. N. (1993). Tumour progression locus-2 (*Tpl*-2) encodes a protein kinase involved in the progression of rodent T cell lymphomas and in T cell activation. *Proceedings of the National Academy of Sciences, USA*, **90**, 2251–5.

Shiraishi, Y., Taguchi, T., Kubonishi, I., Taguchi, H. & Miyoshi, I. (1985). Chromosome abnormalities, sister chromatid exchanges, and cell cycle analysis in phytohemagglutinin-stimulated adult T cell leukemia lymphocytes. *Cancer Genetics and Cytogenetics*, **15**, 65–77.

Tsichlis, P. N. & Bear, S. E. (1991). Infection by mink cell focus-forming (MCF) viruses confers interleukin 2 (IL-2) independence to an IL-2 dependent rat T-cell lymphoma line. *Proceedings of the National Academy of Sciences, USA*, **88**, 4611–15.

Tsichlis, P. N. & Lazo, P. A. (1991). Virus–host-interactions and the pathogenesis of murine and human oncogenic retroviruses. In *Current Topics in Microbiology and Immunology. Retroviral Insertion and Oncogene Activation*, vol. 171, ed. H. J. Kung and P. K. Vogt, pp. 95–179. Springer-Verlag: Berlin, Heidelberg.

Yarden, O., Plamann, M., Ebbole, D. J. & Yanofsky, C. (1992). cot-1, a gene required for hyphal elongation in Neurospora crassa, encodes a protein kinase. *EMBO Journal*, **11**, 2159–66.

Yeatman, T. J. & Nicholson, G. L. (1993). Molecular basis of tumour progression: mechanisms of organ-specific tumour metastasis. *Seminars in Surgical Oncology*, **9**, 256–63.

Yeung, R. S., Taguchi, T., Patriotis, C. *et al.* (1993). New markers, *D16FC1* and *Tpl*-2, differentiate between rat chromosomes 16 and 17. *Cytogenetics and Cell Genetics*, **62**, 149–52.

LYMPHOPROLIFERATION AS A PRECURSOR TO NEOPLASIA: WHAT IS A LYMPHOMA?

H. C. MORSE III, J. W. HARTLEY, Y. TANG, S. K. CHATTOPADHYAY, N. GIESE AND T. N. FREDRICKSON

Laboratory of Immunopathology, National Institute of Allergy and Infectious Diseases, and Registry of Experimental Cancers, National Cancer Institute, National Institutes of Health, Bethesda, Maryland 20892 USA.

INTRODUCTION

The term lymphoma has classically been used to describe lymphoid tumours with characteristic clinical and histological features known to be associated with malignant behaviour. In recent years, it has become apparent that morphologically indistinguishable forms of lymphoma can be subdivided on the basis of genetically determined features. These include chromosomal gene rearrangements resulting in receptor gene expression associated with lineage commitment that can occur in both normal and malignant cells and more specific abnormalities, chromosomal translocations, deletions and mutations, found, with rare exceptions, only in malignancies. It is now generally accepted that the gradual accumulation of these genetic errors in a appearclone of lymphoid cells results in alterations of growth control and survival that give rise to neoplastic behaviour. There is, however, another aspect of lymphomagenesis which is becoming increasingly recognized and that is immunological control exerted by the host over the mutant clones. This aspect of lymphoma development has been revealed in immunosuppressed individuals, and in experimental induction of lymphomas in mice.

The advent of new classes of immunosuppressive drugs such as Cyclosporin A and FK506 for management of transplant recipients, evolving strategies for treatment of cancer patients that often include bone marrow transplantation and, perhaps most profoundly, the dramatic epidemic of immunodeficiency secondary to infection with HIV have greatly increased the population at risk for developing lymphomas. Similar risk is also seen in humans with congenital or age-related immunodeficiencies, with a variety of autoimmune diseases or with angioblastic lymphoproliferative disorders. This has led to the designation of lymphomas arising in these settings as immunodeficiency associated lymphomas (IAL) (German, 1983; Ziegler *et al.*, 1984; Penn, 1986; Knowles *et al.*, 1988; Kaplan *et al.*, 1989; Filipovich *et al.*, 1990).

Although the clinical and morphological features of IAL are often indistinguishable from malignant lymphomas developing in immunocompetent individuals, a fundamental difference in their pathobiology is suggested by several observations. First, cessation of immunosuppressive therapy in transplant patients is sometimes associated with spontaneous tumour regression. Secondly, there is evidence that most IAL present as polyclonal or oligoclonal lymphoproliferative processes rather than the monoclonal lesions as in 'true' lymphomas even though some progress to monoclonal disorders. A third confounding feature is exemplified by the changing clonalities and spontaneous remissions described for lymphomas in patients with Wiskott–Aldrich syndrome.

Thus, under conditions of immunodeficiency, clonality is not the *sine qua non* of malignant lymphoma, and other features must distinguish IAL as tumours that remain susceptible to immune regulation from the immune resistant lymphomas of immunocompetent people. Either immunodeficiency itself is the initiating factor in IAL or chronic inflammatory or infectious conditions resulting in persisting lymphoproliferation would appear to be required conditioning events for development of IAL. These hypotheses cannot be evaluated easily in man and, unfortunately, relatively few animal lymphoma systems seem relevant. Perhaps the most promising model in which immunodeficiency and chronic lymphoproliferation precede the development of true lymphoid neoplasms is the murine leukaemia virus (MuLV)-induced acquired immunodeficiency syndrome of mice designated mouse AIDS (MAIDS; for review see Morse *et al.*, 1992). Mice with this syndrome exhibit lymphoproliferation within 2 to 3 weeks of infection and some mice that live a long time with the disease develop aggressive lymphomas that can invade the central nervous system.

It should be mentioned at the outset that malignant lymphomas are an extremely common occurrence in many strains of mice particularly those expressing high levels of ecotropic MuLV from early in life. They become clinically evident as mice age with the first cases, usually thymic in origin, appearing after six months in strains such as AKR or C58 (Rowe, 1973). Other high virus strains, such as CWD and NFS congenic for AKR ecotropic virus, develop primarily nonthymic B cell lineage lymphomas that start to appear at about 8 months of age and increase steadily thereafter (Fredrickson, Morse & Rowe, 1984; Angel & Bedigian, 1984). Spontaneous tumour incidence as high as 30% to 40% have also been reported in BALB/c (Pattengale & Frith, 1983) and (B6 × C3H)F$_1$ hybrids that express much lower levels of ecotropic viruses. Thus, lymphomas are one of the chief causes of death in old mice.

A long-term, ongoing study of NFS strains congenic for ecotropic MuLV genes from AKR, C58 or C3H/Fg mice has established that mouse lymphomas, which occur in high incidence in these strains (Fredrickson *et al.*, 1985), are much like those occurring in humans. They are generally mono- or biclonal, of B cell origin, at a mature stage of differentiation and are morphologically lymphoblastic or centrocytic. Infiltration of non-lymphoid and distant lymphoid tissues usually occurs as lymphomas spread from sites of origin in the lymph nodes or spleen, giving these lymphomas every appearance of aggressive malignant tumours. Despite their resemblance to human lymphomas, those developing in mice have not been used to any extent as a model system to explore successive steps leading from hyperplasia to malignant neoplasms. Several factors have contributed to their limited use. First, is the mistaken impression that early thymic T cell neoplasms are the primary type of lymphoma in mice. Secondly, it has been difficult to define the role of MuLV in inducing these neoplasms. Perhaps, most importantly, the long latency and slowly progressive incidence curve for development of these tumours present real problems in gaining information about sequential changes required for lymphoma development.

In contrast, the rapidity with which lymphoproliferation develops in MAIDS and its presumed role in the development of lymphomas provide an unusual opportunity to sort out the dynamics of this stage of lymphomagenesis. The possible role of antigen-driven proliferation, the contributions of lymphocyte subsets and of cytokines to growth control and factors contributing to the emergence of dominant clones can be studied during the life-span of approximately six months after infection.

We would stress that MAIDS has a limitation as a model in that mice with the disease usually die prior to development of aggressive lymphomas. The reason for death is unclear, but it is probably associated with the advanced enlargement of all lymph nodes, particularly those in the mediastinum, along with pulmonary compromise secondary to lymphoid infiltration. None the less, this syndrome appears to provide a system for the study of antigen-driven lymphoproliferation which leads to outgrowth of clones in the setting of host immunodeficiency. Additional aspects of clonal evolution can be studied using splenic or nodal transplants in either SCID or nude mice. Accordingly, this chapter will focus on three aspects of MAIDS as they relate to the question posed in the title 'What is a lymphoma?'. The first relates to the histopathological aspects of MAIDS and the phenotypic and functional characteristics of the proliferative tissues. The second concerns cellular interactions and patterns of cytokine expression as they fuel the proliferative phase of this disorder. Finally, features of clonal evolution will be described in mice with MAIDS and in SCIDs or nudes transplanted with tissues from such mice.

CHARACTERISTICS OF MuLV INVOLVED IN MAIDS

The MuLV preparation causative of MAIDS was initially recovered from a radiation-induced tumour of a C57BL (B6) mouse (Laterjet & Duplan, 1962). A derivative of the original preparation used extensively in many laboratories was termed LP-BM5 MuLV (Mosier, Yetter & Morse, 1985). These virus preparations include a mixture of replication competent eco-tropic and mink cell focus-inducing (MCF) MuLV and a number of repli-cation defective viruses (Haas & Reshef, 1980; Legrand *et al.*, 1982; Yetter *et al.*, 1988, Hartley *et al.*, 1989; Chattopadhyay *et al.*, 1989; Aziz, Hanna & Jolicoeur, 1989). The replication competent viruses did not induce MAIDS when inoculated alone or in combination into neonatal or adult mice (Yetter *et al.*, 1988; Hartley *et al.*, 1989), but specific defective viruses induced disease when pseudotyped with ecotropic or amphotropic viruses (Aziz *et al.*, 1989; Chattopadhyay *et al.*, 1991; Huang, Simard & Jolicoeur 1992; Chattopadhyay & Hartley, unpublished observations) or when inoculated helper free (Huang *et al.*, 1989; Hartley & Chattopadhyay, unpublished observations).

Two biologically active defective genomes of approximately 4.9 kb, Du5H (Aziz *et al.*, 1989) and BM5def (Chattopadhyay *et al.*, 1991), were molecularly cloned and sequenced. They were found to be highly similar with markedly truncated *env* and *pol* genes and a single large open reading frame encoded by a relatively well-conserved *gag*. The defective virus gag polyproteins of 60 kD differ from the 65 kD gags of non-pathogenic MuLV primarily in the carboxyterminus of p15 and throughout most of p12. The Pr60^{gag} of the defective virus is normally myristylated and phosphorylated and is expressed at the cell surface in a highly protease resistant form (Huang & Jolicoeur, 1990).

LYMPHOPROLIFERATION AS A CENTRAL FEATURE OF MAIDS

MAIDS susceptible B6 mice rapidly exhibit lymphadenopathy and spleno-megaly following inoculation with LP-BM5 MuLV. This progression is markedly delayed in mice infected with helper free BM5def (unpublished observations); a time course for lymphoproliferation in mice given Du5H helper free has not been published. The progression of disease in mice injected with ecotropic virus pseudotypes of BM5def is accelerated over that of mice-infected helper free (Chattopadhyay *et al.*, 1991), but is still delayed in comparison with mice given mixed virus stocks. Confection of ecotropic virus pseudotypes of BM5def with a BM5 MCF enhances disease pro-gression (Chattopadhyay *et al.*, 1991) by mechanisms not yet understood. However, onset and progression of disease, at least as rapid as that in B6, are seen in some mice, C57L for example, infected with amphotropic pseudo-types of BM5def (Hartley, unpublished observations) indicating that MCF viruses are not required for rapid induction of lymphoproliferation.

An easily measured, reliable method for evaluating the progression of MAIDS is increase in spleen weight which is reflective of accumulation of lymphoid cells in this organ. The data in Fig. 1 show that the spleen size of mice infected with the LP-BM5 mixed virus preparation (BM5mix) in-

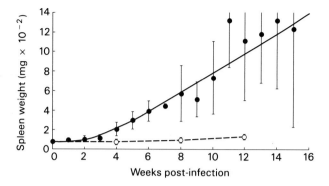

Fig. 1. Changes in spleen weights of mice infected with the pathogenic mixture of LP-BM5 viruses (closed circles) or non-pathogenic LP-BM5 ecotropic virus alone (open circles). Points for the mixed virus infections are means for 8 to 20 mice per time point and those for the ecotropic infection are for four mice per point. Bars represent one standard deviation.

creases slowly during the first two weeks after infection but then mounts rapidly in a linear manner through 15 weeks. Mouse to mouse differences in spleen size, as reflected in the standard deviation, increase progressively with increasing time after infection indicating the high degree of variability associated with progression of MAIDS. Generalized lymphadenopathy is also progressive with the weight of submandibular nodes, normally less than 20 mg, sometimes reaching more than 1 g by 12 weeks after infection.

The MCF viruses recovered from the LP-BM5 mixture replicate relatively poorly when used to infect adult B6 mice and no overt evidence of lymphoproliferation or of MAIDS was seen in inoculated mice. However, B6 mice infected with any of three biologically cloned isolates of BM5 ecotropic MuLV(BM5eco) exhibited small, slowly progressive changes in spleen (Fig. 1) and lymph node size (Lee, Yetter, Holmes & Morse, manuscript in preparation). This is due to what appears to be a chronic reactive response to ecotropic viral infection.

HISTOPATHOLOGICAL CHARACTERISTICS OF THE LYMPHOPROLIFERATIVE PHASE OF MAIDS

The initial response to infection of B6 and other susceptible strains of mice to BM5mix is increased activity within germinal centres (Pattengale *et al.*, 1982). This is most clearly seen in the spleen as progressive enlargement within the periarteolar lymphoid sheaths (PALS) (Fig. 2(*a*)). The increased number of mitotic figures and of immunoblasts and macrophages with tingible bodies is evidence of a response to viral infection. BM5 ecotropic MuLV, when infected alone, produces such a response which does not progress beyond this reactive phase. BM5mix, however, causes progressive enlargement of germinal centres (Figs.2(*a*)–(*e*)) so that the PALS now

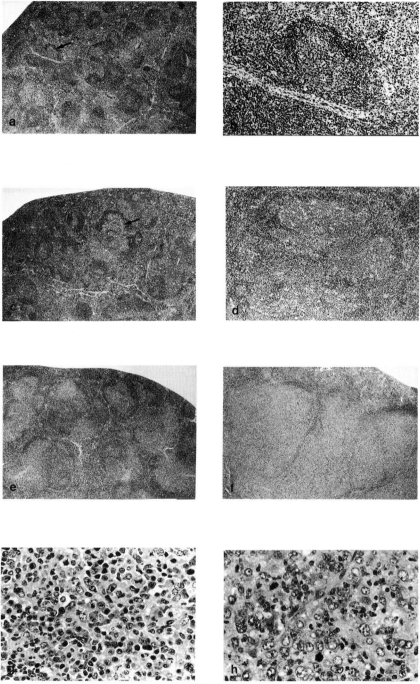

Fig. 2.

become distorted as germinal centre cells encroach on the mantle zone and the T cell domain around the central arteriole. This process continues until each central arteriole is surrounded by a large sheath of germinal centre cells along with plasmablasts, plasmacytoid and plasma cells, which in some areas may represent the majority population. Finally, enlarged PALS merge to form the bulk of cells within the spleen usually sparing only a subcapsular remnant of the red pulp (Fig. 2(f), (g)). Concurrently, but much less clearly defined from a histopathological point of view, cortical germinal centres in lymph nodes enlarge and merge with increasing numbers of immunoblasts and plasma cells within the medullary cords. In advanced MAIDS there is invasion of the lungs, kidney and liver which is clearly perivascular at its onset and may go on to form extensive infiltrates. As MAIDS progresses, a greater proportion of immunoblasts over large and small centrocytes often becomes evident although, in some spleens, and particularly lymph nodes, this can be obscured by an increase in macrophages.

PHENOTYPIC AND FUNCTIONAL CHARACTERISTICS OF THE PROLIFERATING CELLS IN MAIDS

Flow cytometry has been used to identify further the cell populations present in spleens and nodes of mice infected with BM5 mix for varying periods of time. In agreement with the histopathological studies, the major proliferating population is composed of B cells at various stages of differentiation. Within 4 weeks of infection, a high proportion of B cells are activated as shown by decreased levels of surface Ig and the B cell specific

---

Fig. 2(a). Spleen of a B6 mouse given BM5mix showing early reactive white pulp. The PALS are about normal size but contain prominent germinal centres. Note that each centre is surrounded by a mantle of smaller more darkly staining lymphoid cells. H+E ×40.

Fig. 2(b). Enlargement of the PALS designated by the arrow in 2(a). The central arteriole runs diagonally and is surrounded by small lymphocytes. H+E ×200.

Fig. 2(c). Spleen showing moderately more advanced reactive changes particularly in the PALS, indicated by the arrow, which is enlarged and distorted because of proliferation of germinal centre cells. H+E ×40.

Fig. 2(d). Same spleen as in 2(c) showing an enlarged germinal centre with a thinning of the mantle. H+E ×100.

Fig. 2(e). Spleen with early MAIDS; all PALS are enlarged compressing the mantle and, in some cases, proliferative lymphoid cells have encroached into the red pulp. H+E ×40.

Fig. 2(f). Advanced MAIDS with almost complete effacement of splenic architecture; only remnants of mantle zones are squeezed between PALS which are now completely replaced with the typical lymphoid population, some residual red pulp remains next to the splenic capsule. H+E ×600.

Fig. 2(g). Cells from the spleen shown in Fig. 2(f) include immunoblast, plasmacytoid and plasma cells along with scattered small and medium-sized lymphocytes. H+E ×600.

Fig. 2(h). Spleen from a spontaneous case of large cell lymphoma which was shown to be monoclonal and had surface IgK on the large cells with ovoid nuclei containing prominent nucleoli. H+E ×600.

isoform of CD45, B220 (Legrand, Daculsi & Duplan, 1981; Mosier et al., 1985; Klinman & Morse, 1989; Hartley et al., 1989). In addition, they express increased levels of MHC class II molecules and the activation markers CD69 and CD44 (Yetter et al., 1988; Selvey et al., 1993). With time, the proportions of cells with features of presecretory and secretory B cells increases. These phenotypic signs of maturation are associated with functional changes including increased serum IgM, IgG and IgE levels (Pattengale et al., 1982; Mosier et al., 1985; Klinman & Morse, 1989; Aziz et al., 1989; Gazzinelli et al., 1992a). For reasons still not understood, the serum Ig levels increase in parallel with the increase in spleen size through 6 to 8 weeks after infection, but then return to normal levels (Pattengale et al., 1982; Yetter et al., 1988; Basham et al., 1990; Even et al., 1992). Tests of antigen responsiveness revealed that B cells from infected mice rapidly lost their ability to produce specific antibody following immunization with T-dependent or T-independent antigens, were unable to respond to help provided by normal T cells (Mosier et al., 1985, Yetter et al., 1988) and also became unresponsive to mitogenic stimulation (Legrand et al., 1981; Mosier et al., 1985).

B cells are not the only lymphocyte population expanded in spleen of infected mice (Legrand et al., 1981). Although the frequency of CD4[+] T cells decreases slightly, and the percentage of CD8[+] T cells is reduced by more than 70% by 8 weeks after infection, the increase in total lymphoid mass is such that the number of CD4[+] cells is greatly increased while CD8[+] cells are increased several fold (Yetter et al., 1988, Muralidhar et al., 1992). In the first 6 weeks after infection, the expansion of CD4[+] T cells but not CD8[+] cells is biased towards cells expressing $V\beta5$ T cell receptors (TCR) and increased levels of the activation antigens CD69, CD44 and Ly-6C (Holmes et al., 1990; Portnoi et al., 1990; Muralidhar et al., 1992; Selvey et al., 1993). In addition, a high proportion of the CD4[+] cells are Thy-1[−] (Holmes et al., 1990). The abilities of CD4[+] and CD8[+] T cells to function in antigen-specific or non-specific responses are rapidly lost after infection (Mosier et al., 1985; Morse et al., 1989; Cerny et al., 1990a; Rosenberg, Maniero & Morse, 1991). By 8 weeks, both cell subsets are almost completely anergic to any form of activation including stimulation with PMA and the calcium ionophore, ionomycin (N. Giese, manuscript in preparation). Recent studies suggest, however, that responses established prior to infection may be preserved to some extent even when the responses to primary antigenic exposure are lost (M. Doherty, H. Morse & R. Coffman, manuscript in preparation).

Other studies showed increased proportions of CD11b[+] (Mac-1) and NK-1[+] cells in both spleens and peripheral lymph nodes of infected mice (Yetter et al., 1988; Makino et al., 1993; N. Giese et al., submitted for publication). In spleen, the proportions of CD11b[+] cells can be detected as soon as 3 weeks after infection and can reach 35%, reflecting increases in macro-

Table 1. *Changes in cell subsets in B6 mice infected with LP-BM5 MuLV 10 weeks*[a]

Cell type	Fold change	
	Frequency	Total number
CD4+ T	0.2 ↓	8 ↑
CD8+ T	0.8 ↓	2 ↑
B	0.1 ↑	11 ↑
CD11b+ (Mac-1)	4 ↑	40 ↑
NK+	20 ↑	100 ↑

[a] FACS determinations of cell frequency in spleen and lymph node employed monoclonal antibodies to the indicated CD determinants or the NK-1 antigen and for estimating B cell, antibodies to Ig kappa light chain and CD45(B220). The fold changes indicated can only be considered best estimates due to the variability in spleen and lymph node size seen at this stage of disease (Fig. 1).

phages as well as natural killer (NK) cells that coexpress CD11b and NK-1. The increases in NK-1+ cells are less in node than in spleen, but accurate quantitation of the changes of true NK cells in either tissue is not possible as some B cells and T cells can express this marker when activated (Makino *et al.*, 1993). Macrophage function in late MAIDS is greatly impaired as B6 mice infected with LP-BM5 MuLV for more than 8 weeks die following infection with *Leishmania major*, or *Listeria monocytogenes*, organisms that grow exclusively in macrophages (M. Doherty, H. C. Morse & R. L. Coffman, manuscript in preparation; A. Hügin, A. Cerny & H. Morse, submitted). Peritoneal macrophages also exhibit alterations in cytokine expression (Cheung *et al.*, 1991*a*). Increases in the frequency of splenic NK-1+ cell early in MAIDS are paralleled by increased cytotoxicity for NK cell targets including YAK cells, but NK function then drops off rapidly in spite of the mounting numbers of NK-1+ cells (Makino *et al.*, 1993; N. Giese *et al.*, submitted).

To summarize the phenotypic studies (Table 1), the lymphoproliferation of mice with MAIDS involves an increase in total cell mass due mainly to B cells, macrophages and NK cells and to a lesser extent CD4+ and CD8+ T cells. These alterations are associated with a generalized, profound state of immunodeficiency by 10 weeks post-infection.

CELLULAR INTERACTIONS IN MAIDS

Studies of the responses to infection with LP-BM5 MuLV of a large number of mouse strains have shown that immunodeficiency develops in all strains that exhibit lymphoproliferation and, conversely, that resistance to induction of lymphoproliferation is associated with normal or somewhat heightened immune function (Hamelin-Bourassa, Skamene & Gervais, 1989; Makino *et al.*, 1990, 1991, 1992; Huang *et al.*, 1992). This suggests that the same mechanisms contribute to the development of both lymphoproliferation and immunodeficiency or that one outcome of infection, proliferation or immune defects, leads irrevocably to the other.

Two major hypotheses have been offered to explain this observation. The first is that immunodeficiency is a paraneoplastic syndrome with a malignant expansion of transformed B lineage cells being the central defect in MAIDS (Huang, Simard & Jolicoeur, 1989; Jolicoeur, 1991). A crucial feature of this model is the postulate that the defective virus is a transforming agent. The mechanisms by which the Pr60^{gag} of the defective virus might affect transformation in this scenario have not been defined although several theories have been put forward (Jolicoeur, 1991).

The second model postulates that the defective virus gag behaves as an antigen, quite possibly a superantigen, that drives a remarkably strong proliferative response by CD4$^+$ T cells (Hügin, Vacchio & Morse, 1991). This response is associated with production of cytokines that nonspecifically activate other components of the immune system (Pitha *et al.*, 1988; Cerny *et al.*, 1991; Cheung *et al.*, 1991*b*; Gazzinelli *et al.*, 1992*a,b*; Muralidhar *et al.*, 1992). The consequences of cytokine expression are proliferation of activated B and T cells and a deficit in small resting cells capable of participating in cognate immune responses (Morse *et al.*, 1992).

To understand this second model and how it fits with current observations, it will be necessary to summarize the interactions between normal cells responding to immunization or infection and the roles played by different cytokines (Fig. 3). In this Figure, solid lines indicate positive or stimulatory influences of a cytokine directed to a particular cell type whereas the light lines indicate inhibitory or negative influences.

The central players in directing the immune response to most antigens are CD4$^+$ T cells which respond to antigen in the context of MHC class II molecules on antigen presenting cells (APC) including B cells, macrophages, and dendritic cells. CD4$^+$ cells have been divided into two subsets, designated Th1 and Th2, which are distinguished by their patterns of cytokine production (Mosmann *et al.*, 1986; Street & Mosmann *et al.*, 1991; Coffman *et al.*, 1991). Th1 cells produce IL-2 and IFNγ which favour the development of cell-mediated immunity, but do not produce IL-4, -5 or -10. Conversely, Th2 cells produce IL-4, -5 and -10 which bias responses towards the humoral arm of the immune response, but do not produce IL-2 or IFN-γ. The cytokines that characterize these subsets affect cells of other lineages, but also participate in what is known as Th1/Th2 cross-regulation in which factors produced by either Th subset can antagonize the development or function of the other (Coffman *et al.*, 1991; Street & Mosmann, 1991; Sher *et*

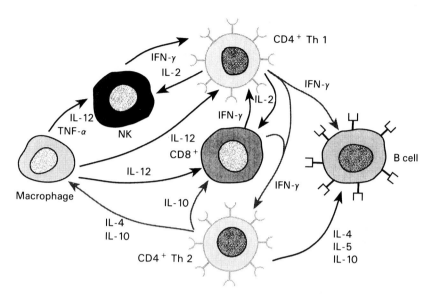

Fig. 3. Cellular interactions among normal cells as they relate to the pathogenesis of MAIDS. The cytokines with positive influences on the growth or differentiation of another cell type are shown in solid lines while cytokines with inhibitory influences are shown in light lines. The inhibitory effects of IL-10 on Th1 CD4$^+$ T cells are mediated through the effects of this cytokine on macrophages.

al., 1992). Thus, IFN-γ inhibits the proliferation of Th2 T cell clones and IL-4 and IL-10 act synergistically through macrophages to inhibit the activation of Th1 cells.

Another factor of importance to this scheme is IL-12 (Kobayashi *et al.*, 1989; Stern *et al.*, 1990; Wolf *et al.*, 1991; Schoenhaut *et al.*, 1992). This cytokine, previously designated natural killer cell stimulating factor or cytotoxic lymphocyte maturation factor, is produced by macrophages/monocytes and possibly some B cells. It favours the induction of Th1 over Th2 cells (Sypek *et al.*, 1993, Manetti *et al.*, 1993; Hsieh *et al.*, 1993) and, in synergy with TNF-α, stimulates production of IFN-γ by NK cells (Tripp, Wolf & Unanue, 1993) as well as promoting their growth. IFN-γ itself, as noted above, is an important contributor to the balance between Th1 and Th2 cells in response to antigen by virtue of its cross regulatory properties. In early studies, it was shown that T cells were required for the development of lymphoproliferation in MAIDS as nude mice did not develop lymphadenopathy, splenomegaly or hypergammaglobulinaemia after infection with LP-BM5 MuLV (Mosier, Yetter & Morse, 1987). Subsequently, it was found that CD4$^+$ cells were required for lymphoproliferation as mice depleted of this T cell subset by chronic treatment with anti-CD4 antibody

Table 2. *Effects of B cell abnormalities on development of MAIDS*

Mouse	Weeks post-infection	Spleen weight mg	MAIDS
B6	Uninfected	80	—
B6	10	730	Advanced
B6.UMT[a]	15	50	None
B6.*xid*[b]	13	80	None

[a] B6 mice with a targeted disruption of the membrane exon of the IgM heavy chain (Kitamura *et al.*, 1991).
[b] B6 mice congenic for the xid mutation.

exhibited no signs of MAIDS following infection (Yetter *et al.*, 1988). This observation was not unexpected, as B cells are normally stimulated to divide and differentiate as a consequence of interaction with CD4[+] T cells.

There are no cell surface markers that reliably distinguish the Th subsets of CD4[+] cells. To determine if one or both of these subsets is activated in MAIDS, it was necessary to evaluate cells for their ability to express cytokines. This was done by testing supernatants of stimulated spleen cells for induced proteins (Gazzinelli *et al.*, 1992a; Muralidhar *et al.*, 1992) and later by PCR analyses of cytokine transcripts (N. Giese, J. Actor, H. C. Morse & R. Gazzinelli, submitted for publication). The results of both types of assay of cells from infected mice were in agreement, and showed that initiation of the infection was associated with a mixed cytokine profile characteristic of Th1 and Th2 activation while the later stages of disease were characterized by a dominant Th2 profile featuring IL-4 and IL-10.

This pattern of cytokine expression represents more than an epiphenomenon of MAIDS as pharmacologics that inhibit cytokine expression, such as cyclosporin A, greatly retard the progression of lymphoproliferation (Cerny *et al.*, 1991, 1992). Together, these studies indicate that lymphoproliferation, due in large part to B cells, is dependent on the activation of the Th2 subset of CD4[+] T cells and the production of characteristic cytokines.

B cells have proven to be a focal point for understanding this disease for several reasons. First, mice depleted of B cells by treatment from birth with anti-IgM antibodies do not develop any signs of T cell activation or defective T cell function following infection with LP-BM5 MuLV (Cerny *et al.*, 1990b). Recently, this observation has been confirmed in another system using mice deficient in B cells due to a gene knockout that prevents membrane expression of IgM (W.-K. Kim, H. C. Morse, unpublished observations). As shown in Table 2, mice infected for 15 weeks exhibited no splenomegaly and had no evidence of MAIDS by flow cytometric or histopathologic studies. B cells are thus absolutely required for induction of T cell abnormalities.

One reason for the pivotal role played by B cells was revealed in studies showing them to be targets for infection by defective virus genomes. Early on, it was found that clonal populations of B cells develop in mice infected with LP-BM MuLV (Klinken *et al.*, 1988)). Subsequently, it was shown that these clonal expansions were associated with clonal integrations of the defective virus genome (Chattopadhyay *et al.*, 1991; Huang *et al.*, 1991) and that the Pr60^{gag} of the defective virus was expressed at the surface of B cell tissue culture lines which also displayed clonal proviral integrations of BM5def (Hügin *et al.*, 1991). The effects of B cells on induction of MAIDS may be quantitative under certain circumstances. First, as reported by Hitoshi *et al.* (1993) and confirmed elsewhere (Y. Tang, A. Hügin & H. C. Morse, unpublished observations), mice with a reduction in B cell frequency due to the *xid* mutation develop disease with a prolonged time course (Table 2). It has been suggested that this effect may be due primarily to a deficit in CD5$^+$ B cells (Hitoshi *et al.*, 1993). In addition, mice that express high levels of IFN-γ secondary to infection with *T. gondii* (Gazzinelli *et al.*, 1992b) or to treatment with IL-12 (N. Giese, R. Gazzinelli, S. Wolf & H. C. Morse, unpublished observations) have reduced levels of B cells in spleen and exhibit reduced severity of MAIDS after infection with LP-BM5 viruses. Finally, treatment of infected mice with high doses of cyclophosphamide, an agent highly toxic for B cells, greatly reduced the progression of disease (Simard & Jolicoeur, 1991). It may be a great oversimplification to attribute a delayed development of MAIDS to reduced B cell frequencies, as in each of these cases components of the immune system other than B cells are likely to be affected by mutation, cytokines or drugs.

The mechanisms by which infected B cells interact with T cells to induce their activation is under investigation. B cell-derived cytokines that stimulate T cells have not been described. However, studies now in progress strongly suggest that CD4$^+$ cells are activated by viral antigen presented in the context of MHC class II molecules on B cells. This view is based on evaluations of MHC class II knockout mice. These mice lack MHC class II expression on B cells, macrophages and other class II antigen presenting cells as the result of a genetically engineered mutation in the *Ab* gene (Grusby *et al.*, 1991). Since class II molecules are needed for positive selection of CD4$^+$ T cells in the thymus, class II deficient mice also lack CD4$^+$ cells. It was found that infected knockout mice do not develop MAIDS, an unremarkable result considering they are CD4$^+$ deficient. However, preliminary results show that no disease develops even after reconstitution with CD4$^+$ cells (N. Giese, H. C. Morse, unpublished results). As a control for this experiment, nude mice, which are also CD4 deficient but express class II molecules, were reconstituted with CD4$^+$ T cells and were shown to develop MAIDS after infection with LP-BM5

MuLV (N. Giese & H. C. Morse, unpublished observations). It is not known if B cells influence T cells in MAIDS in ways other than to present viral antigen.

CD8$^+$ T cells also play a role in MAIDS, but as a population that restricts development of lymphoproliferation. This effect was revealed in a strain, A/J, that normally is resistant to induction of MAIDS and to spread of BM5def. Depletion of CD8$^+$ T cells by long-term treatment with anti-CD8 antibody resulted in development of splenomegaly and lymphadenopathy and BM5def was easily detectable in spleen and lymph nodes of treated mice (Makino *et al.*, 1992).

DEVELOPMENT OF CLONAL LYMPHOID POPULATIONS IN MAIDS

The initial phase of lymphoproliferation has been shown to be truly polyclonal for both B cells and T cells. The evidence is best for B cells, as analyses of B cell antibody production showed that there was a proportional expansion of the normally expressed repertoire during the first 8 to 12 weeks after infection (Klinman & Morse, 1989). In addition, Southern blot hybridization studies of Ig gene organization in mice infected for 12 weeks or less have revealed only a rare clonal expansion of B cells (Klinken *et al.*, 1988). Less detailed analyses of the T cell repertoire have demonstrated a proportional expansion of the Vβ sequences utilized by CD8$^+$ cells and a biased increase among CD4s in the utilization of Vβ5 (Selvey *et al.*, 1993) although agreement on this point is not universal (Muralidhar *et al.*, 1992). During the first 12 weeks of infection, clonal rearrangements of the TCR β locus were observed with the same frequency as those of Ig genes but were still uncommon (Klinken *et al.*, 1988).

In B6 and (B6 × CBA)F$_1$ mice infected for 12 or more weeks, oligoclonal rearrangements of Ig genes were observed in spleen and/or lymph nodes of almost all mice examined (Klinken *et al.*, 1988). Sometimes, a rearranged band was shown to be present in spleen and one or more lymph nodes while other rearrangements were unique to one anatomic site (Klinken *et al.*, 1988). It was estimated that as many as eight different clones were distributed among the tissues of a single mouse. Clonal rearrangements of TCRβ genes were detected less frequently in studies of the same DNA preparations examined for Ig gene organization and multiple rearrangements were not observed (Klinken *et al.*, 1988). Parallel studies of the same samples for newly acquired proviral integrations showed one or more clonal copies of the defective virus were present in tissues with one or more clonal Ig gene rearrangements (Chattopadhyay *et al.*, 1991, Huang *et al.*, 1991). Clonal integrations of helper virus, most often single, were observed at a lower frequency (Klinken *et al.*, 1988; Y. Tang, *et al.*, submitted for publication).

Remarkably, tissues containing oligoclonal or monoclonal populations of

Table 3. *Relations of MAIDS donors to SCID transfers*[a]

	Donor		
Transplant line	Time infected (days)	FACS diagnosis	SCID transfer (FACS diagnosis)
B6-1	174	Pre-B	T (Thy-1$^-$ CD4$^+$ CD8$^-$) Pre-B (Kappa$^-$ B220$^+$)
B6-2	118	MAIDS + T	T (Thy-1$^+$ CD4$^+$ CD8$^-$)
B6-3	124	T	T (Thy-1$^+$ CD4$^+$ CD8$^-$)
B6-4	119	MAIDS + T	T (Thy-1$^-$ CD4$^+$ CD8$^-$) B (Kappa$^+$ B220$^+$)

[a] Spleen cells from donor mice infected for the indicated times were examined by FACS and histopathological techniques. The FACS analyses indicated the presence in each donor tissue of cells with a restricted phenotype consistent with the listed diagnoses. In addition, each mouse had evidence of ongoing MAIDS. Histological changes were consistent only with a diagnosis of MAIDS. The SCID transfers were all diagnosed as lymphoma by histopathology and FACS with the FACS analyses providing the information on cell lineage. The phenotype of each T or B lymphoma and the mixed tumours (B6-1 and B6-4) are given in parentheses.

B cells or T cells were generally indistinguishable histologically from tissues with polyclonal distributions (Klinken *et al.*, 1988; Fredrickson *et al.*, 1993). However, FACS analyses of the same tissues sometimes revealed the presence of discrete populations of cells that differed from normal spleen or lymph node cells in size and intensity of reactivity with different antibodies (Tang *et al.*, 1992; Fredrickson *et al.*, 1993; Y. Tang *et al.*, submitted). The size of these populations ranged from around 15% to as much as 60% of total cells. The patterns of antibody reactivity made it possible to identify these populations as being T, B or pre-B cell in origin (Table 3), and in most cases, the assignment of lineage and state of differentiation was consistent with the rearrangements of TCR or Ig genes seen on Southern analyses of DNA prepared from these tissues (Tang *et al.*, 1992; Fredrickson *et al.*, 1993).

In spite of the high proportion of mice with expanded clones of T cells or B cells detected by FACS and molecular techniques, none of the 17 B6 or (B6 × CBA)F$_1$ animals infected for 9 to 59 weeks had histological features of disease other than MAIDS of varying severity. Specifically, there was no evidence of aggressive infiltration of non-lymphoid organs as seen previously in other mice (Y. Tang *et al.*, submitted).

TRANSFERS OF CLONAL POPULATIONS TO SCID AND NUDE MICE

To gain more information about the clonal populations present in mice with MAIDS, suspensions of spleen cells from mice infected for varying periods were transferred intraperitoneally to mice lacking T or B cells due to the *scid*

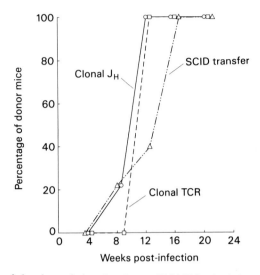

Fig. 4. Evolution of clonal populations in spleens of MAIDS mice in relation to transfers to SCID mice. DNA prepared from donor spleens was tested for the presence of clonal rearrangements of Ig genes by examining the organization of JH genes and for clonal rearrangements of TCR genes by studying the organization of TCRβ locus using Southern blot hybridization. Cells from the same spleens were tested for their ability to transfer to SCID mice. Points indicate the proportions of spleens with rearrangement or undergoing successful transfer among four or five mice per time point.

mutation (Tang *et al.*, 1992), to mice deficient in T cells due to the *nu* (nude) mutation (Kubo *et al.*, 1992) or to normal B6 mice. None of the normal mice developed tumours acutely, and those that exhibited enlarged spleen and nodes after longer periods of time were diagnosed with MAIDS, presumably secondary to transfer of virus with the lymphoid cells (Y. Tang *et al.*, unpublished observations).

Successful transfers to nudes or SCID were evidenced by the development of splenomegaly (200 to 900 mg) and equally prominent lymphadenopathy (Kubo *et al.*, 1992; Tang *et al.*, submitted). Infiltration of abdominal organs, particularly the liver, was seen and some mice had ascites and multiple peritoneal implants. The frequencies of transfer to SCIDs increased with time after infection (Tang *et al.*, submitted). None was obtained with mice infected for 4 weeks but all transfers from mice infected 118 days or more were successful. In trying to relate transfer to the presence of clonal populations in the donor tissue (Fig. 4), it is of interest that clonal TCR or Ig rearrangements were not observed in the spleen of the single successful donor infected for 7 weeks and that only two or five transfers were successful from 83 day infected mice, all of which had clonal rearrangements of both TCR and Ig genes. Transfers to nudes were obtained in 5 of 7 mice infected for 81 to 105 days (Kubo *et al.*, 1992).

The tumours appearing in nude mice and SCID recipients were examined histologically and by FACS, and DNA prepared from them was examined for TCR and Ig gene rearrangements and, in some cases, integrations of BM5def and ecotropic proviruses. All of the transfers to nudes were clonal populations of Thy-1$^+$ CD4$^+$ T cells that contained and expressed the defective virus genome (Kubo *et al.*, 1992). No B cell tumours were recovered. Comparisons of the attributes of four characteristic donor mice and SCID transfers are given in Table 3 and histological studies of tissues from some of these mice are shown in Fig. 5. First, FACS analyses of the transfers showed that they included T cells, confirming the Kubo study, but also B cells (not shown) and remarkably, mixtures of T cells and B lineage cells (Table 3). The T cell transplants were either Thy-1$^+$ CD4$^+$CD8$^-$ or Thy-1$^-$ CD4$^+$ CD8$^-$. Analyses of the organization of TCR and Ig genes demonstrated that the T cell, B cell or mixed cell populations detected in the individual recipients by FACS were monoclonal. In addition, it was found that the T and B cell subsets detected in individual mice could be physically separated and transferred independently to naive SCIDs (Y. Tang *et al.*, submitted).

Finally, in one unusual case, the donor and early transplant generations of one transplant line were dominated by a monoclonal population of T cells. However, later transplants contained a dominant B cell population. FACS and molecular analyses of a cell line grown *in vitro* from the first generation transplant showed that it was of B cell origin and had the same Ig rearrangement as that in the late transplant *in vivo* line, indicating that a 'covert' B lineage lymphoma had overgrown a T cell tumour during serial passage (Y. Tang *et al.*, submitted).

All donor mice were evaluated by histological examination of tissues including material used for transplantation. Typical lesions are illustrated in the section of spleen (Fig. 5(*a*)) from the donor yielding the transplant line B6-3. Infiltration of non-lymphoid tissue, especially the lung (Fig. 5(*b*)) was severe in this advanced case of MAIDS.

Comparisons between diagnoses of donor material established by FACS and by histopathology indicated that the cell subpopulations revealed by FACS as possible lymphoma were usually not distinguishable by histopathology. For example, the FACS-diagnosed subpopulations of T cells present in the donor spleen of the B6-2 transplant line and the predominant pre-B population present in the donor spleen of the B6-1 line did not reveal histopathological characteristics other than advanced MAIDS (Fig. 5(*c*) and Fig. 5(*d*), respectively).

Tissues from implanted SCID mice that contained fairly uniform populations of T cells or B lineage cells which were easily defined by FACS could not be diagnosed by histopathology. For example, the two independent G1 transplants of spleen and lymph node of the B6-3 donor were both judged by FACS to be composed of T cells but were diagnosed as lympho-

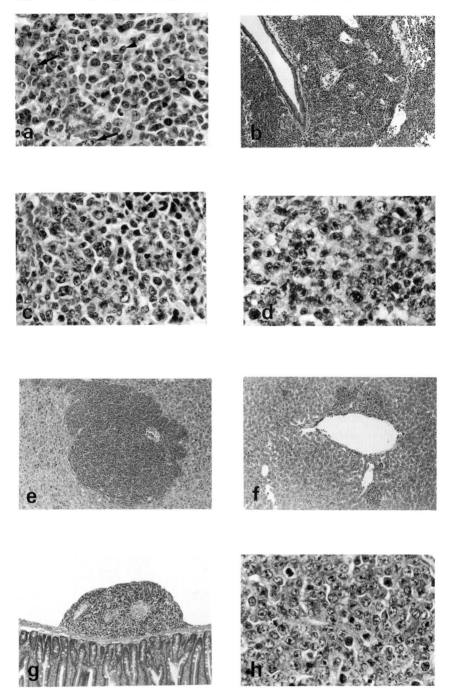

Fig. 5.

blastic or immunoblastic lymphoma, respectively (Fig. 5(*e*) and Fig. 5(*i*)). Immunoblastic lymphomas of T cell origin are uncommon among mouse tumours but have been observed previously. Even advanced passages remained more heterogeneous in cellular morphology (Fig. 5(*h*)) than might be inferred from FACS phenotype, and transplants with mixed T and B subpopulations were indistinguishable histologically from tumours of T or B origin alone (Fig. 5(*j*)). There was, however, a predilection for B cell lineage tumours to show less parenchymal infiltration (Fig. 5(*f*)) than for transplants composed of T cells (Fig. 5(*e*)) and to grow locally within the peritoneal cavity attached to serosal surfaces of visceral organs (Fig. 5(*g*)). It must be taken into account, however, that mouse B cell lymphomas, even those which are monoclonal, often present a heterogeneous population of cells including macrophages and reactive T cells (Fig. 5(*h*)). Therefore, the heterogeneity of cell types characteristic of MAIDS tumours may reflect a response seen frequently in mouse lymphomagenesis.

Fig. 5(*a*). Spleen of donor of B6-3 showing lymphoid cells including immunoblasts (arrow) and plasmacytoid cells (arrow heads) that make up part of the heterogeneous population of the white pulp. Histologically an aggressive stage 3 MAIDS, it was diagnosed as a T cell lymphoma by FACS (Table 3). H+E ×600.

Fig. 5(*b*). Lung from the same mouse as in Fig. 5a showing unusually extensive infiltration that is generally much more restricted to perivascular areas in MAIDS. H+E ×100.

Fig. 5(*c*). Spleen of donor of B6-2 showing variation in cell type with a moderate increase in the number of immunoblasts. Diagnosed by FACS as MAIDS but with a distinct population of large cells with a T cell phenotype. H+E ×600.

Fig. 5(*d*). Spleen of donor of B6-1 showing increase in the proportion of cells with immunoblastic features. Diagnosed by FACS as a pre-B cell lymphoma. H+E ×600.

Fig. 5(*e*). Liver of B6-3 G1 SCID mouse receiving mesenteric lymph node cells 18 days previously from the mouse depicted in Figs. 5(*a*) and (*b*). The pattern of extensive hepatic infiltration in periportal areas and to a lesser extent within sinusoids is typical for transplants diagnosed, as in this case, by FACS as T cell type. H+E ×100.

Fig. 5(*f*). Liver of B6-3 G6 SCID mouse receiving spleen cells of the transplantation series derived from the mouse in Figs. 5(*a*) and (*b*). The relatively restricted periportal infiltration was typical for transplants diagnosed as B cell type by FACS. H+E ×100.

Fig. 5(*g*). Early growth of transplanted cells on the serosa of the intestine in the same mouse as depicted in Fig. 5(*f*). Extensive proliferation within the abdominal cavity as discrete solid growths was typical for B cell transplants which often presented as solid tumour masses. H+E ×100.

Fig. 5(*h*). Spleen of mouse depicted in Fig. 5(*f*). The cells are fairly uniform with relatively small nuclei containing one or two nucleoli and diffuse chromatin and are diagnosed as lymphoblastic. H+E ×600.

Fig. 5(*i*). Spleen of B6-3 G1 SCID mouse inoculated with lymph node cells of the mouse depicted in Figs. 5(*a*) and (*b*). The large round and vesicular nuclei containing a single, prominent central nucleolus along with the well-defined cytoplasm are typical of immunoblasts. H+E ×600.

Fig. 5(*j*). Transplant series B6-4; spleen of B6 mouse (G6 in SCID followed by G2 in B6) with a FACS profile that revealed distinct populations of B and T cells. The cellular size and morphology varies, with a predominance of lymphoblasts. H+E ×600.

GENETIC ABNORMALITIES IN MAIDS LYMPHOPROLIFERATION/ LYMPHOMA

As noted earlier, the development of lymphomas is thought to reflect the accumulation in a single cell of genetic changes that result in altered proliferation or increased life-span. In mouse lymphomas, many of these changes have been found to be caused by proviral insertions in the proximity of cellular protooncogenes (P. Tsichlis, this volume). DNAs prepared from tissues of mice with advanced MAIDS (Klinken *et al.*, 1988; Huang *et al.*, 1989; Huang *et al.*, 1991) or from SCIDs bearing tumour transplants (Y. Tang *et al.*, submitted) were examined for newly acquired insertions of BM5def or ecotropic virus and alterations in the structure of genes known to influence lymphoma development including *myb, myc, Pim-1, Pvt, Gfi-1, Pal-1, Evi-1* and *Evi-5*, a newly defined proviral integration site on mouse chromosome 5 (X. Liao, N. Jenkins and N. Copeland, unpublished observations) closely linked to *Pal-1* (van Lohuizen *et al.*, 1991) and *Gfi-1* (Gilks *et al.*, 1993). Almost all DNA preparations from mice with advanced MAIDS contained one or more clonal integrations of the defective virus and in mice infected with BM5mix, most contained a single clonal ecotropic insertion as well. An altered *myc* was detected in the spleen of one mouse among 10 long-term infected animals in a study that did not examine cell transfers (Klinken *et al.*, 1988). In the SCID transfer study, another example of an altered *myc* locus was observed among 19 clonal populations tested (Y. Tang *et al.*, submitted). It is not known if these changes in *myc* are due to proviral integrations.

Further analyses of DNA from five of these mice and five additional mice not transferred but with longstanding MAIDS revealed structural alterations near the *Evi-5* locus in two animals (X. Liao, Y. Tang, unpublished observations). In both cases, the structural changes are caused by proviral insertion, but it is not yet known if the viruses involved are ecotropic, MCF or defective or if they change the expression of genes in this region. These results suggest that alterations in genes known or thought to affect growth control and growth factor dependence may contribute to clonal expansion in mice with MAIDS and the outgrowth of selected clones following transfer to SCID.

TRANSFER OF SCID LYMPHOMAS TO IMMUNOCOMPETENT MICE

The clones from mice with MAIDS that grew out as aggressive lymphomas in SCID did not grow on primary transfer to normal mice, indicating that they were susceptible to immune elimination. Failure to grow readily in immunocompetent mice was also a feature of SCID tumours, even after as many as eight serial passages (Y. Tang, unpublished observations). When some tumours did transfer, they were found to differ from the SCID

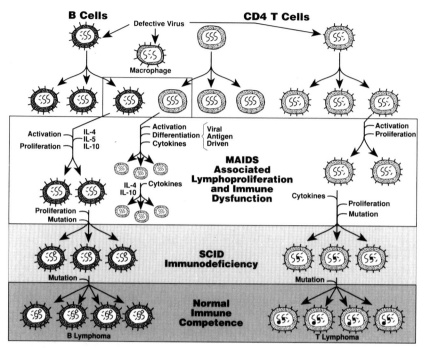

Fig. 6. Progression of lymphoma in mice with MAIDS and following transfers to SCID and then to adult, immunocompetent mice. Lines on the cell surface indicate expression Pr60gag following infection with the defective virus. Waved lines within the cells represent chromosomes. Interruptions in one chromosome indicate proviral insertion(s). Additional mutations that accrue with continued proliferation of infected cells are indicated by closed or open figures affecting other chromosomes. The crucial initiation events in the initiation of B cell lymphomagenesis occur in the top small box in which infected B cells stimulate CD4[+] responsive to the Pr60[gag]. This stimulus may also be the precipitating event for development of T cell lymphomas. The increasingly grey divisions moving down the Figure represent the cumulative effects of mutations on the cells to enhance survival and proliferative advantage.

tumours of origin in TCR or Ig gene organization or proviral insertions, indicating that acquisition of extended growth potential was associated with additional genetic changes.

HYPOTHESIS

Current observations of the MAIDS system can be integrated in a hypothetical scheme of lymphomagenesis shown in Fig. 6. The pseudotyped defective virus infects B cells, macrophages and CD4[+] T cells. Chromosomal integrations of the defective virus as well as ecotropic virus cause mutations, but these are insufficient in the absence of other influences, to provide a prominent growth advantage to infected cells, as nude or CD4 depleted mice do not develop B cell abnormalities and mice without B cells apparently do not develop T cell derangements. These changes may be considered sub-neoplastic, similar to the finding of *bcl-2* rearrangements in hyperplastic but non-lymphomatous tissues of humans (Limpens *et al.*, 1991).

The mutations induced by the initial round of proviral insertions may become important only after cells are driven into a state of perpetual proliferation which serves to increase the likelihood of additional genetic changes. The drive for this proliferative state is provided by $CD4^+$ T cells responding to BM5def-encoded antigen expressed on B cells and is fuelled by Th2 cytokines produced in a polarized CD4 response. The environment generated by this widespread activation of T and B cells is associated with a deficiency in specific immune responsiveness to new antigens although immune responses established early in infection may persist (M. Doherty, H. Morse & R. L. Coffman, manuscript in preparation). Antiviral immune responses induced at the time of infection may wane but could be sufficient to limit the outgrowth of clonal populations that express viral antigens at the cell surface.

It is only after this limited but effective immune response is negated by transfer to the even more immune deficient SCID that the growth stimulus provided by the mutations is readily evident in the unrestrained prolifer-ation and metastases of the clones. None the less, the limited success obtained in transfers from SCID to immunocompetent adult mice strongly indicates that these clones are still susceptible to immune surveillance. It is apparently only after additional mutations are acquired that the ability to exhibit unchecked growth in an immunocompetent environment is ob-tained.

Continued use of this system should prove to be a powerful tool for developing further understanding of lymphoma development in the context of immunodeficiency. It should be possible now to devise strategies to interrupt the complex but reasonably well-defined cellular interactions that drive the polyclonal phase of this disorder. This model should also provide an important opportunity to evaluate the nature of residual immune func-tion in MAIDS that is capable of restricting tumour growth to prevent most mice from developing overt lymphoma. Finally, initial efforts suggest it will be possible to gradually dissect the series of mutations that favour progres-sively more autonomous growth and finally permit full-blown tumour growth in the immunocompetent host.

REFERENCES

Angel, J. M. & Bedigian, H. R. (1984). Expression of murine leukaemia viruses in B-cell lymphomas of CWD/Agl mice. *Journal of Virology*, **52**, 6914.

Aziz, D. C., Hanna, Z. & Jolicoeur, P. (1989). Severe immunodeficiency disease induced by a defective murine leukemia virus. *Nature*, **338**, 505–8.

Basham, T., Rios, C. D., Holdener, T. & Merigan, T. C. (1990). Zidovudine (AZT) reduces virus titer, retards immune dysfunction, and prolongs survival in the LP-

BM5 murine induced immunodeficiency model. *Journal of Infectious Diseases,* **161**, 1006–9.

Cerny, A., Hügin, A. W., Holmes, K. L. & Morse, H. C. III. (1990*a*). CD4$^+$ T cells in murine acquired immunodeficiency syndrome: evidence for an intrinsic defect in the proliferative response to soluble antigen. *European Journal of Immunology,* **20**, 1577–81.

Cerny, A., Hügin, A. W., Hardy, R. R. *et al.* (1990*b*). B cells are required for induction of T cell abnormalities in a murine retrovirus-induced immunodeficiency syndrome. *Journal of Experimental Medicine,* **171**, 315–20.

Cerny, A., Merino, R., Makino, M., Waldvogel, F. A., Morse, H. C. III. & Izui, S. (1991). Protective effect of cyclosporin A on immune abnormalities observed in the murine acquired immunodeficiency syndrome. *European Journal of Immunology,* **21**, 1747–50.

Cerny, A., Merino, R., Fossati, L. *et al.* (1992). Effect of cyclosporin A and zidovudine on immune abnormalities observed in the murine acquired immunodeficiency syndrome. *Journal of Infectious Diseases,* **166**, 285–90.

Chattopadhyay, S. K., Morse, H. C. III, Makino, M., Ruscetti, S. K. & Hartley, J. W. (1989). Defective virus is associated with induction of murine retrovirus-induced immunodeficiency syndrome. *Proceedings of the National Academy of Sciences, USA,* **86**, 3862–6.

Chattopadhyay, S. K., Sengupta, D. N., Fredrickson, T. N., Morse, H. C. III & Hartley, J. W. (1991). Characteristics and contributions of defective, ecotropic, and mink cell focus-inducing viruses involved in a retrovirus-induced immunodeficiency syndrome of mice. *Journal of Virology,* **65**, 4232–41.

Cheung, S. C., Chattopadhyay, S. K., Hartley, J. W., Morse, H. C. III & Pitha, P. M. (1991*a*). Aberrant expression of cytokine genes in peritoneal macrophages from mice infected with LP-BM5 MuLV, a murine model of AIDS. *Journal of Immunology,* **146**, 121–7.

Cheung, S. C., Chattopadhyay, S. K., Morse, H. C. III & Pitha, P. M. (1991*b*). Expression of defective virus and cytokine genes in murine AIDS. *Journal of Virology,* **65**, 823–8.

Coffman, R. L., Varkila, K., Scott, P. & Chatelain, R. (1991). Role of cytokines in the differentiation of CD4$^+$ T-cell subsets *in vivo*. *Immunological Reviews,* **123**, 189–207.

Even, C., Hu, B., Erickson, L. & Plagemann, P. G. (1992). Correlation between levels of immunoglobulins and immune complexes in plasma of C57BL/6 and C57L/J mice infected with MAIDS retrovirus. *Viral Immunology,* **5**, 39–50.

Filipovich, A. H., Shapiro, R., Robison, L., Mettens, A. & Frizzera, G. (1990). Lymphoproliferative disorders associated with immunodeficiency. In *The Non-Hodgkin's Lymphomas*, ed. J. T. Magrath, pp. 135–154, London: Arnold.

Fredrickson, T. N., Morse, H. C. III & Rowe, W. P. (1984). Spontaneous tumours of NFS mice congenic for ecotropic murine leukaemia virus induction loci. *Journal of the National Cancer Institute,* **73**, 521–4.

Fredrickson, T. N., Morse, H. C. III, Yetter, R. A., Rowe, W. P., Hartley, J. W. & Pattengale, P. K. (1985). Multiparameter analyses of spontaneous nonthymic lymphomas occurring in NFS/N mice congenic for ecotropic murine leukaemia viruses. *American Journal of Pathology,* **121**, 349–60.

Fredrickson, T. N., Tang, Y., Chattopadhyay, S. K., Morse, H. C. III & Hartley, J. W. (1993). Retrovirus-induced lymphoproliferation as a model for developing diagnostic criteria for malignant lymphoma in mice. *Journal of Toxicologic Pathology*, in press.

Gazzinelli, R. T., Makino, M., Chattopadhyay, S. K. *et al.* (1992*a*). CD4$^+$ subset regulation in viral infection. Preferential activation of Th2 cells during progression of retrovirus-induced immunodeficiency in mice. *Journal of Immunology,* **148**, 182–8.

Gazzinelli, R. T., Hartley, J. W., Fredrickson, T. N., Chattopadhyay, S. K., Sher, A., Morse, H. C. III. (1992*b*). Opportunistic infections and retrovirus-induced immunodeficiency: studies of acute and chronic infections with *Toxoplasma gondii* in mice infected with LP-BM5 murine leukaemia viruses. *Infection and Immunity,* **60**, 4394–401.

German, J. (1983). Patterns of neoplasia associated with chromosome breakage syndromes. In *Chromosomes, Mutation and Neoplasia*, ed. J. German, pp. 97–134, New York: Alan R. Liss.

Gilks, C. B., Bear, S. E., Grimes, H. L. & Tsichlis, P. N. (1993). Progression of interleukin-2 (IL-2)-dependent rat T cell lymphoma lines to IL-2-independent growth following activation of a gene (*Gfi-1*) encoding a novel zinc finger protein. *Molecular and Cellular Biology,* **13**, 1759–68.

Grusby, M. J., Johnson, R. S., Papaioannou, V. E. & Glimcher, L. H. (1991). Depletion of CD4$^+$ T cells in major histocompatibility complex class II-deficient mice. *Science,* **253**, 1417–20.

Haas, M. & Reshef, T. (1980). Non-thymic malignant lymphomas induced in C57BL/6 mice by cloned dualtropic viruses isolated from hematopoietic stromal cell lines. *European Journal of Cancer,* **16**, 909–17.

Hamelin-Bourassa, D., Skamene, E. & Gervais, F. (1989). Susceptibility to a mouse acquired immunodeficiency syndrome is influenced by the H-2. *Immunogenetics,* **30**, 266–272.

Hartley, J. W., Fredrickson, T. N., Yetter, R. A., Makino, M. & Morse, H. C. III. (1989). Retrovirus-induced murine acquired immunodeficiency syndrome: natural history of infection and differing susceptibility of inbred mouse strains. *Journal of Virology,* **63**, 1223–31.

Hitoshi, Y., Okada, Y., Sonoda, E. *et al.* (1993). Delayed progression of a murine retrovirus-induced acquired immunodeficiency syndrome in X-linked immunodeficient mice. *Journal of Experimental Medicine,* **177**, 621–6.

Holmes, K. L., Morse, H. C. III., Makino, M., Hardy, R. R. & Hayakawa, K. (1990). A unique subset of normal murine CD4$^+$ T cells lacking Thy-1 is expanded in a murine retrovirus-induced immunodeficiency syndrome, MAIDS. *European Journal of Immunology,* **20**, 2783–7.

Hsieh, C-S., Macatonia, S. E., Tripp, C. S., Wolf, S. F., O'Garra, A. & Murphy, K. M. (1993). Development of T_H1 CD4$^+$ T cells through IL-12 produced by *Listeria*-induced macrophages. *Science,* **260**, 547–9.

Huang, M., Simard, C. & Jolicoeur, P. (1989). Immunodeficiency and clonal growth of target cells induced by helper-free defective retrovirus. *Science,* **246**, 1614–17.

Huang, M. & Jolicoeur, P. (1990). Characterization of the gag/fusion protein encoded by the defective Duplan retrovirus inducing murine acquired immunodeficiency syndrome. *Journal of Virology,* **64**, 5764–72.

Huang, M., Simard, C., Kay, D. G. & Jolicoeur, P. (1991). The majority of cells infected with the defective murine AIDS virus belong to the B-cell lineage. *Journal of Virology,* **65**, 6562–71.

Huang, M., Simard, C. & Jolicoeur, P. (1992). Susceptibility of inbred strains of mice to murine AIDS (MAIDS) correlates with target cell expansion and high expression of defective MAIDS virus. *Journal of Virology,* **66**, 2398–406.

Hügin, A. W., Vacchio, M. S. & Morse, H. C. III. (1991). A virus-encoded 'superantigen' in a retrovirus-induced immunodeficiency syndrome of mice. *Science,* **252**, 424–7.

Jolicoeur, P. (1991). Murine acquired immunodeficiency syndrome (MAIDS): an animal model to study the AIDS pathogenesis. *Federation of American Societies for Experimental Biology Journal,* **5**, 2398–405.

Kaplan, L., Abrams, D., Feigal, E. *et al.* (1989). AIDS-associated non-Hodgkin's lymphoma in San Francisco. *Journal of American Medical Association,* **261**, 719–24.

Kitamura, D., Roes, J., Kühn, R. & Rajewsky, K. (1991). Disruption of the membrane exon of the immunoglobulin μ chain gene. *Nature,* **350**, 423–6.

Klinken, S. P., Fredrickson, T. N., Hartley, J. W., Yetter, R. A. & Morse, H. C. III. (1988). Evolution of B cell lineage lymphomas in mice with a retrovirus-induced immunodeficiency syndrome, MAIDS. *Journal of Immunology,* **140**, 1123–31.

Klinman, D. M. & Morse, H. C. III. (1989). Characteristics of B cell proliferation and activation in murine AIDS. *Journal of Immunology,* **142**, 1144–9.

Knowles, D., Chamulak, G., Subon, M. *et al.* (1988). Lymphoid neoplasia associated with the acquired immunodeficiency syndrome (AIDS). *Annals of Internal Medicine,* **108**, 744–53.

Kobayashi, M., Ritz, L., Ryan, M. *et al.*, (1989). Identification and purification of natural killer cell stimulatory factor (NKSF), a cytokine with multiple biologic effects on human lymphocytes. *Journal of Experimental Medicine,* **170**, 827–45.

Kubo, Y., Nakagawa, Y., Kakimi, K. *et al.* (1992). Presence of transplantable T-lymphoid cells in C57BL/6 mice infected with murine AIDS virus. *Journal of Virology,* **66**, 5691–5.

Laterjet, R. & Duplan, J-F. (1962). Experiment and discussion on leukaemogenesis by cell-free extracts of radiation-induced leukaemia in mice. *International Journal of Radiation Biology,* **5**, 339–44.

Legrand, E., Daculsi, R. & Duplan, J. F. (1981). Characteristics of the cell populations involved in extra-thymic lymphosarcoma induced in C57BL/6 mice by RadLV-Rs. *Leukaemia Research,* **5**, 223–33.

Legrand, E., Guillemain, B., Daculsi, R. & Laigret, F. (1982). Leukemogenic activity of B-ecotropic C-type retroviruses isolated from tumours induced by radiation leukaemia virus (RADLV-RS) in C57BL/6 mice. *International Journal of Cancer,* **30**, 241–7.

Limpens, J., de Jong, D., van Krieken, J. H. *et al.* (1991). Bcl-2/JH rearrangements in benign lymphoid tissues with follicular hyperplasia. *Oncogene,* **6**, 2271–6.

Makino, M., Morse, H. C. III, Fredrickson, T. N. & Hartley, J. W. (1990). H-2-associated and background genes influence the development of a murine retrovirus-induced immunodeficiency syndrome. *Journal of Immunology,* **144**, 4347–55.

Makino, M., Davidson, W. F., Fredrickson, T. N., Hartley, J. W. & Morse, H. C. III. (1991). Effects of non-MHC loci on resistance to retrovirus-induced immunodeficiency in mice. *Immunogenetics,* **33**, 345–51.

Makino, M., Chattopadhyay, S. K., Hartley, J. W. & Morse, H. C. (1992). Analysis of role of $CD8^{+}$ T cells in resistance to murine AIDS in A/J mice. *Journal of Immunology,* **149**, 1702–6.

Makino, M., Winkler, D. F., Wunderlich, J., Hartley, J. W., Morse, H. C. III & Holmes, K. L. (1993). High expression of NK-1.1 antigen is induced by infection with murine AIDS virus. *Immunology,* in press.

Manetti, R., Parronchi, P., Giudizi, M. G. *et al.* (1993). Natural killer cell stimulatory factor (interleukin 12 {IL-12}0 induces T helper type 1 (Th1)-specific immune responses and inhibits the development of IL-4-producing Th cells. *Journal of Experimental Medicine*, **177**, 1199–204.

Morse, H. C., Yetter, R. A., Via, C. S. (1989). Functional and phenotypic alterations in T cell subsets during the course of MAIDS, a murine retrovirus-induced immunodeficiency syndrome. *Journal of Immunology*, **143**, 844–50.

Morse, H. C., Chattopadhyay, S. K., Makino, M., Fredrickson, T. N., Hügin, A. W. & Hartley, J. W. (1992). Retrovirus-induced immunodeficiency in the mouse: MAIDS as a model for AIDS. *AIDS*, **6**, 607–21.

Mosier, D. E., Yetter, R. A. & Morse, H. C. III. (1985). Retroviral induction of acute lymphoproliferative disease and profound immunosuppression in adult C57BL/6 mice. *Journal of Experimental Medicine*, **161**, 766–84.

Mosier, D. E., Yetter, R. A. & Morse, H. C. (1987). Functional T lymphocytes are required for a murine retrovirus-induced immunodeficiency disease (MAIDS). *Journal of Experimental Medicine*, **165**, 1737–42.

Mosmann, T. R., Cherwinski, H., Bond, M. B., Giedlin, M. A. & Coffman, R. L. (1986). Two types of murine helper T cell clone. I. Definition according to profiles of lymphokine activities and secreted proteins. *Journal of Immunology*, **136**, 2348–57.

Muralidhar, G., Koch, S., Haas, M. & Swain, S. L. (1992). CD4 T cells in murine acquired immunodeficiency syndrome: polyclonal progression to anergy. *Journal of Experimental Medicine*, **175**, 1589–99.

Pattengale, P. K., Taylor, C. R., Twomey, P. *et al.* (1982). Immunopathology of B-cell lymphomas induced in C57BL/6 mice by dualtropic murine leukaemia virus (MuLV). *American Journal of Pathology*, **107**, 362–77.

Pattengale, P. K. & Frith, C. H. (1983). Immunomorphologic classification of spontaneous lymphoid cell neoplasms occurring in female BALB/c mice. *Journal of the National Cancer Institute*, **70**, 169–9.

Penn, I. (1986). The occurrence of malignant tumours in immunosuppressed states. *Progress in Allergy*, **37**, 259–300.

Pitha, P. M., Biegel, D., Yetter, R. A. & Morse, H. C. III. (1988). Abnormal regulation of IFN-alpha, -beta, and -gamma expression in MAIDS, a murine retrovirus-induced immunodeficiency syndrome. *Journal of Immunology*, **141**, 3611–16.

Portnoi, D., Stall, A. M., Schwartz, D., Merigan, T. C., Herzenberg, L. A. & Basham, T. (1990). Zidovudine (Azido Dideoxythymidine) inhibits characteristic early alterations of lymphoid cell populations in retrovirus-induced murine AIDS. *Journal of Immunology*, **144**, 1705–10.

Rosenberg, A. S., Maniero, T. G. & Morse, H. C. III. (1991). *In vivo* immunologic deficits in mice with MAIDS and effect of LP-BM5 infection on rejection of skin from infected mice. *Transplantation Proceedings*, **23**, 167–9.

Rowe, W. P. (1973). Genetic factors in the natural history of murine leukaemia virus infection: G.H.A. Clowes Memorial Lecture. *Cancer Research*, **33**, 3061–8.

Schoenhaut, D. S., Chua, A. O., Wolitzky, A. G. *et al.* (1992). Cloning and expression of murine IL-12. *Journal of Immunology*, **148**, 3433–40.

Selvey, L. A., Morse, H. C. III, Granger, L. G. & Hodes, R. J. (1993). Preferential expansion and activation of $V\beta5^+$ $CD4^+$ T cells in murine acquired immunodeficiency syndrome. *Journal of Immunology*, **151**, 1712–22.

Sher, A., Gazzinelli, R. T., Oswald, I. P. *et al.* (1992). Role of T-cell derived cytokines in the downregulation of immune responses in parasitic and retroviral infection. *Immunology Review*, **127**, 183–204.

Simard, C. & Jolicoeur, P. (1991). The effect of anti-neoplastic drugs on murine acquired immunodeficiency syndrome. *Science*, **251**, 305–8.

Stern, A. S., Podlaski, F. J., Hulmes, J. D. *et al.* (1990). Purification to homogeneity and partial characterization of cytotoxic lymphocyte maturation factor from human B-lymphoblastoid cells. *Proceedings of the National Academy of Sciences, USA,* **87**, 6808–12.

Street, N. E. & Mosmann, T. R. (1991). Functional diversity of T lymphocytes due to secretion of different cytokine patterns. *Federation of American Societies for Experimental Biology Journal,* **5**, 171–7.

Sypek, J. P., Chung, C. L., Mayor, S. E. H. *et al.* (1993). Resolution of cutaneous leishmaniasis: interleukin 12 initiates a protective T helper type 1 immune response. *Journal of Experimental Medicine,* **177**, 1797–802.

Tang, Y., Fredrickson, T. N., Chattopadhyay, S. K., Hartley, J. W. & Morse, H. C. (1992). Lymphomas in mice with retrovirus-induced immunodeficiency. *Current Topics in Microbiology and Immunology,* **182**, 395–8.

Tripp, C. S., Wolf, S. F. & Unanue, E. R. (1993). Interleukin 12 and tumour necrosis factor α are costimulators of interferon γ production by natural killer cells in severe combined immunodeficiency mice with listeriosis, and interleukin 10 is a physiologic antagonist. *Proceedings of the National Academy of Sciences, USA,* **90**, 3725–9.

van Lohuizen, M., Verbeek, S., Scheijen, S., Wientjens, E., van der Gulden, H. & Berns, A. (1991). Identification of cooperating oncogenes in Eμ-myc transgenic mice by provirus tagging. *Cell,* **65**, 737–52.

Wolf, S. F., Temple, P. A., Kobayashi, M. *et al.* (1991). Cloning of cDNA for natural killer cell stimulatory factor, a heterodimeric cytokine with multiple biologic effects on T and natural killer cells. *Journal of Immunology,* **146**, 3074–81.

Yetter, R. A., Buller, R. M., Lee, J. S. *et al.* (1988). CD4$^+$ T cells are required for development of a murine retrovirus-induced immunodeficiency syndrome (MAIDS). *Journal of Experimental Medicine,* **168**, 623–35.

Ziegler, J. L., Beckstead, J. A., Volberding, P. A. *et al.* (1984). Non-Hodgkin's lymphoma in 90 homosexual men: Relation to generalized lymphadenopathy and the acquired immunodeficiency syndrome. *New England Journal of Medicine,* **311**, 565–70.

HIV AND PREDISPOSITION TO CANCER

I. V. D. WELLER

Academic Department of Genito Urinary Medicine, University College London Medical School, London, UK.

INTRODUCTION

An increased predisposition to cancer has been well documented in patients with naturally occurring immunodeficiency states involving various abnormalities of cellular or humoral immunity or both. The highest incidence is seen in patients with Wiskott–Aldrich syndrome and ataxia telangiectasia (Penn, 1981). In the late 1970s, the first cases of tumours arising in transplant patients were recorded and an increased predisposition has also been seen in patients with a variety of autoimmune diseases. The World Health Organization estimates that, as of mid 1993, over 13 million adults have become infected with HIV globally since the start of the pandemic. HIV infection is associated with a progressive/depletion of CD4 positive T-lymphocytes. As well as reduced numbers, there is also impaired function. The potential mechanisms involved are numerous and have been recently reviewed (Pantaleo, Graziosi & Fauci 1993). Two tumours are clearly associated with the immunosuppression of HIV infection, namely, Kaposi's sarcoma and non-Hodgkin's lymphoma. There is increasing interest in the association between HIV infection and ano-genital neoplasia. Although a large number of other tumours have been described in patients with HIV infection there is currently no evidence that HIV infection increases the risk for these.

KAPOSI'S SARCOMA

Epidemiology

The epidemiology of Kaposi's sarcoma has been recently reviewed (Beral, 1991). Before the HIV epidemic in North America and Europe, Kaposi's sarcoma was a rare tumour with an annual incidence of 0.02–0.06/100 000. It appeared to occur mainly in men over 50 years of age and of Eastern or Mediterranean origin. It was usually confined to the lower limbs, ran an indolent course, and responded well to radiotherapy or chemotherapy (Volberding *et al.*, 1983; Safai & Weiss, 1984).

Kaposi's sarcoma was also endemic in parts of Central and Eastern Africa. In some areas it accounted for 10% of all malignancies. The majority of cases described were similar to the classical form seen in North America and Europe but a lymphadenopathic form was described in children (Taylor *et al.*, 1971; Templeton & Bhana, 1975).

Table 1. *Subjects with AIDS in various countries who were also reported to have Kaposi's sarcoma, by HIV transmission group*

	Percentage of AIDS subjects who have Kaposi's sarcoma (reported number of subjects with AIDS)		
HIV transmission group	USA (90 990)	UK (2830)	Spain (1074)
Homosexual or bisexual men	21%	23%	36%
Heterosexuals	3%	10%	0%
Intravenous drug users	2%	0%	2%
Transfusion recipients	3%	0%	0%
Haemophiliacs	1%	0%	0%
Children infected by perinatal transmission	1%	0%	0%

Modified from Beral, 1991.

Until recent years, Kaposi's sarcoma alone was the presenting feature in around 25% of patients with AIDS in the USA. It is the second most common symptom of first AIDS diagnosis after pneumocystis carinii pneumonia.

The tumour was described in the 1970s in immunocompromised patients, particularly renal allograft recipients. The patients described were often of African, Mediterranean or Middle Eastern origin (Harwood, Osoba & Hofstader, 1979; Penn, 1979). Indeed in these regions it appeared to be a common complication of renal transplantation (Quinibi *et al.*, 1988). However, Kaposi's sarcoma in HIV infection occurs 300 times more commonly than in other immunosuppressed states and is 20 000 times more common than in the general population (Beral *et al.*, 1990). Kaposi's sarcoma has been described in HIV-negative homosexual men (Kitchen, French & Dawkins, 1990) and this description with others in the 1980s has led to the suggestion that this might be a knock-on effect from an epidemic caused by an as yet unidentified causal agent in HIV-positive individuals (Beral, 1991).

Kaposi's sarcoma is far more common in homosexual/bisexual men than in other risk groups (Table 1).

Clinical manifestations

The disease in AIDS is characterized by widespread skin, mucous membrane, visceral and lymph node involvement (Friedman-Kien *et al.*, 1982; Hymes, Cheung & Greene, 1981). Skin lesions are the most common presenting complaint. They appear as pink or red macules or purple plaques and nodules on the face, trunk or limbs. Lesions may arise at the site of trauma (Koebner's phenomenon). The lesions may be surrounded by oedema or be associated more with lymphoedema, particularly affecting the legs and face. The early skin lesions of Kaposi's sarcoma may be very difficult to differentiate from other common skin conditions, including bruises, naevi, dermatofibromata, secondary syphilis or lichen planus.

Histologically, the tumours are thought to be of vascular or lymphatic endothelial cell origin, and consist of spindle-shaped cells arranged in broad bands with vascular slits between and extravasation of erythrocytes between the spindle cells. The histological appearances of epidemic and classical Kaposi's sarcoma are indistinguishable.

The gastrointestinal tract is one of the commonest internal organs to be involved. About 50% of patients have lymph node and/or gut involvement (Safai & Weiss, 1984; Friedman-Kien et al., 1982). Involvement of the oropharynx, oesophagus, stomach, duodenum and colon has been demonstrated. Lesions resemble the range seen in the skin from small, flat telangiectatic lesions not well demonstrated by contrast studies and only seen at endoscopy, to larger nodular or polypoid lesions. The median survival in AIDS patients with Kaposi's sarcoma alone is about 18–24 months and although haemorrhage may occur from lesions opportunistic infection is the most common terminal event.

Aetiology

The increased incidence in homosexual/bisexual men has led to the suggestion that Kaposi's sarcoma, at least in part, may be caused by a sexually transmitted agent. In a study which assessed the social and demographic characteristics, including sexual behaviour, of 65 homosexual or bisexual men with AIDS from London, sexual practices in which there was contact with partners' faeces, before AIDS developed, were the main determinants of Kaposi's sarcoma risk (Beral et al., 1992). Kaposi's sarcoma developed in four out of 22 (18%) of men who reported never having practised insertive rimming (oral–anal sex) compared with 50–75% of those who practised it for less than once a month to more than once a week (Table 2). This sexual practice, more than any other, would facilitate the direct faecal oral transmission of an infectious agent.

All but two of six other study populations, in which the sexual behaviour of homosexual men who had Kaposi's sarcoma was compared with that of men who had other features of AIDS, found an association with faecal contact. In several of these studies, insertive rimming was associated with an increased risk of Kaposi's sarcoma (Jacobsen et al., 1990; Darrow et al., 1992), but the numbers in some of these studies were small. In one study, insertive fisting, but not rimming, was associated with an increased risk of Kaposi's sarcoma (Archibald et al., 1990). In one study from the US and one from Australia, no association was found (Lifson et al., 1990; Elford, Tindall & Schirkey, 1992).

An early study suggested that inhaled nitrites (poppers) used as a sexual

Table 2. *Risk of Kaposi's sarcoma by frequency of insertive rimming (oral anal sex) according to whether subjects were interviewed before or after AIDS developed*

Frequency of insertive rimming during the five years before interview	Percentage of men with Kaposi's sarcoma		
	Interviewed before AIDS ($n = 45$)	Interviewed after AIDS ($n = 20$)	Total ($n = 65$)
Never	25% (4/16)	0% (0/6)	18% (4/22)
<once/month	53% (9/17)	43% (3/7)	50% (12/24)
≥once/month but <once/wk	63% (5/8)	100% (3/3)	73% (8/11)
≥once/wk	75% (3/4)	75% (3/4)	75% (6/8)
Two-sided exact p value for trend	0.04	0.01	<0.001

Modified from Beral *et al.*, 1992.

stimulant was associated with an increased risk of Kaposi's sarcoma (Haverkos *et al.*, 1985). However, re-analysis of this data has shown that the men with Kaposi's sarcoma practised insertive rimming significantly more often than men with opportunistic infections, and no significant association was found with the frequency of popper use (Darrow *et al.*, 1992).

The decrease in proportion of homosexual men with AIDS and Kaposi's sarcoma described in recent years may be due to changes in sexual practices and reduced exposure to a causal agent. Certainly, the frequency of insertive rimming has declined in cohorts of HIV-positive homosexual men in the United States (Beral, 1991).

In the developing world, inadequate sanitation might be an important mode of transmission. One does not necessarily have to postulate that blood-borne transmission of a putative agent occurs to account for the cases in blood transfusion recipients, haemophiliacs, intravenous drug users and heterosexuals, because oral–genital contact may also increase the risk of exposure to a faecally excreted agent.

An early study showed a higher frequency of HLA-DR5 in patients with classical and epidemic Kaposi's sarcoma compared to controls. A higher frequency of this HLA phenotype is also observed in Italians and Jews (Friedman-Kien *et al.*, 1982). Another controlled study suggested that HLA-DR5 positive individuals may acquire the disease at lesser degrees of immunodeficiency (Marmor *et al.*, 1984). In a more recent controlled study, the frequency of HLA DR5 was not increased in Kaposi's sarcoma patients, but these patients had higher frequencies of HLA B35, C4, DR1 and DQ1 and lower frequencies of HLA C5 and DR3 compared to controls (Mann *et al.*, 1990). This group also summarized the antigen frequencies in Kaposi's sarcoma patients in previous studies compared to those in the controlled populations where the studies were carried out. The most consistent finding across the studies was that there was a decreased frequency of HLA DR3 in patients with Kaposi's sarcoma compared to controls, although HLA B35 and C4 were increased in all five other studies and HLA DR1 in two of three.

In early studies, herpes-like virus particles and CMV RNA and DNA were demonstrated in Kaposi's sarcoma tissue. Cytomegalovirus infection is particularly common amongst homosexual men and renal allograft recipients. Early sero-epidemiological studies also showed an association with classical Kaposi's sarcoma. However, more recent studies challenge this suggestion, and the finding of CMV DNA in Kaposi's sarcoma DNA has not been confirmed. The role, if any, of CMV in the aetiology of the tumour is unclear, but it would seem that the virus is more likely to be a passenger.

As with other tumours, there has been an increasing interest in the role of angiogenesis factors and cytokines in the aetiology of Kaposi's sarcoma. Long-term cultures of Kaposi's sarcoma cells from patients established with the addition of a medium from HTLV-II transformed T-cells do produce growth factors which may stimulate their own growth (autocrine function) and the growth of other cells (paracrine) including endothelial cells and fibroblasts. The Kaposi's sarcoma cells also had potent angiogenic properties with extensive vascularization occurring in chorioallantoic membrane. Furthermore, inoculation of cells into nude mice produced lesions similar to those of KS lesions (Salahuddin et al., 1988; Ensoli et al., 1989). These findings led these authors to suggest various stages of the Kaposi's sarcoma development with the induction of a Kaposi's sarcoma precursor cell produced by growth factors and cytokines from HIV infected lymphocytes or macrophages. These cells then produce factors such as interleukin-I or basic fibroblast growth factor (bFGF) which stimulate both their own growth and the growth of other cells. Specific probes for several human retro and DNA viruses including CMV failed to find evidence for any viral nucleotide sequences in these cells.

There is also interest in the role of HIV proteins in terms of their possible oncogenic potential. The HIV *tat* gene is one of the regulatory genes of the virus and a potent transactivator which up-regulates virus expression in cultured cells. The *tat* gene together with the viral regulatory region (LTR) has been introduced into mouse germ cells. The resultant transgenic mice develop skin lesions characterized by an increased cellularity of the dermis consisting largely of spindle-shaped cells at around four months, and at 12–18 months male mice developed nodular multifocal tumours resembling Kaposi's sarcoma. The *tat* gene was expressed in infected skin and tumours but not in the tumour cells derived from the tumours, which suggest that the *tat* message was coming from cells other than tumour cells in surrounding skin. The mechanism of tumourigenesis in this model induced by the tat protein has not been fully elucidated (Vogel et al., 1988).

LYMPHOMA

Epidemiology

Non-Hodgkin's B cell lymphoma is the second most common tumour occurring in AIDS. It is 60 times more common than in the general population. The overall risk is similar to that in transplant recipients. It has been estimated to occur in 3% of AIDS cases, although this may be an underestimate (Beral *et al.*, 1991). Following the first description of the AIDS epidemic, early studies began to identify small numbers of homosexual men with non-Hodgkin's lymphoma developing in a setting of a persistent generalized lymphadenopathy, opportunistic infections or Kaposi's sarcoma (Ziegler *et al.*, 1982; Levine *et al.*, 1984). Originally, lymphomas other than primary cerebral lymphomas were not included in the AIDS definition because lymphomas were known to be associated with immunosuppression. However, cancer registry data in San Francisco and Los Angeles soon indicated up to a three-fold rise in high grade lymphoma in young, never married men in 1983 and the characteristics of 90 such cases were reviewed in 1984 (Ziegler *et al.*, 1984). The median age distribution was identical to that of the AIDS cases that had been reported to the Centers for Disease Control and once the anti-HIV test became routinely available, the link was more firmly made.

Non-Hodgkin's lymphomas are usually classified as intermediate grade (diffuse large cell) or high grade either small non-cleaved lymphoma (Burkitt's or Burkitt's-like) or large cell immunoblastic lymphoma. More than 60% of the non-Hodgkin's lymphomas described in HIV infection are of the high grade type, with large cell immunoblastic being more common than the Burkitt's or Burkitt's-like lymphoma. The epidemiology of HIV-related non-Hodgkin's lymphoma was reviewed recently by studying the 2824 cases from 97 258 AIDS cases, in the USA, reported to the Centers for Disease Control up to 1989 (Beral *et al.*, 1991). The immunoblastic tumour increases in incidence with increasing age. Burkitt's lymphoma shows a peak incidence between the ages of 10 and 19, similar to that seen in a general population. Unlike Kaposi's sarcoma, the frequency of lymphoma seems to be fairly evenly distributed amongst all AIDS risk groups. Burkitt's lymphoma accounts for 20% of AIDS associated non-Hodgkin's lymphomas compared to <1% in other immunosuppressed states (Beral *et al.*, 1991). It is 1000 times more common in AIDS than the general population. Why it should be so common in AIDS is unclear. Primary lymphoma of the brain is several thousand times more common in AIDS patients than the general population, and occurs with equal frequency at all ages rather than increasing in incidence with increasing age.

The disease may present *de novo* or in a setting of prodromal lymphadenopathy, opportunistic infection or Kaposi's sarcoma. Most cases present

Table 3. *Extranodal sites of non-Hodgkin's lymphoma in 88 of 90 patients*

CNS	Brain mass	21
	Other	24
Bone marrow		30
GI tract		22
Lung		8
Liver		8
Skin		7
Other		7

Modified from Ziegler *et al.*, 1984.

with extra-nodal involvement predominantly in the central nervous system, bone marrow and gut (Ziegler *et al.*, 1984, Table 3).

Aetiology

The appearance of the different types of lymphoma, namely, Burkitt's or immunoblastic large cell may depend on the level of immunosuppression. In a recent study, Burkitt's lymphoma appeared to occur at higher CD4 counts than the immunoblastic large cell tumour (Roithman, Tourani & Andrieu, 1991).

The immune dysregulation induced by HIV is associated with polyclonal B cell activation and proliferation leading to hypergammaglobulinaemia and a B cell hyperplasia in lymph nodes. A chronic persistent lymphadenopathy is a common finding in the chronic clinically latent phase of HIV infection. The mechanism for the polyclonal activation of B cells is unclear. Reactivation of infection with EBV is common in both patients with lymphadenopathy and healthy homosexual men.

The progressive depletion in CD4 lymphocytes in HIV infection leads to reactivation of other herpes viruses, recurrent oral or genital ulceration with herpes simplex virus, shingles caused by herpes zoster virus and an increased frequency of CMV shedding and later CMV associated disease.

Evidence for EBV reactivation is found in increased anti-EBV titres and increased EBV shedding in saliva. EBV has also been implicated in the development of oral hairy leucoplakia, a condition that appears as raised corrugated white areas usually on the lateral border of the tongue and is also seen in other immunosuppressed states. It is described as hairy because of the histological appearances of fine keratin projections. Active EBV replication has been demonstrated in these lesions (Niedobitek *et al.*, 1991). EBV has also been suggested to be associated with similar oesophageal lesions which ulcerate (Kitchen *et al.*, 1990).

The B cell proliferation and activation and immune dysregulation may lead to the transition to a monoclonal B cell proliferation and EBV was thought to be involved. However, early studies did not suggest this since, when B cells were polyclonally activated by EBV *in vitro*, 72%–100% of immunoglobulin producing cells were also positive for EBV nuclear antigen (EBNA) and the predominant cytoplasmic immunoglobulin was IgM. In patients with persistent generalized lymphadenopathy with hypergamma-globulinaemia, the predominant cytoplasmic immunoglobulin was found to be IgG and cells were often negative for EBNA even after seven days culture (Crawford *et al.*, 1984).

Studies in which EBV DNA has been examined in non-Hodgkin's lymphoma tissue from AIDS patients are confusing. The EBV genome has been found in up to 50% of peripheral lymphomas by *in situ* hybridization (Hamilton-Dutoit *et al.*, 1991; Meeker *et al.*, 1991), whereas the majority of B cell lymphomas in allograft recipients contain EBV DNA. However, EBV DNA has been found by Southern blotting in the majority of primary cerebral lymphomas in AIDS. Burkitt's lymphoma in Africa is almost always associated with EBV, but EBV DNA has only been found in a proportion of Burkitt's lymphomas in AIDS. It would therefore seem that EBV DNA is found more commonly in primary cerebral lymphoma than in peripheral lymphoma in AIDS and is found less commonly in Burkitt's lymphoma than might be expected. In Burkitt's lymphoma, a reciprocal translocation occurs between the long arm of chromosome 8 and either chromosomes 14, 2 or 22. These arrangements place the c-myc oncogene close to the immunoglobulin heavy chain gene or kappa or lambda light chain genes. C-myc gene rearrangements have been detected in a pro-portion of AIDS non-Hodgkin's lymphomas (Meeker *et al.*, 1991).

In a study from the United States, eight out of 55 patients who were involved in the very early phase I studies of zidovudine developed high grade lymphomas (14.5%) in a setting of severe CD4 lymphopenia. So far there have been no reports of an increase in the incidence of lymphoma in controlled studies, but close longer-term follow-up of participants is necess-ary. So far, no firm relationship has been documented between the use of zidovudine or other nucleoside analogues and the later development of lymphoma.

Hodgkin's disease

Early epidemiological studies did not suggest that there was a significant increase in the incidence of Hodgkin's disease in HIV-infected patients (Biggar *et al.*, 1987). However, some studies did suggest that the clinical presentation was different. Patients presented with more advanced disease and unfavourable histology with an excess of the mixed cellularity or lymphocyte depletion types of tumour (Tirelli *et al.*, 1988). More recently, in a cohort study from San Francisco, an increased incidence of non-Hodgkin's lymphoma was observed in HIV-infected patients in the San Francisco Bay area compared to the general population base but the number of cases was small (Hessol *et al.*, 1992). Currently, Hodgkin's disease is not yet recognized as an HIV-related malignancy.

CERVICAL CARCINOMA

Ten years ago, a high incidence of cervical warts and cervical neoplasia were described in renal transplant patients. Of 132 female patients, 11 developed cervical warts and, of these, six cervical neoplasia, on average 22–38 months, respectively, post-transplant (Schneider, Kay & Lee, 1983). Case control studies in female renal allograft recipients have also shown a high prevalence of cervical intraepithelium neoplasia (CIN) (Alloub *et al.*, 1989). CIN and invasive carcinoma, are associated with human papilloma virus (HPV) infection particularly types 16 and 18 (MacNab *et al.*, 1986). Case control studies and cross-sectional studies have shown the association between HPV 16 and 18 and more severe grades of CIN, i.e. 2 and 3. In a cohort of women initially with normal cervical smears, HPV infection was the most important determinant of the risk for the development of CIN 2/3, with over half associated with HPV types 16 and 18 (Koutsky *et al.*, 1992), although other factors such as age at first intercourse and evidence of previous sexually transmitted diseases were also associated.

There have been a large number of reports of CIN and cervical carcinoma in HIV positive women, some with very rapid progression of disease (Monfardini *et al.*, 1989). In 1990, the Centres for Disease Control reviewed four studies from New York (CDC, 1990). These showed that CIN and human papilloma virus infection are more common in HIV positive compared to HIV negative women. The CIN lesions were of a more severe grade and extent in terms of the area of the cervix affected and involvement of other sites of the lower genital tract and perianal area. However, in New York, where the prevalence of HIV infection in child-bearing women is very high (12.5/1000 in 1987–88), the incidence of cervical cancer did not increase from 1981 to 1986. Obviously, further vigilance is required (CDC, 1990). More recently, 21 published reports or meeting abstracts on the relationship between HIV infection and cervical neoplasia, published between 1986 and 1990, were reviewed (Mandelbatt *et al.*, 1992). Only five included a comparison group of HIV negative women. The sample sizes of the studies were between 43 and 209 patients, and cervical neoplasia was defined as intraepithelial neoplasia or invasive lesions. Differences found between the cases and comparison groups in the studies included ethnicity, history of sexually transmitted disease, use of barrier contraceptives and higher rates of human papilloma virus infection. However, a summary odds ratio indicated that the odds of an HIV positive woman having cervical neoplasia was 4.9 times that of an HIV-negative woman (95% confidence limit 3.8–8.2).

It is clear that there is an association between cervical disease and infection with HIV. However, many other variables, which have been difficult to control for in the studies published so far, may be important such as: the number of sexual partners and the increased risk for acquisition of human papilloma virus infection. Other co-factors such as smoking and infection with other viruses may also play a role. The co-factor role for herpes simplex virus was put forward ten years ago but evidence is still lacking (zur Hausen, 1982). Studies that can adjust for these variables are required, but currently it is recommended that HIV-positive women have annual pap smears. In spite of the lack of definitive data, invasive cervical carcinoma has been included in the expanded surveillance case definition for AIDS in the United States (CDC, 1993).

ANAL CANCER

Early in the 1980s an association was found between anal cancer in men, positive syphilis serology and never having been married (Daling et al., 1982). More recently a case control study comparing patients with anal cancer and controls with colon cancer found that a history of receptive anal intercourse, particularly in men, and a history of genital warts was associated with the development of anal cancer. Certain other genital infections and cigarette smoking were also associated (Daling et al., 1987). The parallels between squamous cell carcinoma of the anus and that of the cervix have been drawn, and human papilloma virus DNA was found in anal cancer tissue in the mid 1980s. Over the last ten years the percentage of anal cancers containing HPV type 16 gene has increased (Scholefield et al., 1990). Earlier epidemiological studies did not suggest that the incidence of anal cancer was increasing as a result of the HIV epidemic, but more recent data may suggest that the incidence is increasing in the San Francisco area (Palefsky et al., 1992).

Anal intraepithelial neoplasia (AIN) like CIN is defined as nuclear abnormalities in the anal epithelium in the absence of inflammation, without breach of the basement membrane, and is graded from 1 to 3. The most frequent site is the anal transitional zone which is an area of transitional epithelium between the columnar epithelium above and the squamous epithelium below. Cytological studies have shown that human papilloma virus infection and dysplasia are common in HIV-positive homosexual men (Frazer et al., 1986). In a cross-sectional study of 97 homosexual men with advanced HIV disease, 39% had anal cytological abnormalities and 15% had AIN. There was a high rate of human papilloma virus infection, particularly with the human papilloma virus genotypes 16 and 18. A longitudinal study in 37 of these patients, with a mean follow-up at 17

Table 4. *HPV screening and typing in anogenital lesions from men seroposit-ive or seronegative for HIV*

	HPV screening			
	HIV+		HIV−	
HPV typing	HPV− $n = 18$ (33.3%)	HPV+ $n = 36$ (66.6%)	HPV− $n = 25$ (46.4%)	HPV+ $n = 29$ (53.6%)
HPV 6/11		6 (16.6%)		18 (62.1%)
HPV 16/18 and/or 31/35/51		30 (83.4%)		11 (37.9%)

Modified from Bernard *et al.*, 1992.

months, showed that the proportion of patients with anal intraepithelial neoplasia increased and there was a trend towards higher grades of AIN over time. However, this study was small, not controlled and there was a drop-out of patients particularly with symptomatic HIV disease.

In a small comparison group study, comparing HIV-positive and HIV-negative homosexual men with anogenital warts, HIV-positive men had a higher frequency of AIN of a more severe grade and were more frequently infected with HPV types 16 and 18. The HIV-negative men had lower grade AIN lesions and were infected more frequently with types 6 and 11 (Bernard *et al.*, 1992; Table 4).

In summary, there is insufficient evidence at the moment to suggest that HIV infection is associated with a higher incidence of anal carcinoma and furthermore, that HIV hastens the progression from human papilloma virus infection through increasingly severe grades of AIN to anal cancer. However, the evidence that does exist is suggestive.

REFERENCES

Alloub, M. I., Barr, B. B., McLaren, K. M., Smith, I. W., Bunney, M. H. & Smart, G. E. (1989). Human papillomavirus infection and cervical intraepithelial neoplasia in women with renal allografts. *British Medical Journal*, **298**, 153–6.

Archibald, C. P., Schechter, M. T., Craib, K. J. P., *et al.* (1990). Risk factors for Kaposi's sarcoma in the Vancouver lymphadenopathy AIDS study. *Journal of Acquired Immune Deficiency Syndromes*, **3**, 18–23.

Beral, V. (1991). Epidemiology of Kaposi's sarcoma. Cancer HIV and AIDS (Imperial Cancer Research Fund). *Cancer Surveys*, **10**, 5–22.

Beral, V., Peterman, T., Berkelman, R. & Jaffe, H. (1991). AIDS-associated non-Hodgkin's lymphoma. *Lancet*, **337**, 805–9.

Beral, V., Peterman, T. A., Berkelman, R. C. & Jaffe, H. W. (1990). Kaposi's sarcoma among persons with AIDS: a sexually transmitted infection? *Lancet*, **335**, 123–8.

Beral, V., Bull, D. & Darby, S. *et al.* (1992). Risk of Kaposi's sarcoma and sexual practices associated with faecal contact in homosexual or bisexual men with AIDS. *Lancet*, **339**, 632–5.

Bernard, C., Mougin, C., Madoz, L. *et al.* (1992). Viral co-infections in human papillomavirus-associated anogenital lesions according to the serostatus for the human immunodeficiency virus. *International Journal of Cancer*, **52**, 731–7.

Biggar, R. J., Horm, J., Goedert, J. J. & Melbye, M. (1987). Cancer in a group at risk of acquired immunodeficiency syndrome (AIDS) through 1984. *American Journal of Epidemiology*, **126**, 578–86.

CDC (1990). Risk for cervical disease in HIV-infected women – New York City. *MMWR*, **39**, 846–9.

CDC (1993). Revised classification system for HIV infection and expanded surveillance case definition for AIDS among adolescents and adults. *MMWR*, **41**, 1–19.

Crawford, D., Weller, I., Iliescu, V., *et al.* (1984). Epstein–Barr (EB) virus infection in homosexual men in London. *British Journal of Venereal Diseases*, **60**, 258–64.

Daling, J. R., Weiss, N. S., Hislop, T. G. *et al.* (1987). Sexual practices, sexually transmitted diseases, and the incidence of anal cancer. *New England Journal of Medicine*, **317**, 973–7.

Daling, J. R., Weiss, N. S., Klopfenstein, L. C., Cochran, L. E., Chou, W. H. & Daifuku, R. (1982). Correlates of homosexual behaviour and the incidence of anal cancer. *Journal of the American Medical Association*, **247**, 1988–90.

Darrow, W. W., Peterman, T. A., Jaffe, H. W., Rogers, M. F., Curran, J. W. & Beral, V. (1992). Kaposi's sarcoma and exposure to faeces. *Lancet*, **339**, 685.

Elford, J., Tindall, B. & Schirkey, T. (1992). Kaposi's sarcoma and insertive rimming. (letter). *Lancet*, **339**, 938.

Ensoli, B., Nakamura, S., Salahuddin, Z. *et al.* (1989). AIDS Kaposi's sarcoma derived cells express cytokines with autocrine and paracrine growth effects. *Science*, **223**, 243.

Frazer, I. H., Crapper, R. M., Medley, G., Brown, T. C. & Mackay, I. R. (1986). Association between anorectal dysplasia, human papillomavirus, and human immunodeficiency virus infection in homosexual men. *Lancet*, **ii**, 657–60.

Friedman-Kien, A., Laubenstein, L. J., Rubinstein, P. *et al.* (1982). Disseminated Kaposi's sarcoma in homosexual men. *Annals of Internal Medicine*, **96**, 693–700.

Hamilton-Dutoit, S. J., Pettesen, G., Franzmann, M. E. *et al.* (1991). AIDS related lymphoma. Histopathology immunophenotype and association with Epstein–Barr virus as demonstrated by in situ nucleic acid hybridisation. *American Journal of Pathology*, **138**, 149–63.

Harwood, A. R., Osoba, D. & Hofstader, S. L. (1979). Kaposi's sarcoma in recipients of renal transplants. *American Journal of Medicine*, **67**, 759–65.

Haverkos, H. W., Pinsky, P. F., Drotman, D. P. & Bregman, D. J. (1985). Disease manifestation among homosexual men with acquired immunodeficiency syndrome: a possible role of nitrites in Kaposi's sarcoma. *Sexually Transmitted Diseases*, **12**, 203–8.

Hessol, N. A., Katz, M. H., Liu, J. Y., Buchbinder, S. P., Rubino, C. J. & Holmberg, S. D. (1992). Increased incidence of Hodgkin disease in homosexual men with HIV infection. *Annals of Internal Medicine*, **117**, 309–11.

Hymes, K. B., Cheung, T. & Greene, J. B. (1981). Kaposi's sarcoma in homosexual men – a report of eight cases. *Lancet*, **ii**, 598–600.

Jacobsen, L. P., Munoz, A., Fox, R. *et al.* (1990). Incidence of Kaposi's sarcoma in a cohort of homosexual men infected with the human immunodeficiency virus type I. *Journal of Acquired Immune Deficiency Syndromes*, **3**, 24–31.

Kitchen, V. S., French, M. A. H. & Dawkins, R. C. (1990). Transmissible agent of Kaposi's sarcoma. *Lancet*, **i**, 797–8.

Kitchen, V. S., Helbert, M., Francis, N. D. *et al.* (1990). Epstein–Barr virus associated oesophageal ulcers in AIDS. *Gut*, **31**, 1223–5.

Koutsky, L. A., Holmes, K. K., Critchlow, C. W. *et al.* (1992). A cohort study of the risk of cervical intraepithelial neoplasia grade 2 or 3 in relation to papillomavirus infection. *New England Journal of Medicine*, **327**, 1272–8.

Levine, A. M., Meyer, P. R., Begandy, M. K. *et al.* (1984). Development of B-cell lymphoma in homosexual men. *Annals of Internal Medicine*, **100**, 7–13.

Lifson, A. R., Darrow, W. W., Hessol, N. A. *et al.* (1990). Kaposi's sarcoma in a cohort of homosexual and bisexual men. *American Journal of Epidemiology*, **131**, 221–31.

MacNab, J. C. M., Walkinshaw, S. A., Cordiner, J. W. & Clements, J. B. (1986). Human papillomavirus in clinically and histologically normal tissue of patients with genital cancer. *New England Journal of Medicine*, **315**, 1052–8.

Mandelbatt, J. S., Fahs, M., Garibaldi, K., Senie, R. T. & Peterson, H. B. (1992). Association between HIV infection and cervical neoplasia: implications for clinical care of women at risk for both conditions. *AIDS*, **6**, 173–8.

Mann, D. L., Murray, C., O'Donnell, M., Blattner, W. A. & Goedert, J. J. (1990). HLA antigen frequencies in HIV-I related Kaposi's sarcoma. *Journal of Acquired Immune Deficiency Syndromes*, **3**, 51–5.

Marmor, M., Friedman-Kien, A. E., Zolla-Pazner, S. *et al.* (1984). Kaposi's sarcoma in homosexual men. A seroepidemiological case-control study. *Annals of Internal Medicine*, **100**, 809–15.

Meeker, T. C., Shiramizu, B., Kaplan, L. *et al.* (1991). Evidence for molecular subtypes of HIV-associated lymphoma: division into peripheral monoclonal, polyclonal and central nervous system lymphoma. *AIDS*, **5**, 669–74.

Monfardini, S., Vaccher, E., Pizzocaro, G. *et al.* (1989). Unusual malignant tumours in 49 patients with HIV infection. *AIDS*, **3**, 449–52.

Niedobitek, G., Young, L. S., Lau, R. *et al.* (1991). Epstein–Barr virus infection in oral hairy leukoplakia: virus replication in the absence of a detectable latent phase. *Journal of General Virology*, **72**, 3035–46.

Palefsky, J. M., Holly, E. A., Gonzales, J., Lamborn, K. & Hollander, H. (1992). Natural history of anal cytologic abnormalities and papillomavirus infection among homosexual men with group IV HIV disease. *Journal of Acquired Immune Deficiency Syndromes*, **5**, 1258–65.

Pantaleo, G., Graziosi, C. & Fauci, A. S. (1993). The immunopathogenesis of human immunodeficiency virus infection. *New England Journal of Medicine*, **328**, 327–35.

Penn, I. (1979). Kaposi's sarcoma in organ transplant recipients. Report of 20 cases. *Transplantation*, **27**, 8–11.

Penn, I. (1981). Depressed immunity and the development of cancer. *Clinical Experimental Immunology*, **46**, 459–74.

Quinibi, W., Akhtar, M., Sheth, K. *et al.* (1988). Kaposi's sarcoma: The most common tumour after renal transplantation in Saudi Arabia. *American Journal of Medicine*, **84**, 225–32.

Roithman, S., Tourani, J. M., Andrieu, J. M. (1991). AIDS associated non-Hodgkin lymphoma. *Lancet*, **338**, 884–5.

Safai, B. & Weiss, H. (1984). Clinical Manifestations of Kaposi's sarcoma. In *AIDS and Infections of Homosexual Men*, ed. P. Ma. & D. Armstrong, pp. 211–214, Yorke Medical Books.

Salahuddin, S. Z., Biberfield, P., Kaplan, M. H., Markham, P. D., Larsson, L. & Gallo, R. C. (1988). Angiogenic properties of Kaposi's sarcoma derived cells after long-term culture *in vitro. Science*, **242**, 430–33.

Schneider, V., Kay, S. & Lee, H. M. (1983). Immunosuppression as a high-risk factor in the development of condyloma acuminatum and squamous neoplasia of the cervix. *Acta Cytologica*, **27**, 220–4.

Scholefield, J. H., McIntyre, P., Palmer, J. G., Coates, P. J., Shepherd, N. A. & Northover, J. M. A. (1990). DNA hybridisation of routinely processed tissue for detecting HPV DNA in anal squamous cell carcinomas over 40 years. *Journal of Clinical Pathology*, **43**, 133–6.

Taylor, J. F., Templeton, A. C., Vogel, C. L., Ziegler, J. C. & Kyalwazi, S. K. (1971). Kaposi's sarcoma in Uganda: a clinico-pathological study. *International Journal of Cancer*, **8**, 122.

Templeton, A. C. & Bhana, D. (1975). Prognosis in Kaposi's sarcoma. *Journal of the National Cancer Institute*, **55**, 1301–4.

Tirelli, U., Vaccher, E., Rezza, G. *et al.* (1988). Hodgkin disease and infection with the human immunodeficiency virus (HIV) in Italy. *Annals of Internal Medicine*, **108**, 309–10.

Vogel, J., Hinrichs, S. H., Reynolds, R. K., Luciw, P. A. & Jay, G. (1988). The HIV tat gene induces dermal lesions resembling Kaposi's sarcoma in transgenic mice. *Nature*, **335**, 606–11.

Volberding, P., Conant, M. A., Stricker, R. B. & Lewis, B. J. (1983). Chemotherapy in advanced Kaposi's sarcoma. *American Journal of Medicine*, **74**, 652–6.

Ziegler, J. L., Beckstead, J. H., Volberding, P. *et al.* (1984). Non-Hodgkins' lymphoma in 90 homosexual men: relation to generalised lymphadenopathy and the acquired immunodeficiency syndrome. *New England Journal of Medicine*, **311**, 565–70.

Ziegler, J. L., Drew, W. L., Miner, R. C. *et al.* (1982). Outbreak of Burkitt's-like lymphoma in homosexual men. *Lancet*, **ii**, 631–3.

zur Hausen, H. (1982). Human genital cancer: synergism between two virus infections or synergism between a virus infection and initiating events? *Lancet*, **ii**, 1370–2.

INDEX